T0205691

LECTURES ON PHASE TRANSITIONS AND THE RENORMALIZATION GROUP

LECTURES ON PHASE TRANSITIONS AND THE RENORMALIZATION GROUP

Nigel Goldenfeld

University of Illinois at Urbana-Champaign

CRC Press is an imprint of the
Taylor & Francis Group, an **informa** business

Advanced Book Program

First published 1992 by Westview Press

Published 2018 by CRC Press
Taylor & Francis Group
6000 Broken Sound Parkway NW, Suite 300
Boca Raton, FL 33487-2742

ISBN 13: 978-0-201-55408-3 (hbk)
ISBN 13: 978-0-201-55409-0 (pbk)

Visit the Taylor & Francis Web site at
http://www.taylorandfrancis.com

and the CRC Press Web site at
http://www.crcpress.com

Library of Congress Cataloging-in-Publication Data

Goldenfeld, Nigel.
Lectures on phase transitions and the renormalization group/
Nigel Goldenfeld.
p. cm. — (Frontiers in physics ; 85)
Lectures given at the University of Illinois at Urbana-Champaign.
Includes index.
ISBN 0-201-55408-9 — ISBN 0-201-55409-7
1. Phase transformations (Statistical physics) 2. Renormalization group. I. Title. II. Series: Frontiers in physics ; v. 85.
QC175.16.P5G65 1992 530.4'74—dc20 92-17055

This book was typeset by the author, using the TEX typesetting language.

Frontiers in Physics
David Pines, Editor

Volumes of the Series published from 1961 to 1973 are not officially numbered. The parenthetical numbers shown are designed to aid librarians and bibliographers to check the completeness of their holdings.
Titles published in this series prior to 1987 appear under either the W. A. Benjamin or the Benjamin/Cummings imprint; titles published since 1986 appear under the Westview Press imprint.

1.	N. Bloembergen	Nuclear Magnetic Relaxation: A Reprint Volume, 1961
2.	G. F. Chew	S-Matrix Theory of Strong Interactions: A Lecture Note and Reprint Volume, 1961
3.	R. P. Feynman	Quantum Electrodynamics: A Lecture Note and Reprint Volume
4.	R. P. Feynman	The Theory of Fundamental Processes: A Lecture Note Volume, 1961
5.	L. Van Hove, N. M. Hugenholtz L. P. Howland	Problem in Quantum Theory of Many-Particle Systems: A Lecture Note and Reprint Volume, 1961
6.	D. Pines	The Many-Body Problem: A Lecture Note and Reprint Volume, 1961
7.	H. Frauenfelder	The Mössbauer Effect: A Review—With a Collection of Reprints, 1962
8.	L. P. Kadanoff G. Baym	Quantum Statistical Mechanics: Green's Function Methods in Equilibrium and Nonequilibrium Problems, 1962
9.	G. E. Pake	Paramagnetic Resonance: An Introductory Monograph, 1962 [cr. (42)—2nd edition]G. E. Pake
10.	P. W. Anderson	Concepts in Solids: Lectures on the Theory of Solids, 1963
11.	S. C. Frautschi	Regge Poles and S-Matrix Theory, 1963
12.	R. Hofstadter	Electron Scattering and Nuclear and Nucleon Structure: A Collection of Reprints with an Introduction, 1963
13.	A. M. Lane	Nuclear Theory: Pairing Force Correlations to Collective Motion, 1964
14.	R. Omnès	Mandelstam Theory and Regge Poles: An Introduction M. Froissart for Experimentalists, 1963
15.	E. J. Squires	Complex Angular Momenta and Particle Physics: A Lecture Note and Reprint Volume, 1963
16.	H. L. Frisch J. L. Lebowitz	The Equilibrium Theory of Classical Fluids: A Lecture Note and Reprint Volume, 1964
17.	M. Gell-Mann Y. Ne'eman	The Eightfold Way (A Review—With a Collection of Reprints), 1964
18.	M. Jacob G. F. Chew	Strong-Interaction Physics: A Lecture Note Volume, 1964
19.	P. Nozières	Theory of Interacting Fermi Systems, 1964
20.	J. R. Schrieffer	Theory of Superconductivity, 1964 (revised 3rd printing, 1983)
21.	N. Bloembergen	Nonlinear Optics: A Lecture Note and Reprint Volume, 1965

Frontiers in Physics

22.	R. Brout	Phase Transitions, 1965
23.	I. M. Khalatnikov	An Introduction to the Theory of Superfluidity, 1965
24.	P. G. deGennes	Superconductivity of Metals and Alloys, 1966
25.	W. A. Harrison	Pseudopotentials in the Theory of Metals, 1966
26.	V. Barger D. Cline	Phenomenological Theories of High Energy Scattering: An Experimental Evaluation, 1967
27.	P. Choquàrd	The Anharmonic Crystal, 1967
28.	T. Loucks	Augmented Plane Wave Method: A Guide to Performing.Electronic Structure Calculations—A Lecture Note and Reprint Volume, 1967
29.	Y. Ne'eman	Algebraic Theory of Particle Physics: Hadron Dynamics In Terms of Unitary Spin Current, 1967
30.	S. L. Adler R. F. Dashen	Current Algebras and Applications to Particle Physics, 1968
31.	A. B. Migdal	Nuclear Theory: The Quasiparticle Method, 1968
32.	J. J. J. Kokkede	The Quark Model, 1969
33.	A. B. Migdal	Approximation Methods in Quantum Mechanics, 1969
34.	R. Z. Sagdeev	Nonlinear Plasma Theory, 1969
35.	J. Schwinger	Quantum Kinematics and Dynamics, 1970
36.	R. P. Feynman	Statistical Mechanics: A Set of Lectures, 1972
37.	R. P. Feynman	Photon-Hadron Interactions, 1972
38.	E. R. Caianiello	Combinatorics and Renormalization in Quantum Field Theory, 1973
39.	G. B. Field H. Arp J. N. Bahcall	The Redshift Controversy, 1973
40.	D. Horn F. Zachariasen	Hadron Physics at Very High Energies, 1973
41.	S. Ichimaru	Basic Principles of Plasma Physics: A Statistical Approach, 1973 (2nd printing, with revisions, 1980)
42.	G. E. Pake T. L. Estle	The Physical Principles of Electron Paramagnetic Resonance, 2nd Edition, completely revised, enlarged, and reset, 1973 [cf. (9)—1st edition

Volumes published from 1974 onward are being numbered as an integral part of the bibliography.

43.	C. Davidson	Theory of Nonneutral Plasmas, 1974
44.	S. Doniach E. H. Sondheimer	Green's Functions for Solid State Physicists, 1974
45.	P. H. Frampton	Dual Resonance Models, 1974
46.	S. K. Ma	Modern Theory of Critical Phenomena, 1976
47.	D. Forster	Hydrodynamic Fluctuation, Broken Symmetry, and Correlation Functions, 1975
48.	A. B. Migdal	Qualitative Methods in Quantum Theory, 1977
49.	S. W. Lovesey	Condensed Matter Physics: Dynamic Correlations, 1980
50.	L. D. Faddeev A. A. Slavnov	Gauge Fields: Introduction to Quantum Theory, 1980
51.	P. Ramond	Field Theory: A Modern Primer, 1981 [cf. 74—2nd ed.]
52.	R. A. Broglia A. Winther	Heavy Ion Reactions: Lecture Notes Vol. I, Elastic and Inelastic Reactions, 1981
53.	R. A. Broglia A. Winther	Heavy Ion Reactions: Lecture Notes Vol. II, 1990

Frontiers in Physics

No.	Author(s)	Title
54.	H. Georgi	Lie Algebras in Particle Physics: From Isospin to Unified Theories, 1982
55.	P. W. Anderson	Basic Notions of Condensed Matter Physics, 1983
56.	C. Quigg	Gauge Theories of the Strong, Weak, and Electromagnetic Interactions, 1983
57.	S. I. Pekar	Crystal Optics and Additional Light Waves, 1983
58.	S. J. Gates M. T. Grisaru M. Rocek W. Siegel	Superspace or One Thousand and One Lessons in Supersymmetry, 1983
59.	R. N. Cahn	Semi-Simple Lie Algebras and Their Representations, 1984
60.	G. G. Ross	Grand Unified Theories, 1984
61.	S. W. Lovesey	Condensed Matter Physics: Dynamic Correlations, 2nd Edition, 1986.
62.	P. H. Frampton	Gauge Field Theories, 1986
63.	J. I. Katz	High Energy Astrophysics, 1987
64.	T. J. Ferbel	Experimental Techniques in High Energy Physics, 1987
65.	T. Appelquist A. Chodos P. G. O. Freund	Modern Kaluza-Klein Theories, 1987
66.	G. Parisi	Statistical Field Theory, 1988
67.	R. C. Richardson E. N. Smith	Techniques in Low-Temperature Condensed Matter Physics, 1988
68.	J. W. Negele H. Orland	Quantum Many-Particle Systems, 1987
69.	E. W. Kolb M. S. Turner	The Early Universe, 1990
70.	E. W. Kolb M. S. Turner	The Early Universe: Reprints, 1988
71.	V. Barger R. J. N. Phillips	Collider Physics, 1987
72.	T. Tajima	Computational Plasma Physics, 1989
73.	W. Kruer	The Physics of Laser Plasma Interactions, 1988
74.	P. Ramond	Field Theory: A Modern Primer, 2nd edition, 1989 [cf. 51—1st edition]
75.	B. F. Hatfield	Quantum Field Theory of Point Particles and Strings, 1989
76.	P. Sokolsky	Introduction to Ultrahigh Energy Cosmic Ray Physics, 1989
77.	R. Field	Applications of Perturbative QCD, 1989
80.	J. F. Gunion H. E. Haber G. Kane S. Dawson	The Higgs Hunter's Guide, 1990
81.	R. C. Davidson	Physics of Nonneutral Plasmas, 1990
82.	E. Fradkin	Field Theories of Condensed Matter Systems, 1991
83.	L. D. Faddeev A. A. Slavnov	Gauge Fields, 1990
84.	R. Broglia A. Winther	Heavy Ion Reactions, Parts I and II, 1990
85.	N. Goldenfeld	Lectures on Phase Transitions and the Renormalization Group, 1992
86.	R. D. Hazeltine J. D. Meiss	Plasma Confinement, 1992
87.	S. Ichimaru	Statistical Plasma Physics, Volume I: Basic Principles,

		1992
88.	S. Ichimaru	Statistical Plasma Physics, Volume II: Condensed Plasmas, 1994
89.	G. Grüner	Density Waves in Solids, 1994
90.	S. Safran	Statistical Thermodynamics of Surfaces, Interfaces, and Membranes, 1994
91.	B. d'Espagnat	Veiled Reality: An Analysis of Present Day Quantum Mechanical Concepts, 1994
92.	J. Bahcall R. Davis, Jr. P. Parker A. Smirnov R. Ulrich	Solar Neutrinos: The First Thirty Years
93.	R. Feynman F. Morinigo W. Wagner	Feynman Lectures on Gravitation
94.	M. Peskin D. Schroeder	An Introduction to Quantum Field Theory
95.	R. Feynman	Feynman Lectures on Computation
96.	M. Brack R. Bhaduri	Semiclassical Physics
97.	D. Cline	Weak Neutral Currents
98.	T. Tajima K. Shibata	Plasma Astrophysics
99.	J. Rammer	Quantum Transport Theory
100.	R. Hazeltine F. Waelbroeck	The Frameworkof Plasma Physics
101.	P. Ramond	Journeys Beyond the Standard Moel
102.	Nutku Saclioglu Turgut	Conformal Field Theory: New Non-Perturbative Methods in String and Feild Theory
103.	P. Philips	Advanced Solid State Physics

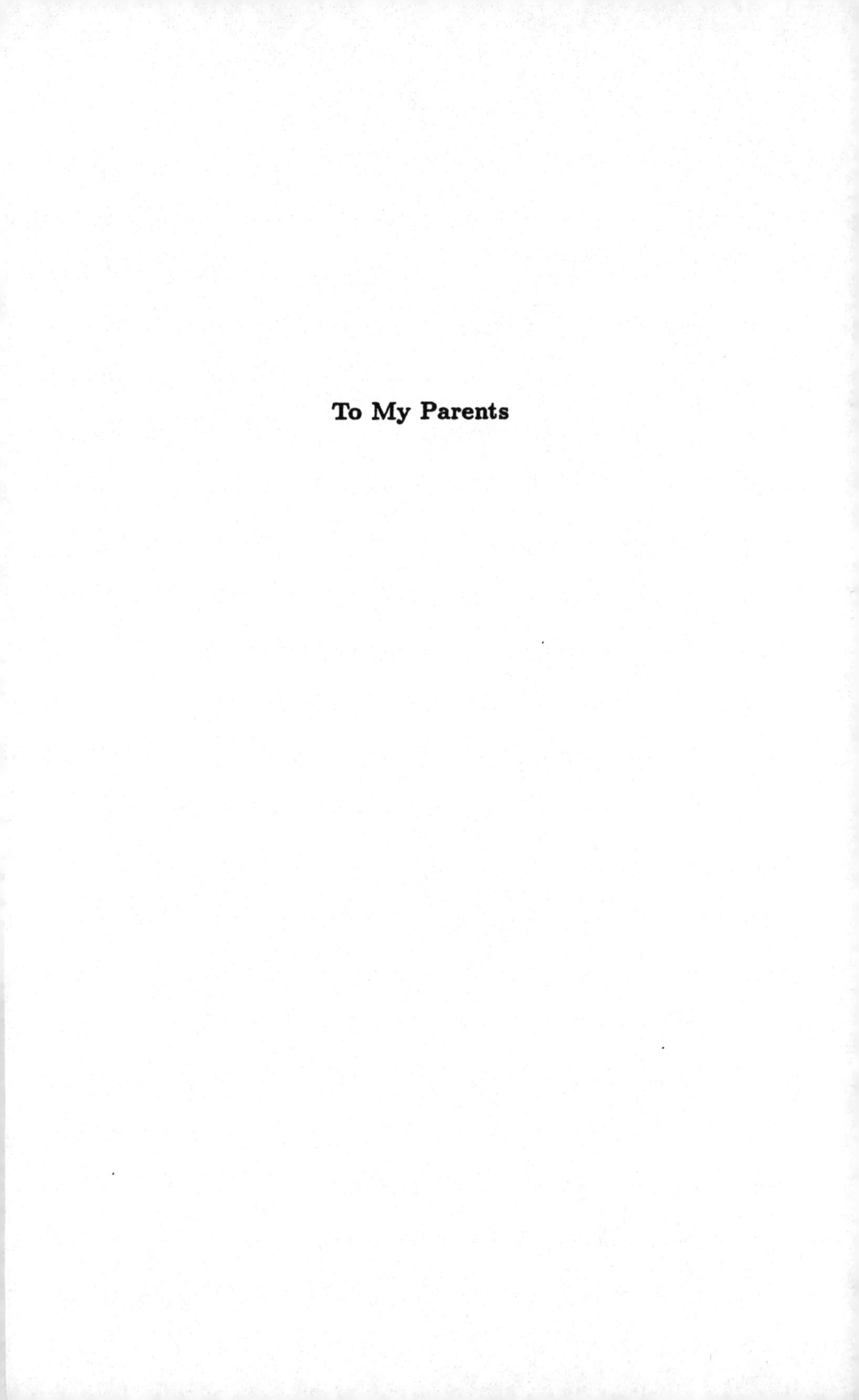

To My Parents

Editor's Foreword

The problem of communicating in a coherent fashion recent developments in the most exciting and active fields of physics continues to be with us. The enormous growth in the number of physicists has tended to make the familiar channels of communication considerably less effective. It has become increasingly difficult for experts in a given field to keep up with the current literature; the novice can only be confused. What is needed is both a consistent account of a field and the presentation of a definite "point of view" concerning it. Formal monographs cannot meet such a need in a rapidly developing field, while the review article seems to have fallen into disfavor. Indeed, it would seem that the people who are most actively engaged in developing a given field are the people least likely to write at length about it.

FRONTIERS IN PHYSICS was conceived in 1961 in an effort to improve the situation in several ways. Leading physicists frequently give a series of lectures, a graduate seminar, or a graduate course in their special fields of interest. Such lectures serve to summarize the present status of a rapidly developing field and may well constitute the only coherent account available at the time. One of the principal purposes of the FRONTIERS IN PHYSICS series is to make notes on such lectures available to the wider physics community.

As FRONTIERS IN PHYSICS has evolved, a second category of book, the informal text/monograph, an intermediate step between lecture notes and formal texts or monographs, has played an increasingly important role in the series. In an informal text or monograph an author has reworked his or her lecture notes to the point at which the manuscript represnts a coherent summation of a newly-developed field, complete with references and problems, suitable for either classroom teaching or individual study.

The modern theory of second order phase transitions and the renormalization group, which was developed some twenty years ago by Kenneth Wilson, represents one of the great accomplishments in theoretical physics in the latter part of this century. It has had a profound influence on subsequent experimental and theoretical work in statistical physics, and, more generally, in condensed matter physics. In the present lecture note volume, Nigel Goldenfeld, who has played a seminal role in applying renormalization group concepts to non-equilibrium phenomena, presents a remarkably lucid elementary account of phase transitions and the renormalization group, an account which is intended for both beginning graduate students and research scientists working in other fields. He draws heavily on experimental examples and elementary physical ideas to discuss how phase transitions occur both in principle and in practice. He shows how scaling ideas and renormalization group concepts may be applied not only to "classical" systems, such as ferromagnets, but also to phenomena which are far from equilibrium, such as non-linear diffusion in fluid dynamics. Systems which display broken ergodicity, such as spin glasses and rubber make their debut in this volume. His carefully reasoned arguments take the reader from elementary scaling arguments to sophisticated renormalization group results. It gives me great pleasure to welcome him as a contributor to FRONTIERS IN PHYSICS.

Contents

LECTURES ON PHASE TRANSITIONS AND THE RENORMALIZATION GROUP

Preface

During the Winter of 1983-1984, it occured to me that certain non-equilibrium phenomena could be analysed using the renormalisation group. At that time, my only encounter with the renormalisation group had been in the contexts of critical phenomena and field theory, and like most condensed matter physicists, I thought that the renormalisation group was, at heart, concerned with "integrating out degrees of freedom." It was difficult to see the relevance of this to those non-equilibrium systems, where there is no partition function or generating functional to compute. Nevertheless, the feeling persisted that there should be a more general way of thinking about renormalisation group transformations, and eventually it became possible to put these ideas into practice. In writing these lecture notes, I have attempted to convey a perspective broader than that usually found in the literature.

The volume before you is based upon a set of lectures that I have given to the physics graduate students at the University of Illinois at Urbana-Champaign. They form a one semester course on the topic of phase transitions and an introduction to the renormalisation group. In giving these lectures, I was guided by two objectives. The first was to explain as simply and as clearly as possible exactly why phase transitions occur, and to cut through some of the exaggerated awe that the subject inspires in the minds of many students. In particular, I wanted to avoid technicalities which hide the crux of the matter. My second objective was to show that there is nothing mysterious about the existence of critical exponents with 'anomalous' values. Even though Landau himself considered that the problem of second order phase transitions was one of the most important in theoretical physics, the actual solution to the problem is, in retrospect, really quite simple. I will have succeeded if the reader's response to the *dénouement* is a vague feeling of disappointment!

These lectures cover the *elementary* aspects of the physics of phase transitions and the renormalisation group. They are not intended to be either monographic or encyclopædic. The choice of topics and order of presentation were chosen so as to instill a logical progression of ideas in the minds of the students. I also decided that these notes would show precisely "how things work." Thus, I have eschewed both elegance and style in favour of explicit detail and directness, an approach which has proven to be popular and effective in the lecture room. Indeed, these notes are almost verbatim transcriptions of the original lectures.

A further note about style is appropriate here. I have sprinkled the text with footnotes — a practice that is uncommon nowadays, possibly as a result of Noel Coward's remark that encountering a footnote is like having to go downstairs and answer the front door in the middle of making love. Whilst I would be flattered to think that my prose style stands up to this analogy, I also feel that the alternative of flicking to the end of the chapter (or worse, the end of the book) is an interruption equivalent to searching the rooms of the house for an alarm clock that has just gone off. Generally speaking, the footnotes provide references to detailed expositions or review articles, and occasionally to original results. I make no claim to completeness; in most cases, cited articles were carefully chosen for their scientific or pedagogical value, rather than for the assignment of priorities. Thus I apologise in advance for the many worthy articles that were necessarily omitted, and hope that the reader will have no difficulty in tracking these down from the paper trail that I have given.

A number of exercises have also been provided, which I consider to be an integral part of the original course. In particular, certain topics, where the development is rather technical and best appreciated in private, are not discussed in the text; instead, the reader is led carefully through the salient points in order, it is hoped, to assist the process of self-discovery. The exercises were, in some sense, enjoyed by the classes at Illinois; I strongly urge the serious reader to attempt them.

I have written these notes keeping in mind a reader who wants to understand why things are done, what the results are, and what in principle can go wrong. These are rightly the concerns of both experimentalists and theorists, and the former will, I hope, not find themselves at a disadvantage when using these notes. I have assumed only a prior knowledge of statistical mechanics at the introductory graduate level, so the present treatment is reasonably self-contained. My emphasis and choice of topic is somewhat different from that of other treatments of phase transitions and the renormalisation group, with a number of topics making their *début* in a book at this level; consequently, I hope that even readers with prior exposure to these topics will find something of value in these pages.

I am too young to be amongst those who actually participated in the development of the renormalisation group in the theory of phase transitions. Some of what I know I learned from textbooks, original articles and from talking to friends and colleagues. In particular, I was fortunate enough to attend two lecture courses: one by John Cardy at the University of California at Santa Barbara, and another by Sandy Fetter at Stanford University. In preparing my own lectures, I was conscious of the pedagogical influence of these fine teachers. I owe an even greater debt of gratitude to Michael Wortis, who taught the course at Illinois before me. His style, approach and, occasionally, method of presentation have inevitably found their way into these notes. My own viewpoint on the renormalisation group has been influenced primarily by two people: G.I. Barenblatt and Yoshitsugu Oono. I wish to take this opportunity to thank them for their inspiration and collaboration respectively. It has also been my good fortune to count amongst my friends and collaborators, Paul Goldbart, with whom a number of ideas about disordered systems were developed.

I am grateful to many people who have wittingly or unwittingly helped in the preparation of this book. To the students of Physics 464 for their naïve and hence difficult questions; to Alex Belic, Paul Goldbart, Byungchan Lee, Yoshi Oono, Kieran Mullen, Lin-Yuan Chen, Jing Shi, Fong Liu, Jan Engelbrecht and Martin Tarlie for suggesting improvements on the original manuscript; to David Pines for encouraging me to write up these notes; to Cris Barnhart for braving TeX and typing some of the manuscript; to Larry Vance for rendering the figures; to Ansel Anderson, Head of the Physics Department at the University of Illinois for the opportunity to present this course; to the National Science Foundation for grant NSF-DMR-90-15791, which provided partial support for some of the results described in chapter 10; to the Alfred P. Sloan Foundation for a Fellowship; and last, but not least, to Joan Campagnolo for her patience and understanding.

In my opinion, the renormalisation group is one of the more profound discoveries in science, because it is a theory about theories. It has enabled physicists to become self-conscious about the way in which they construct physical theories, so that Dirac's dictum that we should seek the most beautiful or simplest equations acquires a meaningful and quantifiable aspect. Like thermodynamics, it will survive the post-quantum revolution, whatever and whenever that may be. Perhaps some reader of these notes will fire the first shot.

Urbana, Illinois
January, 1992

Nigel Goldenfeld

CHAPTER 1

Introduction

1.1 SCALING AND DIMENSIONAL ANALYSIS

The phenomena with which we shall be concerned all exhibit **scaling.** In its simplest form, this just means that two measurable quantities depend upon each other in a **power-law** fashion. A familiar example is Kepler's law, relating the radius R of a planet's circular orbit to the period T of the orbit:

$$T \propto R^{3/2}. \tag{1.1}$$

Another example, possibly not so familiar, is the formula for the phase speed c of waves on shallow water of depth h, neglecting surface tension and viscous effects:

$$c^2 = gh, \tag{1.2}$$

where g is the acceleration due to gravity. The scaling law in this example is $c \propto \sqrt{h}$. Formula (1.2) is only valid when the depth is small compared with the wavelength λ, and the more general relation is[1]

$$c^2 = \frac{g\lambda}{2\pi} \tanh\left(\frac{2\pi h}{\lambda}\right), \tag{1.3}$$

---

[1] See virtually any text-book on fluid dynamics. A clear presentation is given by D.J. Acheson, *Elementary Fluid Dynamics* (Clarendon, Oxford, 1990), Chapter 3.

which indeed reduces to eqn. (1.2) when $h \ll \lambda$. In this case, then, scaling occurs only approximately, but becomes more and more accurate in the limit $h/\lambda \to 0$.

Often, scaling laws can easily be deduced from **dimensional analysis**. For example, in the water wave case, the only variables that c may depend upon in principle are g, h, λ and the fluid density ρ. Using the notation [ ] to denote the dimensions of a given quantity, and the system of units mass-length-time MLT, we have: $[g] = LT^{-2}$, $[h] = [\lambda] = L$ and $[\rho] = ML^{-3}$ in three spatial dimensions. Thus

$$c = (gh)^{1/2} f\left(\frac{h}{\lambda}\right), \tag{1.4}$$

where f is a function that cannot be determined by dimensional analysis. In the limit $h/\lambda \to 0$,

$$c \sim (gh)^{1/2} f(0) \propto \sqrt{h}, \tag{1.5}$$

recovering the scaling law, apparently without requiring any detailed knowledge about fluid dynamics!

Actually, this happy state of affairs is an illusion – we made a very strong assumption in going from eqn. (1.4) to eqn. (1.5), namely *that the limit process was regular*. Usually, this can only be justified properly by considerations other than dimensional analysis. This point is by no means obscure mathematical pedantry: in fact, *the cases where the regularity assumption breaks down constitute the central topic of this book.*

The derivation presented above of the shallow water wave speed is somewhat deceptive for another, more mundane and less far-reaching reason: in writing down eqn. (1.4), we were presented with a choice of which length to use in the prefactor of the function f. We could equally well have used λ instead of h, leading to

$$c = (g\lambda)^{1/2} \tilde{f}\left(\frac{h}{\lambda}\right), \tag{1.6}$$

where $\tilde{f}$ is another function to be determined. Now what happens in the limit $h/\lambda \to 0$? It looks as if something has gone wrong! To proceed, recall that our purpose in taking the limit is to remove the dependence of c on λ: our common sense intuition tells us that when λ is "sufficiently large," it should not affect the result. The only way that λ can cancel out of the formula (1.6) is if the function $\tilde{f}$ has a square root behaviour for small values of its argument:

$$\tilde{f}(x) \sim x^{1/2} f(x) \qquad \text{as } x \to 0, \tag{1.7}$$

where the function f is analytic as $x \to 0$ and tends towards some well-defined limit $f(0)$. In fact, f here is the function of eqn. (1.4), as we see by using the approximation (1.7) in eqn. (1.6); we do indeed obtain the correct result (1.5) for the wave speed.

1.2 POWER LAWS IN STATISTICAL PHYSICS

In the above examples, and in many other scaling laws, the power law, or the **exponent** is a rational fraction, often deduced from simple dimensional considerations. This partly accounts for the fact that the phenomena described are so well understood, and are taught in elementary physics courses. However, there is a broader class of phenomena where power-law behaviour occurs, but the exponent is not a simple fraction (as far as is known). This class of phenomena includes, but is by no means restricted to, phase transitions where there is a **critical point**. We shall shortly discuss precisely what we mean by this; but first, let us consider some examples.

1.2.1 Liquid Gas Critical Point

In figure (1.1) is sketched a portion of the phase diagram for a fluid. The axes are the temperature T and the density of the fluid ρ, and the curve is shown in the fixed pressure, P, plane. Below the **critical** or **transition temperature,** T_c, is the coexistence curve. This has the following interpretation. Below T_c, as density is increased at fixed temperature, it is not possible to pass from a gaseous phase to a liquid phase without passing through a regime where the container of the fluid contains a mixture of both gas and liquid. The two-phase region has a manifestation in the thermodynamic properties of the fluid, which we will discuss later. Above the critical point, it is possible to pass continuously from a gas to a liquid as the density is increased at constant temperature. In this case, there is no density at which there is a coexisting mixture of liquid and gas in the container. Note that even starting below T_c it is always possible to pass from a liquid to a gas without passing through any two-phase region: one simply raises the temperature above T_c, reduces the density, and then lowers the temperature below T_c. This suggests that there is no real way to distinguish between a liquid and a gas. In fact, the question of how one identifies different phases of matter is one with which we shall be concerned in later chapters.

Returning to figure (1.1), the interesting question to ask for the purposes of the present discussion is: what is the shape of the coexistence

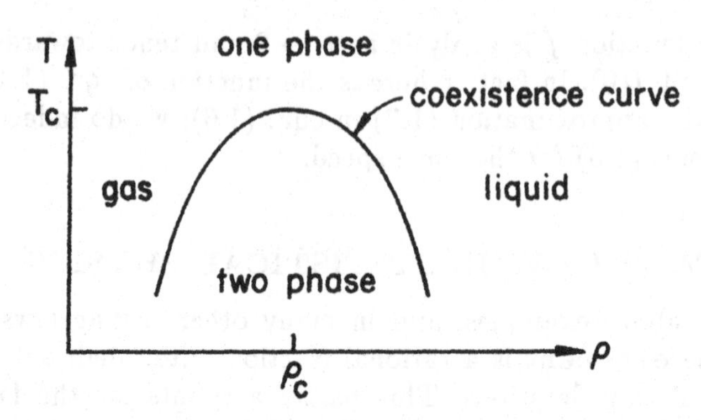

Figure 1.1 Phase diagram of a fluid at fixed pressure.

curve near the critical point? Experimentally, for sulphurhexafluoride, it is found that[2]

$$|\rho_+ - \rho_-| \propto |T - T_c|^{0.327 \pm 0.006} \tag{1.8}$$

is the shape of the curve near the critical point, where $\rho_\pm(T)$ are the values of the density at coexistence on the two branches of the coexistence curve below T_c, as shown in figure (1.1). The number 0.327 ± 0.006 is an example of a **critical exponent**, and does not depend upon the particular fluid system studied. Although you might reasonably have expected that this exponent would be different for the coexistence curve of a different substance, this is not in fact the case! For example, the same measurement on ^{3}He yields a value for the critical exponent[3] of 0.321 ± 0.006. In both of the results quoted, the error bars correspond to two standard deviations.[4] The critical exponent is not obviously a simple rational fraction, and is clearly different from the value 1/2, which, as we will see later, might have been expected from dimensional analysis. In fact, it was the overwhelming experimental evidence that this exponent was different from 1/2 that forced some physicists in the 1930's to realise that there was a deep problem lurking in seemingly unimportant exponents.

---

[2] The data for the liquid gas critical point of sulphurhexafluoride are taken from M. Ley-Koo and M.S. Green, *Phys. Rev. A* **16**, 2483 (1977).

[3] The ^{3}He data are from C. Pittman, T. Doiron and H. Meyer, *Phys. Rev. B* **20**, 3678 (1979).

[4] A useful summary of the experimental situation is given by J.V. Sengers in *Phase Transitions*, Proceedings of the Cargèse Summer School 1980 (Plenum, New York, 1982), p. 95.

1.2.2 Magnetic Critical Point

A second example is the critical point of a ferromagnet. A magnet may be regarded as consisting of a set of magnetic dipoles residing on the vertices of a crystal lattice. We will often refer to the magnetic dipoles as spins. The spins are able to exchange energy through interactions between themselves, as well as between themselves and other degrees of freedom of the crystal lattice (e.g. via spin-orbit coupling). For systems in equilibrium, one can define a temperature T. If one waits a sufficiently long time, equilibrium is established between the lattice and the spins, and both sets of degrees of freedom are described by a single temperature T. On the other hand, the spins can come into equilibrium between themselves well before they come into equilibrium with the lattice: in this case, the spin degrees of freedom and the lattice degrees of freedom may have different temperatures. Here, we ignore such dynamical questions, and assume that we are dealing with a system described by a single temperature T.

At high temperatures and zero external field, the system is in the **paramagnetic** phase: following the time evolution of any spin would reveal that it points in all directions with equal frequency. Thus, no direction is singled out at any given time when considering all of the spins in the system and the net magnetic moment is zero.

Below a **critical temperature**, T_c, however, the spins tend to align along a particular direction in space, even in the absence of an external field. In this case, there is a net **magnetisation**, $M(T)$, and the system is in the **ferromagnetic** phase. The onset of this behaviour is a **continuous phase transition**: the magnetisation rises continuously from zero as the temperature is reduced below T_c, as sketched in figure (1.2). The magnetisation is zero above the transition and is non-zero below the transition temperature. A quantity which varies in this way is referred to as an **order parameter**.

The question naturally arises as to why the system should order along any particular direction: what is special about the direction? This question is far from being naïve, and we shall discuss it in detail later.

In certain systems, the actual dipole interactions between the atoms on the lattice restrict the spins to point parallel or anti-parallel to one particular direction, which we shall take to be the z-axis. In these systems, known as **Ising ferromagnets** , each spin cannot rotate through all possible orientations, but instead can only point along the $+z$ or $-z$ directions. The Ising ferromagnet is therefore relatively simple to study, and we will devote considerable attention to it in these notes. The simplicity is deceptive, however. What may be simple to state may not be simple to solve!

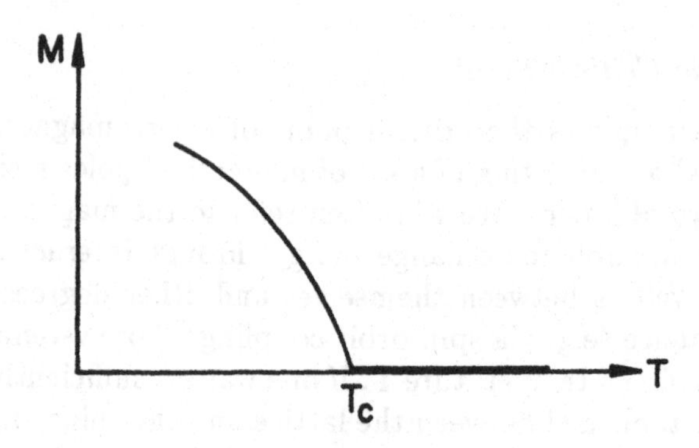

Figure 1.2 Onset of magnetisation in an Ising ferromagnet.

The interaction energy between neighbouring spins in an Ising ferromagnet is lowest when neighbouring spins point in the same direction. However, there is another class of systems, known as **Ising antiferromagnets,** in which the sign of the interaction energy between neighbouring spins is such that the energy is lowered when neighbouring spins point in *opposite* directions. We will later see that the thermodynamics of antiferromagnets with certain crystal lattices in zero applied magnetic field is identical to the thermodynamics of ferromagnets.

The onset of magnetisation in the three dimensional Ising antiferromagnet $DyAlO_3$, in the limit of zero applied magnetic field, exhibits the following behavior experimentally:[5]

$$M \propto (T_c - T)^{0.311 \pm 0.005} \tag{1.9}$$

This result is valid in the limiting case as $T \to T_c$ from below, and is expected to apply to Ising ferromagnets too. As the temperature is reduced below the critical temperature, significant deviations from this result develop. The critical exponent is again not obviously a rational fraction, and furthermore seems to be the same as that for the liquid-gas system, within the experimental precision.

---

[5] The experimental results for the critical point of an antiferromagnet were taken from L.M. Holmes, L.G. Van Uitert and G.W. Hull, *Sol. State Commun.* 9, 1373 (1971). For an exhaustive summary comparing experimental results with predictions based on idealised models of magnetic systems, see L.J. de Jongh and A.R. Miedema, *Adv. Phys.* 23, 1 (1974).

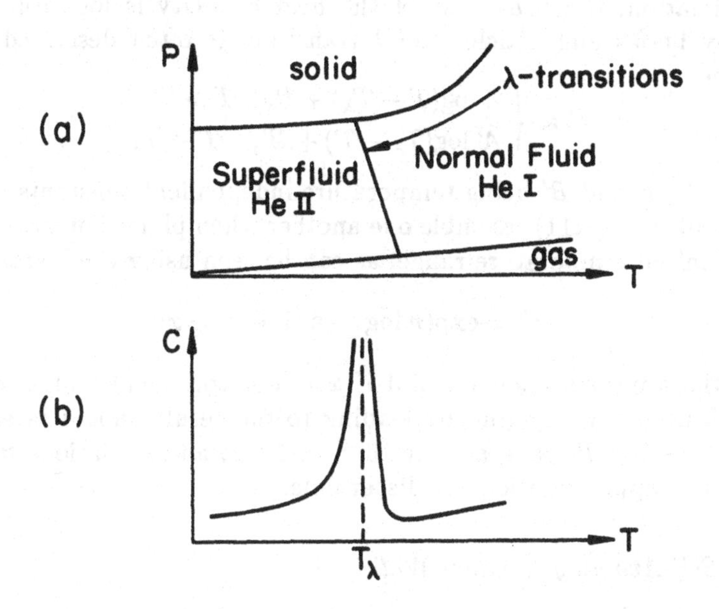

Figure 1.3 (a) Phase diagram of ^{4}He; (b) Heat capacity as a function of temperature at the λ-transition, for fixed pressure.

1.2.3 Superfluid (λ) Transition in ^{4}He

Part of the phase diagram for ^{4}He is sketched in figure (1.3a). For a range of pressures from near zero to about 25 atmospheres, liquid helium undergoes a continuous transition to a superfluid at a temperature of about 2 K. In the superfluid state, ^{4}He exhibits a number of unusual properties, including dissipationless flow through fine capillary tubes. The transition to the superfluid state is sometimes known as the λ-transition, due to the shape of the heat capacity curve, C, as a function of temperature, shown in figure (1.3b). The transition temperature is usually known as T_λ, and its precise value depends upon the pressure.

The best fit to the heat capacity data near the transition[6] is found to be

$$C \propto |T_\lambda - T|^{0.013\pm0.003}. \tag{1.10}$$

This is an experiment where great precision is possible for a variety of technical reasons, and there is little doubt that the critical exponent has the *sign* given. This means that the heat capacity curve is actually a **cusp**, although for many years it was thought that the heat capacity actually exhibited a **divergence** at the λ-transition. Indeed, to a good

---

[6] High resolution experiments on the λ-transition are described in J.A. Lipa and T.C.P. Chiu *Phys. Rev. Lett.* **51**, 2291 (1983).

approximation, the behaviour of the heat capacity is logarithmic, and in many books and articles the λ-transition is often described by the formulae

$$C \approx \begin{cases} A \log(T - T_\lambda) + B, & T > T_\lambda \\ A' \log(T_\lambda - T) + B', & T < T_\lambda \end{cases} \tag{1.11}$$

with A, A', B and B' being temperature independent constants. Expressions (1.10) and (1.11) resemble one another when plotted on graph paper over a limited temperature range, as can be seen using the identity

$$x^n = \exp(n \log x) \approx 1 + n \log x \tag{1.12}$$

where the approximation is valid if x is not too small and $n \ll 1$. In the high accuracy experiments leading to the result quoted in equation (1.10), $|T - T_\lambda|/T_\lambda$ ranges from 10^{-3} to 10^{-8}, and deviations from the logarithmic approximation are discernible.

1.2.4 Self-Avoiding Random Walk

Consider the root mean square distance, R, travelled by a random walker after N steps. By root mean square, we imply that an average has been taken over the probability distribution of the walks. Suppose that we now require that the probability distribution does not permit the walk to intersect itself, but otherwise the walks are random. This is sometimes taken to be a minimal model of a polymer chain in solution, because two molecules making up the polymer cannot occupy the same point in space. Such a walk is called a **self-avoiding walk**. In this case, it is found that in three dimensions the simple scaling law for a random walker is changed from $R \propto \sqrt{N}$, and becomes

$$R \propto N^{0.586 \pm 0.004} \tag{1.13}$$

as $N \to \infty$. The claim is that this formula applies to both a real isolated polymer in solution and a mathematical self-avoiding walk. If this is true, then the molecular structure of the polymer and the various energies of interaction between monomers (repeat units) of the polymer do not seem to influence the scaling behaviour. The exponent quoted in equation (1.13) was obtained from experiment on a dilute polymer solution.[7]

[7] The experimental determination of the scaling of R for polymers is reported in J.P. Cotton, *J. Physique Lett. (Paris)* 41, L231 (1980). The RG calculations for the same quantity were performed by J.C. Le Guillou and J. Zinn-Justin, *J. Physique Lett. (Paris)* 50, 1365 (1989).

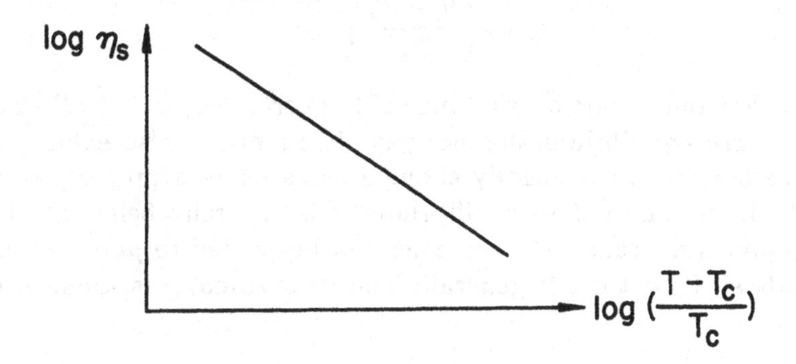

Figure 1.4 Viscosity of a binary fluid near the critical point

Renormalisation group (RG) calculations give a value for the exponent of 0.5880 ± 0.0015, where the uncertainties derive from the mathematical technique used to resum the asymptotic series given by the RG. Indeed, it does seem that the formula is completely independent of the chemistry, and only depends upon the 'spaghetti' nature of the polymer.

1.2.5 Dynamic Critical Phenomena

The examples given above have all exhibited non-trivial power laws in quantities that are unrelated to the time evolution of the physical system in question. For example, magnetisation is a thermodynamic quantity, computed and (in principle) measured in equilibrium. However, non-trivial power laws may also be exhibited by **transport coefficients** in a system near a critical point. A transport coefficient is a phenomenological parameter relating a current to a driving force. For example, Ohm's Law,

$$\mathbf{j} = \sigma \mathbf{E}, \tag{1.14}$$

relates the electric current density $\mathbf{j}$ to the electric field $\mathbf{E}$ which drives the current, through the conductivity tensor σ. Other examples of transport coefficients include diffusion coefficients and viscosity.

Very close to the critical point of a binary fluid mixture, the **shear viscosity**, η_s, is found to diverge weakly with temperature, as sketched in figure (1.4). For example, the shear viscosity of nitroethane-3-methylpentane is found to behave above T_c as[8]

$$\eta_s \propto \left(\frac{T - T_c}{T_c}\right)^{-0.03\pm0.01} \tag{1.15}$$

This behaviour occurs over a range of temperatures $10^{-5} < (T-T_c)/T_c < 10^{-2}$ where equilibrium thermodynamic quantities also exhibit scaling. Although viscosity is usually thought of as being a property of a flowing fluid, *i.e.* one not in equilibrium, it is generally believed that close to equilibrium, transport properties can be related to *purely equilibrium quantities.* This topic is generally known as **linear response theory.**

1.3 SOME IMPORTANT QUESTIONS

These notes are primarily concerned with phase transitions where there is no generation of latent heat. In other words, I shall, for the most part omit discussion of what are called **first order phase transitions** in the **Ehrenfest classification.** Ehrenfest proposed that phase transitions could be classified as 'n^{th} order' if any n^{th} derivative of the free energy with respect to any of its arguments yields a discontinuity at the phase transition. The phenomena which we will describe are often called **critical phenomena** and occur at **second order phase transitions,** although this name is inappropriate, because the Ehrenfest classification is not correct. I prefer the use of the term **continuous phase transition.** Ehrenfest's classification fails, because at the time that it was formulated, it was not known that thermodynamic quantities such as the specific heat actually *diverge* at continuous transitions, rather than exhibiting a simple **discontinuity,** as the Ehrenfest classification implies. We will see in future chapters that this failure is related to the failure of the applicability of **mean field theory,** as exemplified by the **Weiss theory of ferromagnetism** and the **Van der Waals equation** for fluids.

Power law behavior at a critical point, as in examples (1) and (2) of the previous section, is not just restricted to the quantities $|\rho_+ - \rho_-|$ and M. In fact, we shall see that many different observable quantities exhibit scaling behaviour and have associated **critical exponents.** These quantities can be placed in two categories. The first set of quantities are thermodynamic

[8] The divergence of the shear viscosity near the critical point was measured by B.C. Tsai and D. McIntyre, *J. Chem. Phys.* **60**, 937 (1974). Further discussion of this topic may be found in the article by J.V. Sengers mentioned in the footnote on page 4. A review on the topic of dynamic critical phenomena is given by P.C. Hohenberg and B.I. Halperin, *Rev. Mod. Phys.* **49**, 435 (1977).

variables, such as the specific heat. The second set characterise spatial ordering in a system and address the question: how does, for example, the *local* magnetic moment vary from point to point in a ferromagnet? The examples which we have mentioned above prompt us to consider many questions, amongst which the most interesting are perhaps:

Question 0: *Why do phase transitions occur at all?*

In statistical mechanics, thermodynamics arises from the free energy, F, and its derivatives, given by Gibbs' formula

$$e^{-F/k_BT} = \mathrm{Tr}\, e^{-H/k_BT} \tag{1.16}$$

where k_B is Boltzmann's constant and T is the temperature. H is the Hamiltonian, and Tr denotes a sum over all degrees of freedom mentioned in H. Since the Hamiltonian will usually be a non-singular function of the degrees of freedom, the right hand side of equation (1.16) — the **partition function** — is nothing more than a sum of terms, each of which is the exponential of an analytic function of the parameters in the Hamiltonian. How, then, can such a sum give rise to *non-analytic* behaviour, of the sort described in the previous section? Is it even clear *a priori* that a single partition function can describe multiple phases?

Question 1: *How can we calculate the phase diagram of a system as we change the external parameters?*

These external parameters, as well as the temperature, enter the free energy through the Hamiltonian, *via* equation (1.16). Anticipating the answer to the last question posed in the previous paragraph, it is sometimes forgotten that apparently different states of matter, such as liquid and solid, are actually described by the *same* Hamiltonian. Simply changing T in equation (1.16) can cause a change of state.

Question 2: *How do we compute the exponents which are observed at a continuous transition?*

Even when we have understood, in principle, how it is that non-analytic or critical behaviour can arise, the challenge remains to account for the precise values of the critical exponents. Mean field theories, built from the foundations laid by Van der Waals and Weiss, always lead to exponents given by rational fractions. However, the observed values do not seem to agree with these predictions.

The discrepancy between the mean field theory prediction for the shape of the liquid - gas coexistence curve, eqn. (1.8), which gives an exponent of 1/2, and the observed value of about 0.325 is far from insignificant. Usually in physics, we are satisfied with a qualitative understanding of a given phenomenon together with a reasonable estimate of the quantitative consequences, always with the assurance that a more refined calculation would improve the quantitative predictions. Why then is the difference between 0.5 and 0.35 of such apparent significance? The point is that until about twenty-five years ago or so, *it was not possible, even in principle, to account for this discrepancy.* The numerical discrepancy of 30% in a critical exponent is but the tip of a well-hidden iceberg. Classical physics makes an assumption so subtle that it was not even recognised explicitly for many years. It is no exaggeration to say that in solving this problem, a new way of looking at physics emerged, which has infused condensed matter physics and high energy physics. In recognition of this, K.G. Wilson, the principle architect of the renormalisation group approach, was awarded the 1982 Nobel Prize in physics.

We shall shortly see that the critical exponents are often independent of the specific system under consideration. For example, the critical exponent β for the liquid-gas critical point

$$|\rho_+ - \rho_-| \propto |T - T_c|^\beta \tag{1.17}$$

and the exponent β for the ferromagnetic critical point

$$M \propto (T_c - T)^\beta \tag{1.18}$$

are the same within the accuracy of the experiments! The fact that two apparently different physical systems might share precisely the same sets of critical exponents is known as **universality**. Thus, we ask

Question 3: *Why does universality occur, and what are the factors that determine which set of phenomena have the same critical exponents?*

Phenomena with the same set of critical exponents are said to form a **universality class**. The usefulness of the concept of universality class lies in the fact that, in general, members of a universality class have only three things in common: the symmetry group of the Hamiltonian (*not* the lattice, if one is present), the dimensionality, and whether or not the forces are short-ranged.

1.4 HISTORICAL DEVELOPMENT

Complete answers to most of the questions above were not known 20 years ago. In these lecture notes, we shall follow approximately the historical development of the subject, although we will start with a precise statistical mechanical definition of phase transition, emphasising the important notion of the **thermodynamic limit**, in which the system size is taken to be infinitely large, and indicating how, strictly speaking, phase transitions arise only in this limit.

Of central importance to the development of the subject were studies of two physical systems, the liquid-gas system and the ferromagnet, at the *classical* level of description. The term 'classical' is not used to indicate an alternative to quantum theory; instead, it signifies that the theory in question ignores thermal fluctuations, an approximation used in all but a few exactly solvable cases, until the advent of the renormalisation group. In fact, we shall see that these classical theories, due to Van der Waals and Weiss respectively, are actually rather good in many ways, and already exhibit some (but not all) qualitatively correct features of phase transitions. In our presentation, we will also expose some of the similarities between these two apparently different phenomena, showing how the mathematical descriptions of these phenomena become identical near a critical point.

This observation forms the basis of **Landau's theory of phase transitions**, which is the most succinct encapsulation of the classical approach. However, we will present Landau theory in a way which anticipates the developments that follow. In fact, we shall see that classical theories, and thus Landau's theory, are **mean field theories**; a physical variable such as the magnetisation is replaced by its average value, and fluctuations about that value are ignored. It is possible to use Landau theory to estimate the importance of fluctuations, and thus to check on the self-consistency of the theory. However, it is found that near a critical point, fluctuations are not negligible, and thus the theory is not self-consistent. Landau theory contains within it the seeds of its own destruction!

The next significant step was the development of the notion of **scaling laws**, first arrived at on a phenomenological basis by B. Widom. We will see that the equation of state of a physical system near to a critical point obeys what appears to be an analogue of the **law of corresponding states**, encountered in the equation of state (such as that due to Van der Waals) for a fluid. In the latter case, however, the law of corresponding states is always valid, not just near to a critical point. In the context of magnetic systems, the equation of state relates the magnetisation, M, the temperature T, and the external magnetic field, H:

$$H = f(M,t), \qquad t = \left|\frac{T - T_c}{T_c}\right| \tag{1.19}$$

Ostensibly, f is a function of two variables. Nevertheless, Widom discovered that near a critical point, the equation of state may be written as

$$H = M^{\delta}\Phi\left(\frac{t}{M^{1/\beta}}\right) \tag{1.20}$$

with Φ being a function of just *one* variable. The exponents δ and β appear in equation (1.20) in accord with convention, and Widom's striking discovery was that δ, β and T_c can be chosen so that the experimental data from *different* materials (Fe, Ni, ...) all satisfy equation (1.20) *with the same function* Φ! Furthermore, it was noticed that equation (1.20) implies a relationship between the critical exponents for different thermodynamic quantities, such as the specific heat, the susceptibility and the magnetisation.

Where do scaling laws come from? L.P. Kadanoff proposed a simple, intuitive explanation, namely that a relation of the form of (1.20) follows if one assumes that near a critical point, the system 'looks the same on all length scales'. We shall formulate this notion a bit more precisely later and see why it is, in fact, not really quite correct. Kadanoff's argument is important, because it provides the basic physical insight on which the technique of the **renormalisation group** (RG) is built; Kadanoff almost certainly was aware of the flaws in his argument, but nevertheless had the intellectual courage to propose it anyway.

The modern era began with a series of seminal papers by K.G. Wilson in 1971, in which the renormalisation group was developed and explained in the contexts of both condensed matter physics and high energy physics. These and subsequent papers by Wilson initiated an explosion of activity which continued unabated for a decade. Many of the applications of the renormalisation group utilised perturbation expansion techniques, with such small parameters as the variables $\epsilon \equiv 4 - d$, where d is the dimensionality of space and $1/n$, where n is the number of components of the order parameter (*i.e.*, the magnetisation, which is a vector, has 3 components). Today, some of this body of work has become part of the mainstream of physics.

At the time of writing, it is probably fair to say that the frontiers of renormalisation group physics have shifted away from phase transitions and field theory, towards **non-equilibrium phenomena**. One active avenue of research is the study of dynamical phenomena where there is a fluctuating noise source present. An example is the growth of an interface by the random deposition of atoms. Such an interface is rough, and will exhibit height fluctuations about its average position. The RG and other methods, originally developed to treat critical phenomena, have been used to study how these fluctuations scale in both space and time.

Another line of inquiry is the dynamics of systems approaching an equilibrium state, but still not close enough to equilibrium that linear response theory is valid. In these systems, spatial correlations sometimes exhibit scaling behaviour in time. One well-known example is that of the **phase separation** of a binary alloy below the critical point, where the time dependent X-ray scattering intensity at wavenumber k, $S(k,t)$, is found to obey a relation of the form

$$S(k,t) = t^{d\phi} F(kt^{\phi}) \tag{1.21}$$

with ϕ close, if not exactly equal, to 1/3, and d being the spatial dimensionality. It is not yet well-understood how to apply the RG to pattern formation problems of this type.

In another set of problems described by partial differential equations without noise present, a system may be well-described by a **similarity solution** of the form

$$u(x,t) \sim t^{\alpha} f(xt^{\beta}), \qquad \text{as } t \to \infty, \tag{1.22}$$

where u is some observable and x and t are space and time respectively. In some cases, the exponents α and β characterising the similarity solution may not be simply obtained by dimensional analysis; nevertheless, the RG can be successfully used to solve these problems too. Examples of such problems arise in many areas of fluid mechanics, for example, and this new topic will be introduced in chapter 10.

The condensed matter physics literature contains two versions of the RG: the Gell-Mann–Low RG and the Wilson RG. In critical phenomena, the use of the former is based upon perturbation theory, whilst the latter has a direct *geometrical* interpretation and is in principle, non-perturbative. Indeed, it was the introduction of Wilson's method in 1971 which began the modern era of the RG. The connection between the two versions is by no means obvious. We shall present the Gell-Mann–Low RG when we explain renormalisation in chapter 10, in the context of phenomena far from equilibrium; there we will also mention the connection with Wilson's formulation of RG. We will see that renormalisation has a direct physical interpretation, and may be easily understood without the technical complications of quantum field theory.

EXERCISES

Exercise 1–1

Dimensional analysis (DA) is often a powerful tool in physics. This question requires you to use dimensional arguments to solve a couple of interesting problems. DA is usually used in two ways: (1) The fundamental theorem of DA asserts[9] that in any physical problem involving a number of dimensionful quantities, the relationship between them can be expressed by forming all possible independent dimensionless quantities, denoted by $\Pi, \Pi_1, \Pi_2, \ldots, \Pi_n$. Then the solution to the physical problem is of the form $\Pi = f(\Pi_1 \ldots \Pi_n)$, where f is a function of n variables. (2) Sometimes there is only one dimensionless combination of variables relevant to a given problem. Then (1) implies $\Pi =$ constant.

(a) By noting that the area of a right-angled triangle can be expressed in terms of the hypotenuse and (*e.g.*) the smaller of the acute angles, prove Pythagoras' theorem using dimensional analysis. You will find it helpful to construct a well-chosen line in the right-angled triangle. *Note: the whole point of dimensional analysis is that you do NOT need to solve for the functional form of the solution to a given problem. Thus, in this question, you must pretend that you do not know trigonometry.*

(b) Now consider the case of Riemannian or Lobachevskian geometry (*i.e.* the triangle is drawn on a curved surface such as a riding saddle or a football). What happens in this case?

Exercise 1–2

In 1947, a sequence of photographs of the first atomic bomb explosion in New Mexico in 1945 were published in *Life* magazine. The photographs show the expansion of the shock wave caused by the blast at successive times in ms. From the photographs, one can read off the radius of the shock wave as a function of time: the result is shown in the accompanying table. Assuming that the motion of the shock is unaffected by the presence of the ground, and that the motion is determined only by the energy released in the blast E and the density of the undisturbed air into which the shock is propagating, ρ, derive a scaling law for the radius of the fireball as a function of time. Use the data from the photographs to test your scaling law and hence deduce the yield of the blast. *You must test*

[9] First enunciated apparently by E. Buckingham in a delightful paper (see especially the concluding paragraph), *Phys. Rev.* 4, 345 (1914). Fourier is usually attributed with the principle that every term in a physical equation must have the same dimensions.

Table 1.1 RADIUS R OF BLAST WAVE AFTER TIME T

T/msec	R/m
0.10	11.1
0.24	19.9
0.38	25.4
0.52	28.8
0.66	31.9
0.80	34.2
0.94	36.3
1.08	38.9
1.22	41.0
1.36	42.8
1.50	44.4
1.65	46.0
1.79	46.9
1.93	48.7
3.26	59.0
3.53	61.1
3.80	62.9
4.07	64.3
4.34	65.6
4.61	67.3
15.0	106.5
25.0	130.0
34.0	145.0
53.0	175.0
62.0	185.0

your scaling law by plotting a graph. You should consider carefully and then explain what is the most useful graph to plot. You should assume that all numerical factors are of order unity.[10] Although the photographs were declassified in 1947, the yield of the explosion was to remain classified until several years later.

[10] For a detailed analysis and discussion of the data, see G.I. Taylor, *Proc. Roy. Soc. A* **201**, 175 (1950).

CHAPTER 2

How Phase Transitions Occur in Principle

The purpose of this chapter is to explain how phase transitions can occur *in principle.* Thus, we shall define precisely what we mean by a phase transition, and examine how it is possible for a phase transition to occur in a simple model of a magnet – the Ising model. The Ising model is the *drosophila* of statistical mechanics, and we will not shrink from using it as an example to illustrate many of the general ideas that we shall develop. We shall also explore some of the wider and more complex issues associated with phase transitions, using other examples. In particular, we shall focus on ergodicity breaking, using disordered systems as an example. Throughout this chapter, we shall aim to be descriptive: our goal is to examine the possibilities that may occur, rather than to show how they occur in practice. The following chapter will present some calculational techniques whose results illustrate the general considerations given here.

2.1 REVIEW OF STATISTICAL MECHANICS

Let us now review the basic results of statistical mechanics, but in a rather formal way. The formality is actually only cosmetic, but gives us a convenient general notation to cover all sorts of different systems. This will turn out to be essential, because we shall later wish to consider the (formal) space of all possible Hamiltonians. In fact, when we use the RG, we will be lead to consider **dynamical systems** in the space of all possible Hamiltonians.

The goal of statistical mechanics is to compute the partition function Z. We specify the system of interest as some sample region Ω, in which is defined the Hamiltonian H_Ω. The volume of the region is $V(\Omega)$ and the surface area is $S(\Omega)$. Often it will be convenient to think of our system as having some characteristic linear dimension L, so that $V(\Omega) \propto L^d$, $S(\Omega) \propto L^{d-1}$, where d is the dimensionality of the system. Unless otherwise stated, we will only consider classical statistical mechanics here; for reasons that will become clear, quantum statistical mechanics is not required to describe the vicinity of a phase transition, *even if the underlying phenomenon is quantum mechanical* (as in superconductivity, for example).

Usually, there will be **boundary conditions** specified on the boundary of Ω. Often these will be periodic or hard wall (*i.e.* zero flux of particles through a wall). The system may exist as a continuum (*e.g.* a fluid) or on a lattice (*e.g.* a magnet); for now, $V(\Omega)$ is *finite*.

We write the Hamiltonian for the system as

$$-\frac{H_\Omega}{k_B T} = \sum_n K_n \Theta_n \tag{2.1}$$

where K_n are the **coupling constants** and the Θ_n are combinations of the dynamical degrees of freedom, which are summed over in the partition function. We shall sometimes refer to the Θ_n as **local operators**. The so-called coupling constants K_n are the external parameters, such as fields, exchange interaction parameters, temperature ... So, for example, if we are dealing with a magnet, the degrees of freedom are the (vector) spins on the lattice sites $\mathbf{S}_i$, where $1 \leq i \leq N(\Omega)$. Thus the Θ_n are built out of combinations of the $\mathbf{S}_i$: *i.e.* $\Theta_1 = \sum_i \mathbf{S}_i$, $\Theta_2 = \sum_{ij} \mathbf{S}_i \cdot \mathbf{S}_j$, etc. In this context,

$$\mathrm{Tr} \equiv \sum_{\mathbf{S}_1} \sum_{\mathbf{S}_2} \sum_{\mathbf{S}_3} \cdots \sum_{\mathbf{S}_{N(\Omega)}}, \tag{2.2}$$

where each sum is over all possible values that each $\mathbf{S}_i$ can take.

A simple example of a term in H_Ω is that responsible for the **Zeeman effect**. The coupling constant in this context is the external field $\mathbf{H}$, and the corresponding local operator is the magnetic moment at a lattice site i, $\mathbf{S}_i$. Then the contribution of these terms to H_Ω is $-\sum_i \mathbf{H} \cdot \mathbf{S}_i$.

The partition function itself is given by

$$Z[\{K_n\}] \equiv \mathrm{Tr}\ e^{-\beta H_\Omega}, \qquad \text{where } \beta \equiv 1/k_B T \tag{2.3}$$

and the operation Tr means "sum over all degrees of freedom, the sum including every possible value of each degree of freedom." After carrying out the trace, Z depends on all of the K_n as indicated by the notation $[\ldots]$. We shall sometimes indicate the dependence of a quantity on the *entire set* of coupling constants in a compact way: $[K] \equiv [\{K_n\}]$. Thus, for example, the partition function might be written as $Z[K]$. We shall also have be very careful about what Tr means when we discuss **spontaneous symmetry breaking.**

The **free energy** is defined by

$$F_\Omega[\{K_n\}] = -k_B T \log Z_\Omega \tag{2.4}$$

and information on the **thermodynamics** of the system Ω is contained in the derivatives $\partial F_\Omega/\partial K_n, \partial^2 F_\Omega/\partial K_n \partial K_m, \ldots$ which include bulk effects, surface effects, and finite – size effects. However, at this stage, where Ω is finite, there is *no* information about phase transitions or phases.

2.2 THE THERMODYNAMIC LIMIT

Experience tells us that the free energy is extensive for a large system: $F_\Omega \propto V(\Omega)$. Thus, we expect that for a finite system, we can write

$$F_\Omega = V(\Omega) f_b + S(\Omega) f_s + O(L^{d-2}), \tag{2.5}$$

where f_b is the **bulk free energy per unit volume** or **bulk free energy density** and f_s is the **surface free energy per unit area.** We can give a precise definition of these important quantitites as follows:

$$f_b[K] \equiv \lim_{V(\Omega)\to\infty} \frac{F_\Omega[K]}{V(\Omega)} \tag{2.6}$$

when the limit exists and is independent of Ω. For a system defined on a lattice, with $N(\Omega)$ lattice sites, the bulk **free energy per site** is

$$f_b[K] \equiv \lim_{N(\Omega)\to\infty} \frac{F_\Omega[K]}{N(\Omega)} \tag{2.7}$$

when the limit exists and is independent of Ω. The bulk free energy $f_b[K]$ describes extensive thermodynamic behaviour (proportional to $V(\Omega)$) but does not describe surface or finite size behaviour. This information may be computed from the surface free energy

$$f_s[K] \equiv \lim_{S(\Omega)\to\infty} \left\{ \frac{F_\Omega[K] - V(\Omega) f_b[K]}{S(\Omega)} \right\} \tag{2.8}$$

when the limit exists and is independent of Ω.

The limit in eqns. (2.6), (2.7) and (2.8) is known as the **thermodynamic limit.** Sometimes, an auxiliary constraint is imposed simultaneously with the limit: for example, in fluid systems, taking the limit $V(\Omega) \rightarrow \infty$ is senseless unless one simultaneously takes the limit that the number of particles $N(\Omega)$ in the system also tends to infinity, in such a way that the density $N(\Omega)/V(\Omega)$ stays constant. We will see that the concepts of phase and and phase transition are only sharply defined in this limit.

The existence of the thermodynamic limit is not trivial. In order for a uniform bulk behaviour to exist, the forces in the system must satisfy certain properties, and the thermodynamic limit must be taken carefully. It is instructive to consider how the thermodynamic limit can fail to exist, firstly because the physics will eventually give us some insight about phase transitions, and secondly, because we will see a very useful trick, which we will employ many times in these notes.

2.2.1 Thermodynamic Limit in a Charged System

Let us consider the energy at $T = 0$ of a system of uniform charge density ρ in 3 dimensions with an interaction potential between two particles separated by a distance r given by **Coulomb's Law:**

$$U(r) = A/r, \tag{2.9}$$

with A being a constant. For a spherical system of radius R, the energy E is given by

$$E(R) = \int_0^R \left(\frac{4}{3}\pi r^3 \rho\right) \cdot \frac{A}{r} \cdot 4\pi r^2 \rho \, dr. \tag{2.10}$$

In this expression, we have used the fact that in 3 dimensions for **inverse square law** forces, the charge can be considered to reside at the center of the sphere (Gauss' Law) and we have used the fact that charge outside the shell of thickness dr at radius r does not contribute to the electrostatic energy of the shell. Doing the integral we find

$$E(R) = A\frac{(4\pi)^2}{15}\rho^2 R^5 \tag{2.11}$$

and so the energy per unit volume is

$$E_b \equiv \frac{E(R)}{V(R)} = \frac{4\pi A}{5}\rho^2 R^2 \tag{2.12}$$

which diverges as $R \to \infty$. We conclude that inverse square law forces, such as gravity and electrostatics are too *long-ranged* to permit thermodynamic behaviour. This is a consequence of the fact that we have only allowed for charge of one sign. If both positive and negative charges are present, then the force law is no longer long ranged. In fact, if the charges are mobile, then the phenomenon of **screening** occurs, and the interaction potential changes from $1/r$ to $\exp\{-r/\ell_D\}/r$ where ℓ_D is the **Debye screening length**, which depends on the density of positive and negative charge in the system.

In fact, it is not even obvious that a system of an equal number N of oppositely charged point particles is stable against collapse; if this occured, the question of the existence of the thermodynamic limit would be moot. It turns out that to prevent collapse and obtain a sensible thermodynamic limit requires **quantum mechanics**, and furthermore, at least one of the species of particles must be a **fermion**. Without the **exclusion principle**, the ground state energy of the system diverges as $N^{7/5}$ and the thermodynamic limit is not well-defined![1]

2.2.2 *Thermodynamic Limit for Power Law Interactions*

It is instructive to repeat the above calculation, for an interaction potential $U(r) = A/r^{\sigma}$ in d-dimensions. For this more general case, Gauss' Law does not apply, and we make progress by using a simple trick which we will encounter often in the study of critical phenomena. Expression (2.10) becomes

$$E(R) = \frac{1}{2}\int_{\Omega} d^d r d^d r' \rho(\mathbf{r}) U(\mathbf{r}-\mathbf{r}') \rho(\mathbf{r}') \tag{2.13}$$

where $\rho(\mathbf{r})$ is the charge density at $\mathbf{r}$ and Ω is a d-dimensional sphere of radius R. For a uniform system, $\rho(\mathbf{r}) = \rho$ and the integral becomes

$$E(R) = A\frac{\rho^2}{2}\int_{\Omega} d^d r d^d r' \frac{1}{|\mathbf{r}-\mathbf{r}'|^{\sigma}}. \tag{2.14}$$

Make the change of variables $\mathbf{r} = R\mathbf{x}$; $\mathbf{r}' = R\mathbf{y}$ and Ω becomes the unit sphere. Then we can simply extract the dependence on R:

$$\begin{aligned} E(R) &= \frac{1}{2}A\rho^2 \int_{u.s.} R^d d^d x R^d d^d y \frac{1}{|R\mathbf{x} - R\mathbf{y}|^{\sigma}} \\ &= \frac{1}{2}A\rho^2 R^{2d-\sigma} C \end{aligned} \tag{2.15}$$

---

[1] A clear and accessible discussion of the stability of matter may be found in E. Lieb, *Rev. Mod. Phys.* **48**, 553 (1976).

where *u.s.* denotes integration over the unit sphere, and C is a constant independent of R:

$$C = \int_{u.s.} d^d x d^d y \frac{1}{|\mathbf{x} - \mathbf{y}|^\sigma}. \tag{2.16}$$

Thus

$$E_b \equiv \frac{E(R)}{V(R)} = \frac{A\rho^2 C R^{2d-\sigma}}{2V_d R^d} \sim R^{d-\sigma}, \tag{2.17}$$

where V_d is the volume of a unit sphere in d dimensions; we shall also denote the surface area of the unit sphere by S_d. In the limit $R \to \infty$, we see that the thermodynamic limit is only well defined if and only if $\sigma > d$. We shall say more below about the case $\sigma = d$.

In fact, we should be a little more careful, because the integral over the unit sphere represented by C may not converge. The problem arises in principle when $\mathbf{r} \to \mathbf{r}'$ for $\sigma > d$, since the singularity of the integrand of eqn. (2.14) is not integrable. This does not represent a serious problem in practice, because the charges may be considered either to reside on a lattice of spacing a, or if they are mobile in a continuum, to have a hard-core repulsion at a radius a. In either case, the interaction $U(r) \sim r^{-\sigma}$ only applies for $r > a$. The integral for C in eqn. (2.16) can be examined by making the change of variables

$$\mathbf{u} = \mathbf{x} - \mathbf{y} \tag{2.18}$$

$$\mathbf{v} = \frac{1}{2}(\mathbf{x} + \mathbf{y}) \tag{2.19}$$

for which the Jacobian is unity. The integral over $\mathbf{v}$ simply gives V_d, because the integrand is independent of $\mathbf{v}$. Thus we get

$$C = V_d \int_{a/R}^{1} S_d u^{d-1} \frac{du}{u^\sigma} = \frac{V_d S_d}{d - \sigma}\left(1 - \left[\frac{a}{R}\right]^{d-\sigma}\right), \quad \text{for } d \neq \sigma \tag{2.20}$$

and

$$E_b = \frac{A\rho^2 S_d R^{d-\sigma}}{2(d-\sigma)}\left(1 - (a/R)^{d-\sigma}\right), \quad \text{for } d \neq \sigma. \tag{2.21}$$

With this final, improved calculation of the existence of the thermodynamic limit, we can address the question of what happens when $\sigma = d$. In equation (2.20), the inner integral yields $\log(R/a)$, so that as $R \to \infty$ for fixed a, the bulk energy per unit volume E_b diverges. Thus, as advertised, in this example, the thermodynamic limit exists only for $\sigma > d$.

2.3 PHASE BOUNDARIES AND PHASE TRANSITIONS

When $f_b[K]$ exists, then a precise definition of a **phase boundary** follows. Let us suppose that there are D coupling constants. The axes of the **phase diagram** are $K_1, K_2, \ldots, K_D$, and hence the dimension of the phase diagram is D. As a function of $[K]$, $f_b[K]$ is analytic almost everywhere. The possible non-analyticities of $f_b[K]$ are points, lines, planes, hyperplanes, *etc.* in the phase diagram. These **singular loci** have a dimensionality associated with them ($D_s = 0, 1, 2, \ldots$ respectively), and the important **invariant** for each type of singular locus is the **codimension**, C:

$$C \equiv D - D_s. \tag{2.22}$$

This is an invariant in the sense that if we decide to include an extra variable in $\{K_n\}$ (and thus to the phase diagram), both D and D_s increase by 1 so that C remains fixed.

Regions of analyticity of $f_b[K]$ are called **phases.** Loci of codimension $C = 1$ (*i.e.* loci which separate phases), are called **phase boundaries.** Loci of codimension greater than one cannot possibly represent phase boundaries. To give a prosaic example, in order to partition a room into two, it is necessary to build a wall across it — a pole through the centre of the room will not suffice.

2.3.1 Ambiguity in the Definition of Phase Boundary

The definition of **phase** given above is ambiguous. There may exist a path along which $f_b[K]$ is analytic going from one side of a phase boundary to the other. Using our "room analogy", this is like having a wall which does not quite reach the ceiling. At floor level, the room is partitioned, but a flying insect may be able to pass from one side of the room to the other without encountering any impediment to its progress.

An example is the liquid-gas-solid phase diagram, shown in figure (2.1). Although it is not possible to pass from fluid to solid without encountering a phase transition, it is possible to choose a path in p–T space which goes from liquid to gas without encountering any singular behaviour in thermodynamic quantities. We shall see later that this is a reflection of the fact that the liquid and gas states have the same degree of symmetry, whereas a fluid has a higher degree of symmetry than a solid (at least if the solid is crystalline, which is the usual case in equilibrium).[2]

---

[2] Even if the solid were in equilibrium and amorphous, it would still represent a state of broken translational invariance, whereas in a fluid, this symmetry is unbroken. See section 2.10.

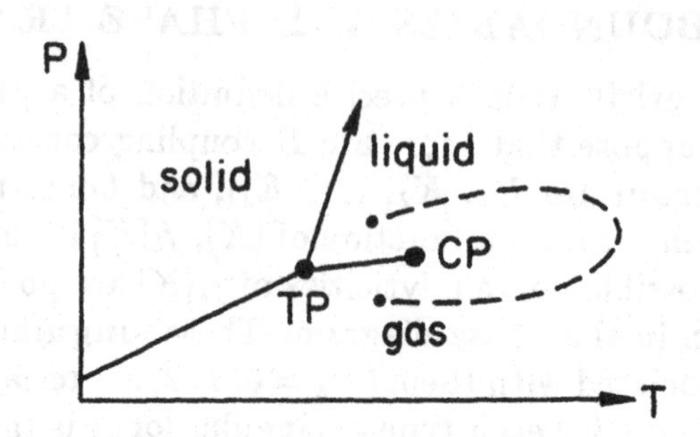

Figure 2.1 Phase diagram of a typical substance: p is pressure, T is temperature. The point TP is a triple point ($C = 2$) and the point CP is a critical point ($C = 2$). The solid–liquid phase boundary extends to arbitrarily high pressure (in principle), whilst the gas–liquid phase boundary ends at CP. The dashed curve represents a trajectory in the phase diagram along which no phase transition is encountered, even though a change of phase has apparently taken place.

2.3.2 *Types of Phase Transition*

We shall show shortly that $f_b[K]$ is everywhere *continuous*: our demonstration will be specific to a particular model system, but the conclusion is true in general. This implies that where phase boundaries exist, they must come in two classes:

(1) $\partial f_b/\partial K_i$ is discontinuous across a phase boundary. It can be either one or more of the $\partial f_b/\partial K_i$ which is discontinuous. If this case occurs, then the transition is said to be a **first-order phase transition.**

(2) The only other remaining possibility for non-analytic behaviour is that all $\partial f_b/\partial K_i$ are continuous across the phase boundary. If this occurs, the transition is said to be a **continous phase transition.** Sometimes, this is referred to as a second order transition too, but for reasons already explained, this terminology is not encouraged.

2.3.3 *Finite–Size Effects and the Correlation Length*

In practice, the thermodynamic limit is never attained: $10^{23} \neq \infty$. Thus it is reasonable to ask whether or not the thermodynamic limit is physically relevant. If there were perfect instrumental resolution, a change in the physical properties in a finite system would not occur over an infinitesimal interval of the relevant coupling constant, but would occur over some range. This phenomenon is an example of a **finite size effect**, and will be discussed in detail later. In practice, however, instrumental resolution is usually the limiting factor.

We can give a simple criterion for when predictions using $f_b[K]$ are not reliable, by introducing the concept of the **correlation length**, usually denoted by ξ. Loosely speaking, the correlation length describes the spatial extent of fluctuations in a physical quantity about the average of that quantity. For example, in a gas, there will be density fluctuations in thermal equilibrium. In a particular region of the sample, the density may be higher than the average density. We can choose to think of such a region as a droplet of near-liquid density floating in the gas. In thermal equilibrium, there is a distribution of droplet sizes, of course, but it turns out that there is a well-defined average size, at least away from the critical point itself. This characteristic size is, roughly speaking, what we mean by the term correlation length. We will see later how to give a more precise definition of this important quantity. It is not unreasonable that the droplet distribution should depend upon the position in the phase diagram: how close the system is to a phase boundary or critical point for example. Thus, we might expect that ξ depends upon the coupling constants, in particular temperature. The correlation length depends strongly on *temperature* near a continuous phase transition, diverging to infinity at the transition itself. This is what gives rise to scaling behaviour, as we will see.

Now we are ready to address the question of when the finite size of a real system is important. In a finite system, the correlation length is not able to diverge to infinity, since it cannot exceed L, the characteristic linear dimension of the system. Thus, for temperatures sufficiently close to T_c that the correlation length of an infinitely large system would exceed L, the behaviour of the finite system departs from the ideal behaviour described by f_b. To make a rough estimate of when this occurs, let us assume that

$$\xi \approx \xi_0 t^{-2/3} \qquad \text{where } t \equiv \frac{T - T_c}{T_c} \tag{2.23}$$

with $\xi_0 \approx 10\text{Å}$ being the correlation length well away from the critical point. This form for ξ is not unrealistic for magnetic or fluid systems: the value for ξ_0 is probably an overestimate in many cases, so our calculation will be quite conservative. For a system with $L = 1$ cm, we find that $\xi = L$ when the **reduced temperature** $t \approx 10^{-11}$. In this situation, finite size effects would be hard to observe.

On the other hand, in computational simulations of critical phenomena, the system size L is usually only a modest multiple of ξ_0, and finite size effects must be taken into account carefully in order to obtain useful results. We will discuss this topic further once we have introduced the RG.

2.4 THE ROLE OF MODELS

The casual reader of any textbook or research paper on phase transitions and statistical mechanics cannot help being struck by the frequency of the term "model." The phase transition literature is replete with models: the **Ising model**, the **Heisenberg model**, the **Potts model**, the **Baxter model**, the **F model**, and even such unlikely sounding names as the **non-linear sigma model**! These "models" are often systems for which it is possible (perhaps only in some limit or special dimension) to compute the partition function exactly, or at least to reduce it to quadrature (*i.e.* one or a finite number of integrals, rather than an infinite number of integrals). Beautiful mathematics has emerged from this enterprise, revealing intriguing connections with solitons, Ramanujan identities, string theory and topology.[3]

We will not, for the most part, deal with these topics here. The techniques are very specialized and only work in certain special cases. For example, although the Ising model in zero external field was solved in a *tour de force* by Lars Onsager in 1944, it still has not been solved in an external magnetic field. Much effort has been expended to solve it in 3 dimensions in zero magnetic field, although a solution of the $d = 2$, $H \neq 0$ case might be more significant. The techniques do not generalize well, and what we really require is an approach which will always work, albeit in some approximation. Nevertheless, the exactly solvable models have been of enormous importance in statistical mechanics, providing a benchmark on which to test approximation methods.

Before we proceed to a discussion of the Ising model, it is appropriate to make a few remarks about the role of models in statistical physics. There are two diametrically opposing views about the way models are used. The "traditional" viewpoint has been to construct a faithful representation of the physical system, including as many of the fine details as possible. In this methodology, when theory is unable to explain the results of an experiment, the response is to fine-tune the parameters of the model, or to add new parameters if necessary. An example of a branch of

[3] R. J. Baxter *Exactly Solved Models in Statistical Mechanics* (Academic, New York, 1989); *Knot Theory and Statistical Mechanics*, V.F.R. Jones, *Scientific American*, Nov. 1990, pp. 98-103

science where this is considered appropriate is quantum chemistry. On the other hand, such fine detail may actually not be needed to describe the particular phenomenon in which one is interested. Many of the parameters may be irrelevant, and even more importantly, the directly measurable quantities may well form dimensionless numbers, or even universal functions, which to a good approximation do not depend on microscopic details. An example of a more "modern" viewpoint is the BCS theory of superconductivity, which predicts a variety of dimensionless ratios (*i.e.* the famous relation between the zero temperature gap and the transition temperature: $2\Delta/k_BT_c \approx 3.5$) and functional forms (*i.e.* specific heat as a function of temperature) for all weak-coupling superconductors. In such a case, it is only important to start with the correct **minimal model**, *i.e.* that model which most economically caricatures the essential physics. The BCS Hamiltonian is the simplest 4 fermion interaction with pairing between time-reversed states that one can write down. In this viewpoint, all of the microscopic physics is subsumed into as *few* parameters, or phenomenological constants, as possible. As we shall see, the existence of such a viewpoint is a consequence of RG arguments. The RG also provides a calculational framework for explicitly identifying and calculating *universal* or model independent observables.[4]

2.5 THE ISING MODEL

This is a model of a ferromagnet or antiferromagnet on a lattice. It was first studied in 1925 by Lenz and Ising, who showed that in dimension $d = 1$ the model does not have a phase transition for $T > 0$. They concluded (incorrectly) that the model does not exhibit a phase transition at a non-zero temperature for $d > 1$, and so could not describe real magnetic systems.

We consider a lattice in d dimensions of sites $\{i\}$ labelled $1 \dots N(\Omega)$, which we will take to be hypercubic, unless otherwise stated. The degrees of freedom are classical spin variables S_i, residing on the vertices of the lattice, which take only two values: up or down, or more usefully,

$$S_i = \pm 1. \tag{2.24}$$

The total number of states of the system is $2^{N(\Omega)}$. The spins interact with an external magnetic field (in principle varying from site-to-site) H_i

[4] A good discussion of this philosophy is to be found in the article by Y. Oono, *Adv. Chem. Phys.* 61, 301 (1985).

and with each other through **exchange interactions** J_{ij}, K_{ijk}, ... which couple two spins, three spins, ...

A general form of the Hamiltonian is

$$-\mathcal{H}_\Omega = \sum_{i\in\Omega} H_i S_i + \sum_{ij} J_{ij} S_i S_j + \sum_{ijk} K_{ijk} S_i S_j S_k + \dots \tag{2.25}$$

although in the following we will neglect three and higher spin interactions represented by K_{ijk} and ... in eqn. (2.25).

For this system, the operation Tr means

$$\sum_{S_1=\pm 1} \sum_{S_2=\pm 1} \cdots \sum_{S_{N(\Omega)}=\pm 1} \equiv \sum_{\{S_i=\pm 1\}} . \tag{2.26}$$

The free energy is given by

$$F_\Omega(T, \{H_i\}, \{J_{ij}\} \dots) = -k_B T \log \text{Tr } e^{-\beta \mathcal{H}_\Omega} \tag{2.27}$$

and thermodynamic properties can be obtained by differentiation, as mentioned after eqn. (2.4). For example, the average value of the magnetisation at the site i is obtained by differentiating with respect to the external magnetic field H_i:

$$\begin{aligned} \frac{\partial F_\Omega}{\partial H_i} &= -k_B T \cdot \frac{1}{\text{Tr } e^{-\beta \mathcal{H}_\Omega}} \cdot \text{Tr } \frac{S_i}{k_B T} e^{-\beta \mathcal{H}_\Omega} \\ &= -\langle S_i \rangle_\Omega . \end{aligned} \tag{2.28}$$

Even if we are interested in the situation where there is no magnetic field, or one that is constant over lattice sites, it is still a useful trick to allow the term $\sum_i H_i S_i$ in $\mathcal{H}_\Omega$ so that formal identities like eqn. (2.28) can be established. At the end of a given calculation, H_i can be set to any given desired value, including zero. We shall use this device, sometimes known as the **method of sources**, often in these notes.

What about the thermodynamic limit for the Ising model? For the finite system, the free energy is an analytic function of each of its arguments, at least in some strip including the real axis, because for each of the $2^{N(\Omega)}$ states of the system, the energy of the nth state, E_n, is simply a linear combination of the coupling constants. Writing

$$Z_\Omega = \sum_{n=1}^{2^{N(\Omega)}} \exp\left(-\beta E_n[K]\right) \tag{2.29}$$

shows that $Z_\Omega[K]$ is analytic.

Phase transitions can only arise in the thermodynamic limit. It can be shown[5] that for the thermodynamic limit to exist it is necessary that the two-spin interaction J_{ij} satisfies

$$\sum_{j(\neq i)} | J_{ij} | < \infty. \tag{2.30}$$

As before, in our discussion of the existence of the thermodynamic limit at $T = 0$, the range of the interaction J_{ij} and the dimensionality determine whether or not this limit exists. For example, let us suppose that for two spins S_i and S_j at lattice sites with positions in space $\mathbf{r}_i$, $\mathbf{r}_j$,

$$J_{ij} = A \left|\mathbf{r}_i - \mathbf{r}_j\right|^{-\sigma}. \tag{2.31}$$

Then, we require $\sigma > d$ for the limit to exist. In $d = 3$, this condition excludes magnetic dipole-dipole interactions, which fall off with distance r as r^{-3}. However, as long as the dipoles are not aligned, it can be shown[5] that the thermodynamic limit does exist and is independent of the shape of Ω.

2.6 ANALYTIC PROPERTIES OF THE ISING MODEL

Assuming the existence of the bulk free energy density $f_b[K]$, let us examine its analytic properties. This is important to do, because we are interested in precisely when and how non-analytic behaviour can arise.

Notation:– We shall sometimes use the abbreviation f for f_b. The notation $< ij >$ means "i and j are nearest neighbour sites".

We consider the nearest neighbour Ising model Hamiltonian

$$-\mathcal{H}_\Omega = H \sum_{i=1}^{N(\Omega)} S_i + J \sum_{<ij>} S_i S_j \tag{2.32}$$

where we have assumed that the external magnetic field H is uniform in space, and that the only interaction between spins is that between *neighbouring* spins, and denoted by J. With a uniform external magnetic field, we can define the magnetisation or magnetic moment per site, M:

$$M \equiv \frac{1}{N(\Omega)} \sum_{i=1}^{N(\Omega)} \langle S_i \rangle . \tag{2.33}$$

[5] A clear presentation is given by R. Griffiths in *Phase Transitions and Critical Phenomena*, vol. 1, C. Domb and M.S. Green (eds.) (Academic Press, New York, 1972).

This can be obtained by differentiating with respect to H:

$$M = -\frac{1}{N(\Omega)}\frac{\partial F_\Omega}{\partial H}. \tag{2.34}$$

The principle analytic properties of f are:

(a) $f < 0$.
(b) $f(H, J, T, \ldots)$ is continuous.
(c) $\partial f/\partial T, \partial f/\partial H, \ldots$ exist almost everywhere. Right and left derivatives exist everywhere and are equal almost everywhere.
(d) The entropy per site $S \equiv -\partial f/\partial T \geq 0$ almost everywhere.
(e) $\partial f/\partial T$ is monotonically non-increasing with T. Thus,

$$\frac{\partial^2 f}{\partial T^2} \leq 0, \tag{2.35}$$

which implies that the specific heat at constant magnetic field $C_H \geq 0$:

$$C_H \equiv T\frac{\partial S}{\partial T}\bigg|_H = -T\frac{\partial^2 F}{\partial T^2}\bigg|_H \geq 0. \tag{2.36}$$

(f) $\partial f/\partial H$ is monotonically non-increasing with H. Thus,

$$\frac{\partial^2 f}{\partial H^2} \leq 0, \tag{2.37}$$

which implies that the isothermal susceptibility $\chi_T \geq 0$, where

$$\chi_T \equiv \frac{\partial M}{\partial H}\bigg|_T = -\frac{\partial^2 f}{\partial H^2}. \tag{2.38}$$

These properties are postulates in thermodynamics, but can be proved in statistical mechanics. The strategy of the proofs is to start with a finite system Ω, and then take the thermodynamic limit. We shall prove (a) and (d) by simple *ad hoc* methods, and show that the other properties follow from the important notion of **convexity**.

Proof of (a): Z_Ω is a sum of $2^{N(\Omega)}$ *positive* terms for any finite T, since $\exp(-x) > 0$ for any finite real x. The free energy density in the finite system is given by

$$f_\Omega \equiv \frac{F_\Omega}{N(\Omega)} = -k_B T \log Z_\Omega. \tag{2.39}$$

If we can show that $Z_\Omega > 1$, then (a) will follow from eqn. (2.39). But this must be true, because there is at least one configuration of the spins, denoted by $\{S_i{}^*\}$, for which H_Ω is negative and hence $\exp(-\beta H_\Omega)$ is *greater* than one. Since all terms in Z_Ω are positive, the existence of this single configuration guarantees that $Z_\Omega > 1$ as desired. What is the configuration $\{S_i^*\}$? For $H < 0$, $J > 0$, $S_i^* = -1$ for all sites i does the trick. A slicker way to see that there must be a configuration $\{S_i^*\}$ is to note that

$$\text{Tr } H_\Omega[S] = 0. \tag{2.40}$$

Since the terms in the trace are non-zero, there must be at least one configuration with H_Ω negative in order that eqn. (2.40) can be satisfied. Finally, since $Z_\Omega > 1$, the desired result (a) follows after taking the thermodynamic limit. Furthermore, since the logarithm function only develops singularities when its argument vanishes, we can conclude that F_Ω is analytic.

Q.E.D.

Proof of (d): Assume existence of derivative for the moment. We shall prove this below. Then

$$\begin{aligned} -\frac{\partial F_\Omega}{\partial T} &= k_B \log \text{Tr } e^{-\beta H_\Omega} + k_B T \cdot \frac{1}{k_B T^2} \cdot \frac{\text{Tr } H_\Omega e^{-\beta H_\Omega}}{\text{Tr } e^{-\beta H_\Omega}} \\ &= k_B \left[\log Z_\Omega + \frac{\text{Tr } \beta H_\Omega \, e^{-\beta H_\Omega}}{Z_\Omega} \right] \\ &= -k_B \, \text{Tr } (\rho_\Omega \log \rho_\Omega) \end{aligned} \tag{2.41}$$

where

$$\rho_\Omega \equiv \frac{\exp(-\beta H_\Omega)}{\text{Tr } e^{-\beta H_\Omega}}. \tag{2.42}$$

But eqn. (2.41) is a sum of positive terms, since $0 < \rho_\Omega < 1$ and thus $\log \rho_\Omega < 0$. Dividing by $N(\Omega)$ and taking the thermodynamic limit we obtain the desired result.

Q.E.D.

The other results listed are a direct consequence of the property of convexity: F_Ω and f are **convex functions** of their arguments T, H and J.

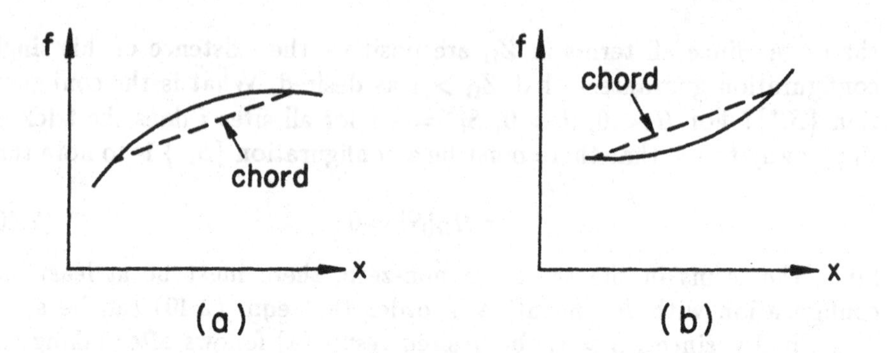

Figure 2.2 (a) A convex up function. (b) A convex down function.

2.6.1 Convex Functions

Definition: A function $f(x)$ is said to be **convex up (down)** in x if and only if for all numbers α_1 and α_2 such that for $0 \leq \alpha_1, \alpha_2 \leq 1$, $\alpha_1 + \alpha_2 = 1$, we have

$$f(\alpha_1 x + \alpha_2 y) \geq \alpha_1 f(x) + \alpha_2 f(y) \qquad \text{(convex up)} \tag{2.43}$$

or

$$f(\alpha_1 x + \alpha_2 y) \leq \alpha_1 f(x) + \alpha_2 f(y) \qquad \text{(convex down).} \tag{2.44}$$

These definitions are best understood graphically, as shown in figure (2.2): **Convex up** means that the function f is always *above* any chord. **Convex down** means that the function f is always *below* any chord.

From this definition it can be proved that if $f(x)$ is bounded and convex up (down) then:

(i) $f(x)$ is continuous.

(ii) $f(x)$ is differentiable almost everywhere.

(iii) df/dx is monotonically non-increasing (non-decreasing).

These theorems are trivial graphically. For example, any discontinuity clearly violates convexity, as shown in figure (2.3a)

Also, $f(x)$ may have a point discontinuity in slope, as shown in figure (2.3b), and still remain convex.

2.6.2 Convexity and the Free Energy Density

Now we will show that the free energy density $f(H, J, T)$ is convex up in H. The convexity in T and J is left as an exercise to the reader. The

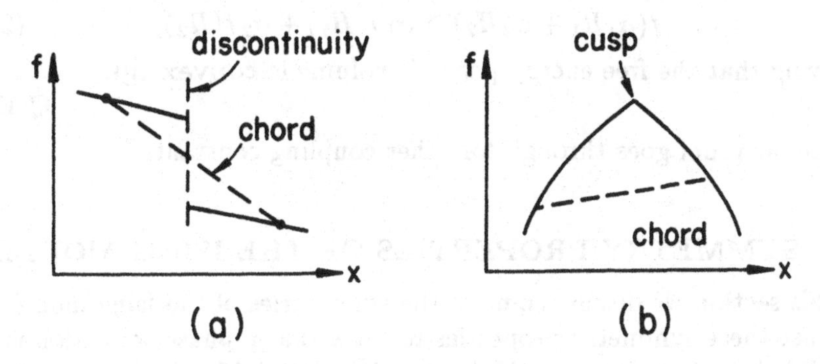

Figure 2.3 (a) $f(x)$ lies both above and below chord. (b) Convex up function with slope discontinuity.

proof of convexity relies on the **Hölder inequality:**[6] given two sequences $\{g_k\}$ and $\{h_k\}$ with g_k and $h_k \geq 0$ for all k, and two real non-negative numbers α and β such that $\alpha + \beta = 1$, then

$$\sum_k \left(g_k\right)^\alpha \left(h_k\right)^\beta \leq \left(\sum_k g_k\right)^\alpha \left(\sum_k h_k\right)^\beta . \tag{2.45}$$

Now consider $Z_\Omega(H)$. We have suppresed all arguments except H, the magnetic field. By definition,

$$Z_\Omega(H) = \mathrm{Tr}\ e^{\beta H \sum_i S_i} \underbrace{\exp\{\beta J \sum_{<ij>} S_i S_j + \ldots\}}_{\mathcal{G}[S]}, \tag{2.46}$$

defining the variable $\mathcal{G}[S]$. Thus

$$\begin{aligned} Z_\Omega(\alpha_1 H_1 + \alpha_2 H_2) &= \mathrm{Tr}\ \exp\left\{\beta\alpha_1 H_1 \sum_i S_i + \beta\alpha_2 H_2 \sum_i S_i\right\} \mathcal{G}[S] \\ &= \mathrm{Tr}\ \left(e^{\beta H_1 \sum_i S_i} \mathcal{G}[S]\right)^{\alpha_1} \left(e^{\beta H_2 \sum_i S_i} \mathcal{G}[S]\right)^{\alpha_2} \end{aligned} \tag{2.47}$$

where we have used $\alpha_1 + \alpha_2 = 1$ to write $\mathcal{G}$ [S] $= \mathcal{G}[S]^{\alpha_1} \mathcal{G}[S]^{\alpha_2}$. Now use the Hölder inequality:

$$\begin{aligned} Z_\Omega(\alpha_1 H_1 + \alpha_2 H_2) &\leq \left(\mathrm{Tr}\ e^{\beta H_1 \sum_i S_i} \mathcal{G}[S]\right)^{\alpha_1} \left(\mathrm{Tr}\ e^{\beta H_2 \sum_i S_i} \mathcal{G}[S]\right)^{\alpha_2} \\ &\leq Z_\Omega(H_1)^{\alpha_1} Z_\Omega(H_2)^{\alpha_2}. \end{aligned} \tag{2.48}$$

---

[6] See, (*e.g.*), L.M. Graves, *The Theory of Functions of Real Variables* (McGraw-Hill, New York, 1946), p. 233.

Taking the logarithm, multiplying by $-k_BT$, dividing by $N(\Omega)$, and taking the thermodynamic limit we find

$$f(\alpha_1 H_1 + \alpha_2 H_2) \geq \alpha_1 f(H_1) + \alpha_2 f(H_2), \tag{2.49}$$

showing that the free energy per unit volume is **convex up**.

Q.E.D.

A similar proof goes through for other coupling constants.

2.7 SYMMETRY PROPERTIES OF THE ISING MODEL

In this section we discuss some of the symmetries of the Ising model. We will use these symmetry properties to show that a phase transition is not possible in a finite system. We begin with a trivial lemma:

Lemma: For any function ϕ which depends on the spin configuration $\{S_i\}$,

$$\sum_{\{S_i=\pm1\}} \phi(\{S_i\}) = \sum_{\{S_i=\pm1\}} \phi(\{-S_i\}). \tag{2.50}$$

Proof: By inspection. Just write out the terms. Every term occurs once in the summation on both sides of the equation.

Q.E.D.

2.7.1 Time-reversal Symmetry

The first symmetry which we discuss is up-down symmetry, sometimes called **time-reversal symmetry** or Z_2 symmetry. The definition of H_Ω in eqn. (2.32) implies that

$$H_\Omega(H, J, \{S_i\}) = H_\Omega(-H, J, \{-S_i\}). \tag{2.51}$$

Thus

$$\begin{aligned} Z_\Omega(-H,J,T) &= \sum_{\{S_i=\pm1\}} \exp\left[-\beta H_\Omega(-H,J,\{S_i\})\right] \\ &= \sum_{\{S_i=\pm1\}} \exp\left[-\beta H_\Omega(-H,J,\{-S_i\})\right] \quad \text{from eqn. (2.50)} \\ &= \sum_{\{S_i=\pm1\}} \exp\left[-\beta H_\Omega(H,J,\{S_i\})\right] \quad \text{from eqn. (2.51)} \\ &= Z_\Omega(H,J,T). \end{aligned} \tag{2.52}$$

Thus the free energy density is even in H:

$$f(H,J,T) = f(-H,J,T). \tag{2.53}$$

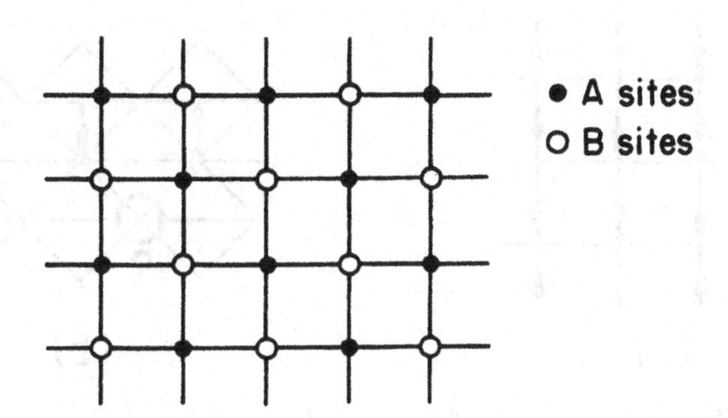

Figure 2.4 Cubic lattice divided into two interpenetrating sub-lattices A and B.

2.7.2 Sub-lattice Symmetry

There is a second symmetry which occurs when $H = 0$. This is called **sub-lattice symmetry**. We divide our hypercubic lattice up into two interpenetrating **sub-lattices**, which we shall call A and B, and label the spins on them accordingly, as shown in figure (2.4). In the Hamiltonian

$$\mathcal{H}_\Omega(0, J, \{S_i\}) = -J \sum_{<ij>} S_i S_j \tag{2.54}$$

the spins on sub-lattice A only interact with spins on sub-lattice B and *vice versa*, so we can trivially re-write the Hamiltonian as

$$\mathcal{H}_\Omega(0, J, \{S_i^A\}, \{S_i^B\}) = -J \sum_{<ij>} S_i^A S_i^B. \tag{2.55}$$

The trace operation can be decomposed into two traces over each sub-lattice:

$$\mathrm{Tr} \equiv \sum_{\{S_i = \pm 1\}} = \sum_{\{S_i^A = \pm 1\}} \sum_{\{S_i^B = \pm 1\}} . \tag{2.56}$$

The Hamiltonian, written as in eqn. (2.55), exhibits the symmetry

$$\begin{aligned} \mathcal{H}_\Omega(0, -J, \{S_i^A\}, \{S_i^B\}) &= \mathcal{H}_\Omega(0, J, \{-S_i^A\}, \{S_i^B\}) \\ &= \mathcal{H}_\Omega(0, J, \{S_i^A\}, \{-S_i^B\}) \end{aligned} \tag{2.57}$$

Now let us examine the implications of this symmetry for the thermodynamics. The partition function in zero field is

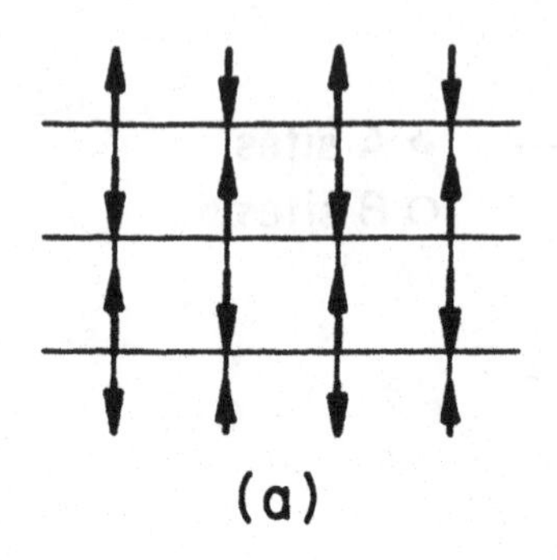

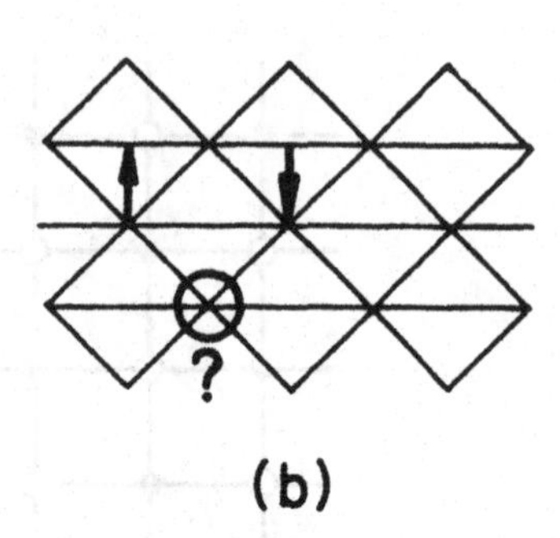

Figure 2.5 Minimum energy configuration of an anti-ferromagnet on (a) Bipartite lattice. (b) Non-bipartite lattice.

$$\begin{aligned}
Z_\Omega(0,-J,T) &= \text{Tr } e^{-\beta H_\Omega(0,-J,T)} \\
&= \sum_{\{S_i^A\}} \sum_{\{S_i^B\}} e^{-\beta H_\Omega(0,-J,\{S^A\}\{S^B\})} \\
&= \sum_{\{S_i^A\}} \sum_{\{S_i^B\}} e^{-\beta H_\Omega(0,J,\{-S^A\}\{S^B\})} \quad \text{from (2.57)} \\
&= \sum_{\{S^A\}} \sum_{\{S^B\}} e^{-\beta H_\Omega(0,J,\{S^A\}\{S^B\})} \quad \text{from (2.50)} \\
&= Z_\Omega(0,J,T). \qquad (2.58)
\end{aligned}$$

Thus, we obtain the second symmetry property of the free energy density:

$$f(0,J,T) = f(0,-J,T). \qquad (2.59)$$

In zero field, the ferromagnetic Ising model ($J > 0$) and the anti-ferromagnetic Ising model ($J < 0$) on a hypercubic lattice have the same thermodynamics.

This conclusion relies on the fact that a hypercubic lattice is **bipartite**: it can be subdivided into two equivalent sub-lattices. The theorem does *not* apply on a triangular lattice, which is not bipartite.

You might also imagine (correctly) that it would be difficult to write down the zero temperature configuration of an Ising model in $d = 2$ on a triangular lattice with $J < 0$. As shown in figure (2.5), it is not possible to simultaneously minimise all the interactions on the triangular lattice, as it is possible to do on a square lattice. This phenomenon is an example of **frustration**.

2.8 EXISTENCE OF PHASE TRANSITIONS

How do we construct the phase diagram of a particular system? One of the most useful techniques, particularly for the purpose of doing rigorous mathematics is the so-called **energy-entropy argument**, which we shall give here. The basic idea is quite simple. At high temperature, the entropy S always dominates the free energy, and the free energy is minimised by maximising S. At low temperature, there is the possibility that the internal energy E dominates TS in the free energy, and the free energy may be minimised by minimising E. If the macroscopic states of the system obtained by these two procedures are different, then we conclude that at least one phase transition has occurred at some intermediate temperature.

Consider the Ising model in d dimensions at $T = 0$. The free energy, then, is just equal to the **internal energy**, E, and the problem of finding the free energy is reduced to that of finding the energy E. This is easy for the Ising model Hamiltonian eqn. (2.32), and for generalisations, such as those involving next nearest neighbour interactions, can be often simplified by using the following lemma:

Lemma: If the energy E is the sum of a number of terms dependent upon the spin configuration $[S]$:

$$E = \sum_{\lambda} E_{\lambda}[S], \tag{2.60}$$

then

$$\min_{[S]} E \geq \sum_{\lambda} \min_{[S]} E_{\lambda}[S]. \tag{2.61}$$

Here "min E" means the minimum value of the function E, found by searching over all the configurations $[S]$. In other words, *if* one can find a configuration which minimises each term in the Hamiltonian separately, then that configuration is a ground state (not necessarily unique).

Proof: $E_{\lambda}[S] \geq \min_{[S]} E_{\lambda}[S]$ by definition. Summing over λ and then taking the min yields the result.

Q.E.D.

The ground state is any state for which the equality is satisfied, again by definition. As the contributions E_{λ} to the energy depend upon the coupling constants, the minimising configuration may be anticipated to depend upon the values of the coupling constants.

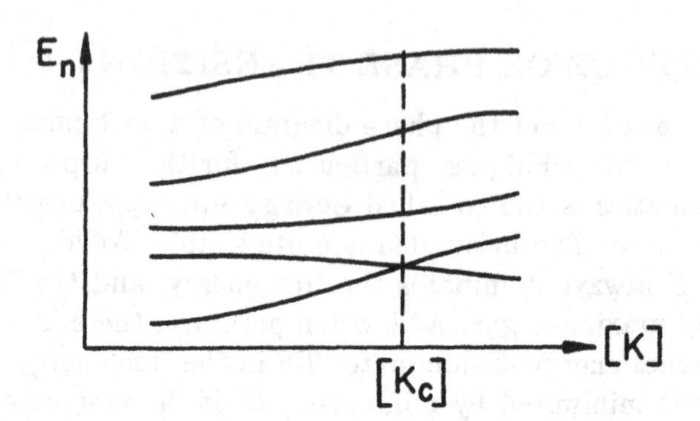

Figure 2.6 Zero temperature phase transition by level crossing.

2.8.1 Zero Temperature Phase Diagram

Let us suppose that, for a given set of values of the coupling constants $[K]$, we have obtained the energy levels of the system: that is, we know the configuration that minimises the energy, the configuration that corresponds to the first excited state of the system *etc.* Now we make a small change in the coupling constants. In general, all the energy levels change by a small amount. Thus, we can map out the energy levels as a function of $[K]$, as sketched above. It may happen that as the coupling constants pass through the set of values denoted by $[K_c]$ the first excited state and the ground state cross: the energy of the system is now minimised by a configuration that previously corresponded to an excited state of the system. This will usually generate a first order transition, because there will be a discontinuity in $\partial E/\partial K_i$, as shown in figure (2.6). This mechanism for a **zero temperature phase transition** is known as **level crossing.** Note that it is *not* necessary to take the thermodynamic limit to achieve a zero temperature phase transition by this mechanism. The non-analytic behaviour is permitted to occur in this case, not because $N \to \infty$, but because $\beta \to \infty$.

Now let us put this into practice for the Ising model with $J > 0$. We observe that the term

$$-J \sum_{<ij>} S_i S_j \tag{2.62}$$

is minimized by $S_i = S_j$, and that the term

$$-H \sum_i S_i$$

is minimized by

$$S_i = \begin{cases} +1 & H > 0; \\ -1 & H < 0. \end{cases} \tag{2.63}$$

Thus the ground state configuration, for each spin S_i, is:

$$S_i = \begin{cases} +1 & H > 0, \quad J > 0 \; ; \\ -1 & H < 0, \quad J > 0 \; . \end{cases} \tag{2.64}$$

Note the dependence on the coupling constant H (external field). For $H > 0$, the **magnetisation**

$$M_\Omega \equiv \frac{1}{N(\Omega)} \sum_{i \in \Omega} S_i = +1 \tag{2.65}$$

and for $H < 0$, $M_\Omega = -1$. Thus the zero temperature phase diagram for $J > 0$ has a phase transition at $H = 0$.

At zero external field H, the cases $S_i = +1$ (spin up) and $S_i = -1$ (spin down) have the same energy. Which is observed in practice must therefore depend on the **initial conditions** in which a given system is prepared. We will say more about this important case below.

2.8.2 Phase Diagram at Non-Zero Temperature: $d = 1$

Phase transitions at $T = 0$ may or may not disappear for $T > 0$. In fact, what happens depends on the **dimensionality** of the system. We will now see, using heuristic arguments due to Landau and Peierls, that in $d = 1$ for $T > 0$ there is no **long range order** (*i.e.* no ferromagnetic state), whereas for $d = 2$, long range order can exist above $T = 0$, with a transition at $T_c > 0$ to a paramagnetic (*i.e.* disordered) state. The dimension above which a given transition occurs for $T > 0$ is often referred to as the **lower critical dimension.** For now, the term "long range order" simply means a state in which the degrees of freedom (spins here) order over arbitrarily long distance, such as the ferromagnetic state. The paramagnetic state does not have long range order: two spins widely separated will not, on the average, point in the same direction, whereas they will do so in the ferromagnetic state. We will define long range order precisely later on. The heuristic arguments given here form the basis of a rigorous proof of these results.[7]

In the preceding section, we found that at zero temperature, there are two possible phases at zero field: all spins up and all spins down. Consider the spin up phase

$$\uparrow\uparrow\uparrow\uparrow\uparrow \cdots \uparrow\uparrow\uparrow\uparrow\uparrow \, .$$

---

[7] A clear account may be found in R. Griffiths in *Phase Transitions and Critical Phenomena*, vol. 1, C. Domb and M.S. Green (eds.) (Academic Press, New York, 1972).

As the temperature is raised above zero, each spin executes a sort of **Brownian motion** by virtue of being in thermal equilibrium. Does this spin flipping destroy the long range order? In the thermodynamic limit, the uncorrelated flipping of a finite number of spins, n, cannot destroy long range order:

$$\lim_{N(\Omega)\to\infty} \frac{N(\Omega)-n}{N(\Omega)} = 1. \tag{2.66}$$

The only fluctuations which can potentially destroy long range order are those involving a thermodynamically large number of spins. In other words, a non-zero *fraction* of spins, f, must be reversed on average:

$$\lim_{N(\Omega)\to\infty} \frac{N(\Omega)-fN(\Omega)}{N(\Omega)} < 1. \tag{2.67}$$

What effect does this have on the thermodynamics? We have already considered the energy of the system; when spins are flipped, there are many ways that this can occur for a given value of the overall magnetisation. Thus, we need to compute the entropy corresponding to a state with given magnetisation. In the ordered state the entropy is zero if all the spins are aligned, so the free energy is just

$$F_N = -NJ, \tag{2.68}$$

where we write, for ease of notation, $N \equiv N(\Omega)$. Now consider the state with 2 domains

$$\uparrow\uparrow\uparrow\uparrow\cdots\uparrow\wr\downarrow\cdots\downarrow\downarrow\downarrow\downarrow$$

The interface $\wr$ between the domains has cost energy. In fact,

$$E_N = -NJ + 2J = -(N-1)J + J. \tag{2.69}$$

What is the entropy S_N? The interface between the two domains can be at any of N sites (assuming periodic boundary conditions, for convenience). Then

$$S_N = k_B \log N. \tag{2.70}$$

Hence, the free energy difference between the state with an interface and the state with all spins up is

$$F_N^{(\text{boundary})} - F_N^{(\text{one phase})} \equiv \Delta F = 2J - k_B T \log N \tag{2.71}$$

For any $T > 0$, $\Delta F \to -\infty$ as $N \to \infty$. The system can lower the free energy by creating a domain wall for *any* temperature $T > 0$. In fact, the free energy can be lowered still further by splitting each domain in

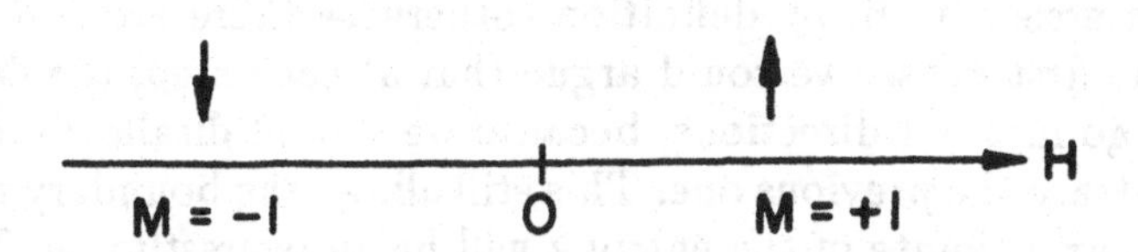

Figure 2.7 Zero temperature phase diagram for the ferromagnetic Ising model in $d = 1$.

two. We may continue this argument until there are simply no domains remaining at all.

Thus, the long range order state is unstable towards thermal fluctuations for $T > 0$: the magnetisation in zero external field for $T > 0$ is strictly zero, whereas for $T = 0$ the magnetisation is either $+1$ or -1. A corollary of our result that thermal fluctuations destroy long range order in this case is that there is no phase transition for $T > 0$, because there are no longer two phases at $H = 0$.

The argument given above assumed that the interaction between the spins was short-ranged: this meant that the energy difference between the spin up system and the two domain system was independent of N. In fact it can be shown that if the interaction J_{ij} between spins at $\mathbf{r}_i$ and $\mathbf{r}_j$ varies as

$$J_{ij} = \frac{J}{|\mathbf{r}_i - \mathbf{r}_j|^\sigma} \tag{2.72}$$

then long range order may persist for $0 < T < T_c$ as long as[8] $1 \leq \sigma \leq 2$. For $\sigma < 1$, the thermodynamic limit does not exist. For $\sigma > 2$, the interaction qualifies as being short-ranged, and the argument for the destruction of long range order for $T > 0$ applies.

2.8.3 Phase Diagram at Non-Zero Temperature: $d = 2$

As above, we consider a domain of flipped spins, in a background of spins with long range order, but now the domain is two-dimensional. Suppose the border between the flipped spins and the up spins contains n bonds. Then the energy difference between the state with a domain and one with complete long range order is $\Delta E \sim 2Jn$. What is the entropy? We can estimate it as follows. Choose a point on the boundary. How many ways are there for the boundary to go from the starting point? If the co-ordination number of the lattice is z, then an upper bound on the

[8] For references and the solution of the difficult case $\sigma = 2$ see J. Fröhlich and T. Spencer, *Commun. Math. Phys.* **84**, 87 (1982).

number of configurations of the boundary or domain wall is z^n. The precise number is less than this, because the boundary of just a single domain cannot intersect itself, by definition (otherwise there are two domains!); as a crude first guess, we could argue that at each step, the domain wall can only go in $z-1$ directions, because we should disallow the step that would re-trace the previous one. This still allows the boundary to intersect itself, so our estimate of the entropy will be an overestimate. Then, as in $d = 1$, we can compute the entropy difference due to the presence of the domain: $\Delta S \sim k_B n \log(z-1)$. Thus, the change in free energy due to a domain whose boundary contains n bonds is

$$\Delta F_n = [2J - (\log(z-1))k_B T]n. \tag{2.73}$$

Using the same argument as in the one-dimensional case, we are only interested in domains which contain a thermodynamically large number of spins (*i.e.* a non-zero *fraction* of the total number of spins). Thus, in the thermodynamic limit, $n \to \infty$. The behaviour of ΔF_n in this limit depends upon the temperature. If

$$T > T_c \equiv \frac{2J}{k_B \log(z-1)} \tag{2.74}$$

then $\Delta F \to -\infty$ as $n \to \infty$ and the system is unstable towards the formation of domains.[9] Accordingly we anticipate a disordered, paramagnetic phase with $M = 0$. For $0 < T < T_c$, however, ΔF is minimized by $n \to 0$, and the state with long range order is stable. Thus, for $0 < T < T_c$, the system exhibits a net magnetisation M, which can be either positive or negative, *in the absence of an applied field.* This magnetisation is often referred to as **spontaneous magnetisation.** In conclusion, **long range order** may exist at sufficiently low but non-zero temperature in the two dimensional Ising model. A corollary of this result is that the zero field two dimensional Ising model exhibits a phase transition at a temperature $T_c > 0$. Note that the transition temperature, as estimated in eqn. (2.74), depends on the coordination number z, and therefore on the type of lattice. The transition temperature is not a universal quantity.

Our results imply that the lower critical dimension is $d = 1$. For $d > 1$, long range order is possible for $T > 0$. In fact, although we have assumed

[9] For a square lattice, with $z = 4$, the estimate yields $T_c \approx 1.82J/k_B$, whereas the exact result is $k_B T_c = 2J/(\sinh^{-1}(1)) \approx 2.27J$, and was obtained by Kramers and Wannier, *Phys. Rev.* **60**, 252 (1941); *ibid* **60**, 263 (1941). They obtained the expression for T_c *without* solving the model! The transition temperature in eqn. (2.74) is too small, because we have overestimated the entropy.

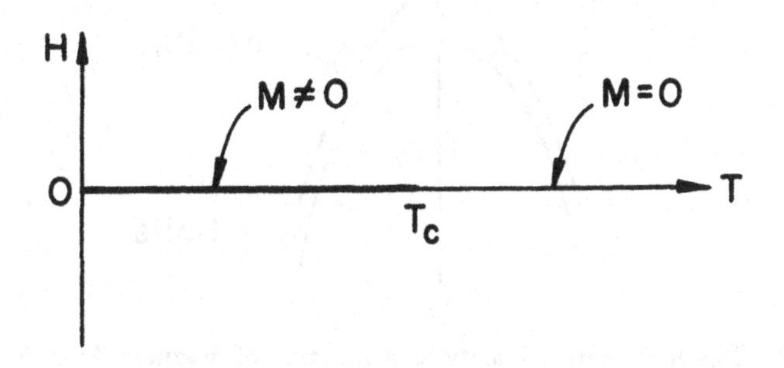

Figure 2.8 Phase diagram of the ferromagnetic Ising model in $d = 2$.

d is an integer in writing these results, the conclusion can also be shown to be correct if *non-integral* values of d are allowed. That is, if we write $d = 1 + \epsilon$, then the critical temperature T_c is $O(\epsilon)$. Finally, in deriving these results, we worked with the Ising model, but our analysis really only used the fact that domain walls exist. Thus our conclusions apply in general for a Hamiltonian with a discrete symmetry.

2.8.4 Impossibility of Phase Transitions

We can easily see that the existence of a phase transition is impossible! From time-reversal symmetry, as expressed by eqn. (2.53), we know that

$$F_\Omega(H, J, T) = F_\Omega(-H, J, T). \tag{2.75}$$

The magnetisation $M(H)$ satisfies

$$\begin{aligned} N(\Omega)M_\Omega(H) &= -\frac{\partial F_\Omega(H)}{\partial H} \\ &= -\frac{\partial F_\Omega(-H)}{\partial H} \\ &= \frac{\partial F_\Omega(-H)}{\partial(-H)} \\ &= -N(\Omega)M_\Omega(-H). \end{aligned} \tag{2.76}$$

Thus

$$M_\Omega(H) = -M_\Omega(-H). \tag{2.77}$$

At $H = 0$ we must have

$$M_\Omega(0) = -M_\Omega(0) = 0. \tag{2.78}$$

This "impossibility theorem" shows that the magnetisation in zero external field must be zero! What has gone wrong?

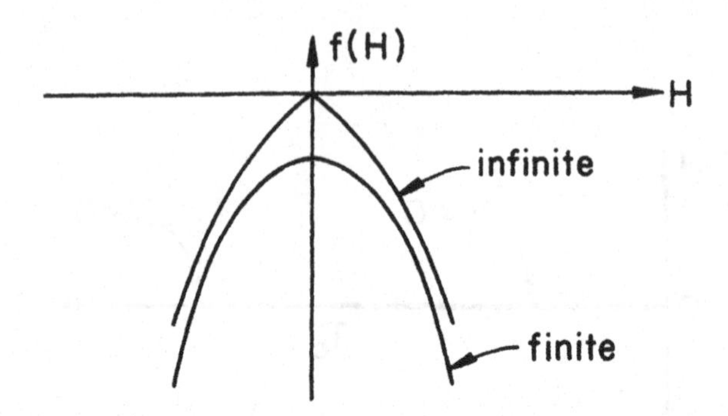

Figure 2.9 The free energy density as a function of magnetic field H for finite and infinite systems, for $T < T_c$.

2.9 SPONTANEOUS SYMMETRY BREAKING

The argument above that $M_\Omega(0) = 0$ is indeed correct *for a finite system.* It fails in the thermodynamic limit, however, because $f(H)$ can develop a discontinuity in its first derivative $\partial f/\partial H$. We know that $f(H)$ is a convex up function. The condition $f(H) = f(-H)$ does *not* imply $M(0) = 0$ unless we make the additional assumption that $f(H)$ is smooth at $H = 0$ and the left and right derivatives are equal. Smoothness follows if

$$f(H) = f(0) + O(H^\sigma) \qquad \sigma > 1 \tag{2.79}$$

and

$$\lim_{\epsilon\to 0} \frac{f(+\epsilon) - f(0)}{\epsilon} = \lim_{\epsilon\to 0} \frac{f(-\epsilon) - f(0)}{\epsilon} = 0. \tag{2.80}$$

However, none of the properties of $f(H)$ guarantee that smoothness occurs. We can evade the consequences of the "impossibility theorem" and still satisfy the analytical properties of the free energy density if the behaviour near $H = 0$ is as sketched in figure (2.9):

$$f(H) = f(0) - M_s|H| + O(H^\sigma), \qquad \sigma > 1. \tag{2.81}$$

This is not differentiable at $H = 0$, but is convex. Eqn. (2.81) implies that

$$\frac{\partial f}{\partial H} = \begin{cases} -M_s + O(H^{\sigma-1}), & H > 0 \\ +M_s + O(H^{\sigma-1}), & H < 0\,. \end{cases} \tag{2.82}$$

As $|H| \to 0$

$$M = -\frac{\partial f}{\partial H} = \begin{cases} M_s & H > 0 \\ -M_s & H < 0\,. \end{cases} \tag{2.83}$$

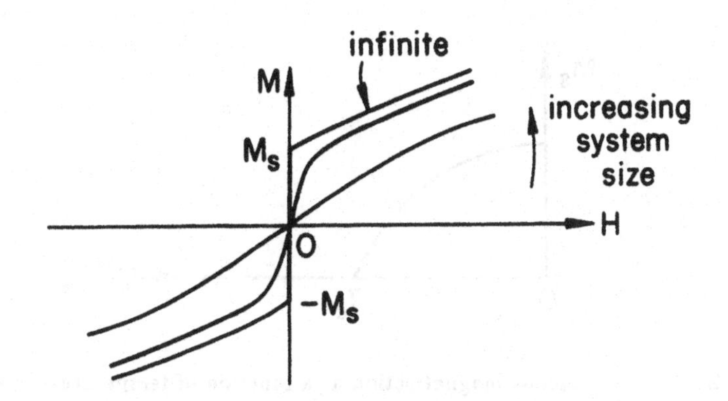

Figure 2.10 Magnetisation M plotted against external field H.

The graph of magnetisation M *versus* external field H is sketched in figure (2.10). We see that M_s, the spontaneous magnetisation is given by

$$M_s = \lim_{H\to 0^+} -\frac{\partial f(H)}{\partial H} \tag{2.84}$$

and

$$-M_s = \lim_{H\to 0^-} -\frac{\partial f(H)}{\partial H}. \tag{2.85}$$

Notice that the limits

(a) $N(\Omega) \to \infty$
(b) $H \to 0$

do *not* commute:

$$\lim_{N(\Omega)\to\infty} \lim_{H\to O} \frac{1}{N(\Omega)} \frac{\partial F_\Omega(H)}{\partial H} = 0 \tag{2.86}$$

whereas

$$\lim_{H\to 0} \lim_{N(\Omega)\to\infty} \frac{1}{N(\Omega)} \frac{\partial F_\Omega(H)}{\partial H} \neq 0. \tag{2.87}$$

Notice also that the value of the spontaneous magnetisation M_s is a function of temperature: at zero temperature, we already argued that M_s should be unity, because the spins will be either all up or all down. As the temperature rises towards T_c, the value of the spontaneous magnetisation is reduced, as an increasingly greater fraction of spins are flipped by thermal fluctuations. At T_c, the spontaneous magnetisation has fallen to zero, as depicted in figure (2.11).

This set of phenomena is referred to as **spontaneous symmetry breaking**. When $H = 0$,

$$\mathcal{H}_\Omega[\{S_i\}] = \mathcal{H}_\Omega[\{-S_i\}]. \tag{2.88}$$

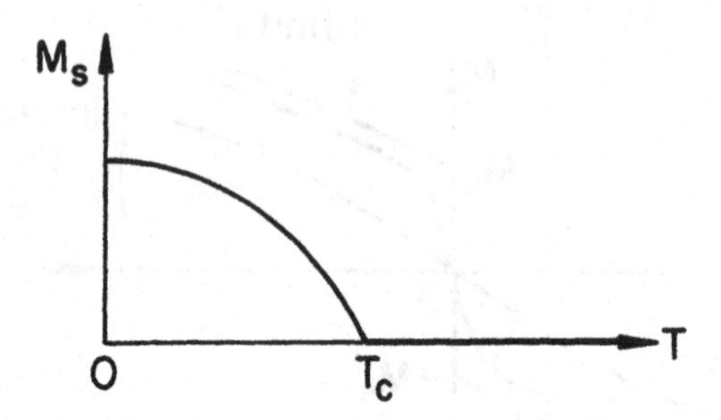

Figure 2.11 Spontaneous magnetisation as a function of temperature.

Even though the Hamiltonian is invariant under $\{S_i\} \to \{-S_i\}$, the statistical expectation values are *not* invariant under time-reversal symmetry: $\langle S_i \rangle \neq 0$ and

$$M = \lim_{N\to\infty} \frac{1}{N(\Omega)} \sum_i \langle S_i \rangle \neq 0. \tag{2.89}$$

The use of the word "spontaneous" in the above definition is to distinguish this phenomenon from the appearance of magnetisation in the presence of an external field $H \neq 0$. This rather mundane symmetry breaking is not really symmetry breaking at all: the symmetry in the Hamiltonian is not there to begin with for $H \neq 0$.

2.9.1 Probability Distribution

Spontaneous symmetry breaking should appear to you paradoxical for another reason. If the probability of finding the system in a state $\{S_i\}$ is given by the **Boltzmann distribution**

$$P_\Omega(\{S_i\}) = \frac{\exp(-\beta H_\Omega(\{S_i\}))}{Z_\Omega}, \tag{2.90}$$

then the fact that H_Ω is time-reversal invariant implies that P_Ω is also time-reversal invariant. Thus

$$\langle S_i \rangle \equiv \mathrm{Tr}\, P_\Omega(\{S_i\}) S_i = 0, \tag{2.91}$$

since $P_\Omega(\{S_i\})$ is even and S_i is odd in the variable S_i. This is, of course, nothing more than a restatement of the "impossibility theorem." This version of the theorem fails; in an infinite system $Z \to \infty$ and we have to be more careful about the probability distribution: in the thermodynamic limit, the probability distribution is *not* necessarily given by the Boltzmann distribution.

Let us explore the consequences of this remark, and obtain an understanding of spontaneous symmetry breaking from the point of view of the probability distribution. Suppose that we consider the Ising model with $H = 0$. Let two configurations of the system labelled by A and B be related by time-reversal symmetry – the spins in the state A are related to those in state B by a simple flip of all the spins. Both states A and B are equally likely, and the magnetisation in state A, M, is minus the magnetisation in state B.

Now apply an external field H. What is the probability of the system being in state A, P_A, as opposed to the probability of being in state B, P_B? Due to the coupling term $-H\sum_i S_i$ in the Ising model Hamiltonian eqn. (2.32), it follows that

$$\frac{P_A}{P_B} = \frac{e^{-\beta(-HN(\Omega)M)}}{e^{-\beta(HN(\Omega)M)}} = e^{2\beta HN(\Omega)M}. \tag{2.92}$$

Taking the thermodynamic limit $N(\Omega) \to \infty$, for $H > 0$, gives

$$\frac{P_B}{P_A} \to 0 \tag{2.93}$$

and thus the system must be in state A with magnetisation $+M$. This is true regardless of the magnitude of H, and in particular applies as $H \to 0^+$. On the other hand, if $H < 0$, taking the thermodynamic limit gives

$$\frac{P_B}{P_A} \to \infty \tag{2.94}$$

and the system must be in state B with magnetisation $-M$. This is also the case as $H \to 0^-$.

Thus the presence of an infinitesimal field $H \to 0^+$ or 0^- *together with the thermodynamic limit* provides a macroscopic weighting of the state with magnetisation $+M$ over the state with magnetisation $-M$ (or *vice versa*). How can we explain spontaneous magnetisation in the thermodynamic limit, in *zero* field, from the probabilistic picture?

The use of the limit $H \to 0^+$ and the thermodynamic limit is equivalent to setting $H = 0$ but using a *restricted ensemble* in which microstates contributing to $-M_s$ are *not* included. Thus, the probability distribution for a system *after* the thermodynamic limit has been taken is *identically zero* for all states whose magnetisation is negative. There is another probability distribution that must be defined, corresponding to the simultaneous use of the $H \to 0^-$ limit and the thermodynamic limit: it gives zero weight to states whose magnetisation is positive![10] In each case, for the states with non-zero weight, the weight is given by the Boltzmann distribution after care has been taken to include the mathematical delicacies of the meaning of probability in an infinite system![11]

2.9.2 Continuous Symmetry

The result that long range order is possible in the Ising model only for $d > 1$ relies on the fact that the symmetry group of H_Ω is Z_2, *i.e.* $H_\Omega[\{S_i\}] = H_\Omega[\{-S_i\}]$. The fact that the Ising model has a **discrete symmetry** means that the width of the domain walls must be *finite*. We can see that easily, because a domain wall separates $\uparrow$ from $\downarrow$, and since those are the only possible states of the spins, the domain wall thickness is one lattice unit.

However, things change if the spins S_i obey a **continuous symmetry** rather than a discrete symmetry. One way for this to occur is if the S_i are *vectors*, allowed to point in *all* directions (4π steradians) rather than just up or down. This occurs in the **Heisenberg ferromagnet** with HamiltonianHeisenberg model

$$H_\Omega[\{\mathbf{S}_i\}] = -\sum_{<ij>} J_{ij}\,\mathbf{S}_i \cdot \mathbf{S}_j - \sum_i \mathbf{H}_i \cdot \mathbf{S}_i. \tag{2.95}$$

In general, this is a more realistic model of a ferromagnet than the Ising model, since there is no preferred direction for the spins to point in. In a real crystal, spin-orbit coupling is present in addition to the exchange interaction already considered. The spins will then couple to the electronic charge density, which reflects the presence and symmetry of the underlying crystal lattice. In the cases where this is significant, the rotational symmetry of H_Ω is broken, and there is a tendency for spins to align along crystallographically preferred directions. In extreme cases, this can be sufficiently strong that a better model Hamiltonian is not the Heisenberg Hamiltonian but the Ising Hamiltonian.

The Heisenberg model has a continuous symmetry. Suppose that

$$\mathbf{S}_i = (S_i^x, S_i^y, S_i^z) \tag{2.96}$$

---

[10] A quantum version of the argument given here may be found in D. Forster, *Hydrodynamic Fluctuations, Broken Symmetry and Correlation Functions* (Benjamin, Reading, 1975), Section 7.2.

[11] A complete discussion of the rigorous mathematics is given in J. Glimm and A. Jaffe, *Quantum Physics: A Functional Integral Point of View* (Springer-Verlag, New York, 1981), especially Chapters 5 and 16.

and $R(\theta)$ is a rotation matrix which rotates a vector in (x, y, z) space by an angle θ about direction y, for example.

Then the energy of eqn. (2.95) is invariant under the simultaneous rotation of *all* the spins by an *arbitrary* angle θ;

$$H_\Omega[\{R(\theta)\mathbf{S}_i\}] = H_\Omega[\{\mathbf{S}_i\}] \tag{2.97}$$

Of course, in general, we do not have to single out the y-axis. R can be an arbitrary rotation about an arbitrary direction. Thus, the Heisenberg model is *rotationally invariant*, and this symmetry is sometimes called $O(3)$ (in three dimensional space). It is very important that you realize that the rotations are of the spins, keeping the lattice fixed in space. The rotation operator $R(\theta)$ acts on the degrees of freedom, not the spatial co-ordinate system.

The continuous rotational symmetry is spontaneously broken in the state of long range order. It is "easier" for thermal fluctuations to destroy long range order when there is a continuous symmetry as opposed to a discrete symmetry, because "there are more directions to point in (in spin space)." This should be intuitively clear, and will made precise in chapter 11. To compensate for the increased entropy due to the larger dimensionality of the order parameter, an increased energy is required if there is to be long range order in the Heisenberg model at $T > 0$. This can be achieved by increasing the dimensionality of the lattice: spins have more neighbours with which to interact. A detailed analysis[12] shows that to have long range order, we need $d > 2$. These results are a consequence of the fact that with a continuous symmetry, the width of a domain wall is the size of the system.

2.10 ERGODICITY BREAKING

Statistical mechanics is intended to represent the actual dynamical behaviour of a system in equilibrium. What are the consequences of spontaneous symmetry breaking for the dynamics? To answer this, let us recall how the actual dynamics of a system enters the statistical mechanics.

Usually, statistical mechanics is justified by identifying time averages with ensemble averages. To be more precise, for any observable $A\{\eta_i\}$, where $\eta_i(t)$ are the dynamical degrees of freedom as a function of time t, (position and momentum for a fluid, for example), the time average is

[12] The Mermin-Wagner theorem, due to N.D. Mermin and H. Wagner, *Phys. Rev. Lett.* **17**, 1133 (1966).

given by

$$\langle A \rangle \equiv \lim_{t\to\infty} \frac{1}{t} \int_0^t A\{\eta_i(t')\}dt' \tag{2.98}$$

and in statistical mechanics it is hypothesised that

$$\langle A \rangle = \int \prod_i d\eta_i P_{eqm}(\{\eta_i\}) A\{\eta_i\}, \tag{2.99}$$

where $P_{eqm}\{\eta_i\}$ is the *equilibrium* probability distribution of the variables $\{\eta_i\}$. These variables constitute **phase space** for the system in question: for a single particle in a three dimensional box, phase space is six-dimensional, the variables being position and momentum.

The basic assumption of statistical mechanics is that these two averages give the same result. This hypothesis is the **ergodic hypothesis**: as $t \to \infty$, $\{\eta_i(t)\}$ comes arbitrarily close to every possible configuration of the $\{\eta_i\}$ allowed by energy conservation.[13]

When $H = 0$ and $T < T_c$, regions of configuration (or phase) space with $M_\Omega = +M$ and $M_\Omega = -M$ are sampled equally. An observer of the system would notice that at first the system had net (*e.g.*) positive magnetisation. After a while a large cluster of down spins might form and then grow, eventually causing the system to have a net negative magnetisation. The rate R at which such a cluster might form is expected to be of the Arrhenius form

$$R \propto e^{-\beta \Delta F}, \tag{2.100}$$

where ΔF is the free energy of the critical cluster: clusters larger than this critical cluster will grow. Since ΔF is proportional to the surface area of the critical cluster, we expect that $\Delta F \propto N^{(d-1)/d}$. Hence, the *lifetime* of a state with a given magnetisation, τ, is roughly $1/R$, *i.e.*

$$\tau \sim \exp[N^{(d-1)/d}]. \tag{2.101}$$

In the thermodynamic limit, the lifetime will rapidly grow very large, so that the *initial condition* determines whether or not the magnetisation is positive or negative. Thus, in the thermodynamic limit, the system is

---

[13] For the Ising model, the Hamiltonian does not generate an equation of motion. The latter must be provided separately, and we shall assume that this has been done in the following.

effectively trapped in one or the other region of configuration or phase space. This is known as **ergodicity breaking**.[14]

An important feature of this form of ergodicity breaking is that the configuration space has been fragmented into two regions corresponding to positive or negative magnetisation. Thus, for every state of the system in one of the regions, there is a corresponding time-reversed state in the other region, *i.e.* there is a one-to-one mapping between states of the system in the two disjoint regions of configuration space. The two regions are said to be related by symmetry, the symmetry in question being the same symmetry that has been spontaneously broken.

We will explore these ideas further below. Note that in discussing ergodicity breaking, we had to invoke the thermodynamic limit. Thus, when we discuss phase space below, we will always assume that the thermodynamic limit has already been taken. The presentation will be partly verbal, although some references to the rigorous mathematics will be provided for the interested reader.

2.10.1 Illustrative Example

Consider a system with degrees of freedom $\{\mathbf{c}_i\}$ ($i = 1, \ldots, N$) and a Hamiltonian $H_\lambda\{\mathbf{c}_i\}$. The degrees of freedom $\mathbf{c}_i$ may be thought of as the position of particle i in a many-body system. Let us suppose that the system is translationally invariant. This means that configurations, which are identical apart from a translation by an arbitrary vector $\mathbf{a}$, have the same energy:

$$H_\lambda\{\mathbf{c}_i\} = H_\lambda\{\mathbf{c}_i + \mathbf{a}\}. \qquad (2.102)$$

The parameter λ is supposed to represent all the parameters or coupling constants in the Hamiltonian; by varying λ, we intend to make the system undergo a phase transition with spontaneous symmetry breaking.

Now suppose that we are interested in computing the expectation value of the function $f_i(\mathbf{k})$, $\mathbf{k} \neq 0$, which we shall take to be

$$f_i(\mathbf{k}) = \exp(i\mathbf{k} \cdot \mathbf{c}_i). \qquad (2.103)$$

The function $f_i(\mathbf{k})$ serves as our order parameter in this example. It has a simple interpretation: when summed over i, it is the Fourier transform

---

[14] A detailed discussion of this topic may be found in the introduction by A.S. Wightman to R.B. Israel, *Convexity and the Theory of Lattice Gases* (Princeton University Press, Princeton, 1978). Less formal accounts are given by R.G. Palmer, *Adv. Phys.* 31, 669 (1982) and by A.C.D. van Enter and J.L. van Hemmen, *Phys. Rev. A* 29, 355 (1984).

of the density

$$\rho(\mathbf{r}) \equiv \sum_i \delta(\mathbf{r} - \mathbf{c}_i). \tag{2.104}$$

Thus, we expect that for $\lambda > \lambda_c$, $\langle f_i(\mathbf{k})\rangle = 0$, whilst for $\lambda < \lambda_c$, $\langle f_i(\mathbf{k})\rangle \neq 0$. To be more precise,

$$\langle f_i(\mathbf{k})\rangle_\sigma = \frac{\mathrm{Tr}_\sigma f_i(\mathbf{k}) \exp(-\beta H_\lambda\{\mathbf{c}_i\})}{\mathrm{Tr}_\sigma \exp(-\beta H_\lambda\{\mathbf{c}_i\})}, \tag{2.105}$$

where the subscript σ denotes the microstates (*i.e.* the values of the vectors $\mathbf{c}_i$) included in the trace operation.

What is the physical interpretation of the phase transition that occurs at $\lambda = \lambda_c$? The fact that $\langle f_i(\mathbf{k})\rangle_\sigma$ becomes non-zero means that the density of the system is not simply a constant in space, as it is for $\lambda > \lambda_c$. Thus the transition is associated with some ordering of the particles in space — in other words, **solidification.** We will now show that in statistical mechanical language, this is just the spontaneous breaking of the symmetries associated with translational (and rotational) invariance.

To proceed, we must first specify which microstates should be included in σ. If all possible microstates are included, then because of the translational invariance expressed in eqn. (2.102),

$$\langle f_i(\mathbf{k})\rangle_\sigma = e^{i\mathbf{k}\cdot\mathbf{a}} \langle f_i(\mathbf{k})\rangle_\sigma, \tag{2.106}$$

and hence $\langle f_i(\mathbf{k})\rangle_\sigma = 0$ regardless of the value of λ. The only way in which $\langle f_i(\mathbf{k})\rangle_\sigma$ can be non-zero (as it should be for $\lambda < \lambda_c$) is if σ does not include all the allowed microstates. However, this statement is not precise enough: if some microstates are omitted from σ, but all the residual microstates remain in σ under the action of the translation operation, then $\langle f_i(\mathbf{k})\rangle_\sigma$ will still vanish. Therefore, it is only if the residual set σ is *not* invariant under the symmetry group of $H_\lambda\{\mathbf{c}_i\}$ that $\langle f_i(\mathbf{k})\rangle_\sigma$ can acquire a non-zero value. The residual set of microstates represents the system subject to the constraint that the centre of mass is at a certain position in space; we will see in the following section that there are infinitely many other residual sets σ' in phase space, each corresponding to a different centre of mass position.

The requirement that the set of microstates included in the trace be restricted for $\lambda < \lambda_c$ is, once again, **ergodicity breaking.** In this example, it is associated with spontaneous symmetry breaking, although this need not be the case: on the coexistence line of the liquid – gas transition, for example, phase space is fragmented into two distinct sets, which are not distinct in terms of their symmetry properties.

In summary, there are two complementary ways of looking at spontaneous symmetry breaking:

(i) Method of small fields: $H \to 0^+$ *etc.*

(ii) Ergodicity breaking and the restricted ensemble.

For simple problems like ferromagnetism, (i) is almost always more convenient. However, for some problems, such as superfluidity, the analogue of the magnetisation is a thermally averaged quantum mechanical operator, for which there is no physical counterpart to the external field H. In this case, it is often convenient to formally invent such a quantity so that one can differentiate the free energy with respect to it (*i.e.* form the analogue of $-\partial f/\partial H = M$). For some complex problems, especially those in which **disorder** plays a crucial role, (ii) has proven to be a more useful formalism.

In the following sections, we will explore in more detail the consequences of spontaneous symmetry breaking and ergodicity breaking. Our goal will be to obtain a qualitative understanding of the structure of phase space. When reading these sections, bear in mind the very important point that the structure of phase space depends upon the values of the parameters λ in the Hamiltonian and on β. Thus, the structure of phase space is a reflection of the state of the system, in other words, where the system is in its phase diagram.

2.10.2 *Symmetry and its Implications for Ergodicity Breaking*

In the preceding section, we showed by example that ergodicity breaking necessarily accompanies spontaneous symmetry breaking, although the converse is not true. In what manner is phase space broken? To answer this, consider the simple example given in the preceding section, for $\lambda < \lambda_c$. We saw there that if the set σ of microstates is the entire phase space, then $\langle f_i(\mathbf{k})\rangle_\sigma = 0$, and furthermore, that this quantity could only be non-zero if σ was a subset of the entire phase space which did *not* transform into itself under the symmetry of the Hamiltonian $T_\mathbf{a} : \mathbf{c}_i \to \mathbf{c}_i + \mathbf{a}$, where $\mathbf{a}$ is arbitrary.

Let us now consider the set σ_1 of microstates in phase space, defined to be the "largest" set for which $\langle f_i(\mathbf{k})\rangle_{\sigma_1} \neq 0$. Is this set unique? We will now show that there are a non-denumerably infinite number of such sets, each generated from σ_1 by the symmetry operations $T_\mathbf{a}$.

Let σ_2 be a set of microstates generated from the set σ_1 by some member T of the transformations $T_\mathbf{a}$: $\sigma_2 = T(\sigma_1)$. By definition, $\sigma_2 \neq \sigma_1$, and $H(T\mathbf{c}_i) = H(\mathbf{c}_i)$. The partition function, taken over the set σ_2 is

$$\begin{aligned} Z_2 &= \sum_{\mathbf{c}_i \in T(\sigma_1)} e^{-\beta H_\lambda\{\mathbf{c}_i\}} \\ &= \sum_{\mathbf{c}_i \in \sigma_1} e^{-\beta H_\lambda\{T\mathbf{c}_i\}} \quad \text{(making a change of dummy variable)} \\ &= \sum_{\mathbf{c}_i \in \sigma_1} e^{-\beta H_\lambda\{\mathbf{c}_i\}} \\ &= Z_1. \end{aligned} \tag{2.107}$$

Thus,

$$\begin{aligned} \langle f_i(\mathbf{k}) \rangle_{\sigma_2} &= \frac{1}{Z_2} \sum_{\mathbf{c}_i \in T(\sigma_1)} e^{i\mathbf{k}\cdot\mathbf{c}_i} e^{-\beta H_\lambda\{\mathbf{c}_i\}} \\ &= \frac{1}{Z_1} \sum_{\mathbf{c}_i \in \sigma_1} e^{i\mathbf{k}\cdot[\mathbf{c}_i+\mathbf{a}]} e^{-\beta H\{\mathbf{c}_i+\mathbf{a}\}} \\ &= e^{i\mathbf{k}\cdot\mathbf{a}} \langle f_i(\mathbf{k}) \rangle_{\sigma_1} \end{aligned} \tag{2.108}$$

In the region of phase space (*i.e.* the set of microstates) σ_2, the order parameter $f_i(\mathbf{k})$ is non-zero. The set σ_2 was obtained from σ_1 by a simple translation by $\mathbf{a}$ *i.e.*, by the translational symmetry which was spontaneously broken for $\lambda < \lambda_c$. Hence, we conclude that if ergodicity is broken so that a certain order parameter is non-zero in some set σ of microstates in phase space, then the order parameter will attain a (possibly different) non-zero value in all other sets of microstates related to σ by applying the symmetry of the Hamiltonian that has been spontaneously broken. In the example here, these other sets of microstates correspond to the system being translated in space from the position it occupies in the microstates in σ. Hence we have justified the statement made in the preceding section that phase space is broken into ergodic regions, each describing the solid with a different centre of mass position.

Although we have shown that spontaneous symmetry breaking generates disjoint ergodic regions in phase space, related by the symmetry that has been broken, our argument does *not* imply that these are the *only* sets of microstates in phase space. That is, we must logically allow, in principle, for the possibility that the symmetry-related ergodic regions in phase space do *not* exhaust phase space. To explain the interpretation of this statement, let us again use our simple example. If an ergodic region σ and its symmetry-related counterparts do exhaust phase space, then apart from simple translations (and rotations), there is only one way for the solid state to exist, corresponding to the microstates contained within σ. Hence, if there are, for example, *two* sets σ_1 and σ_2, each

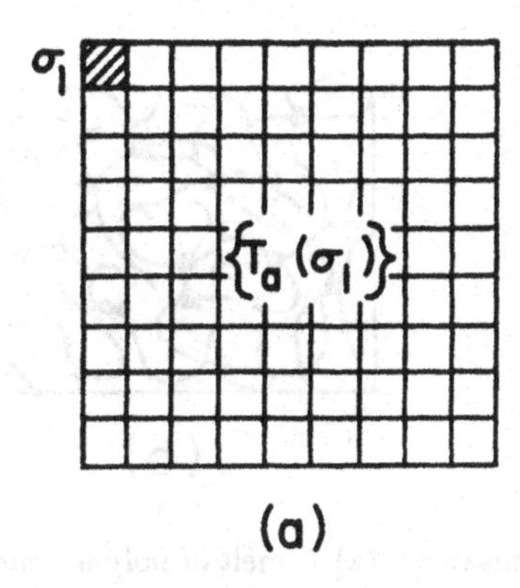

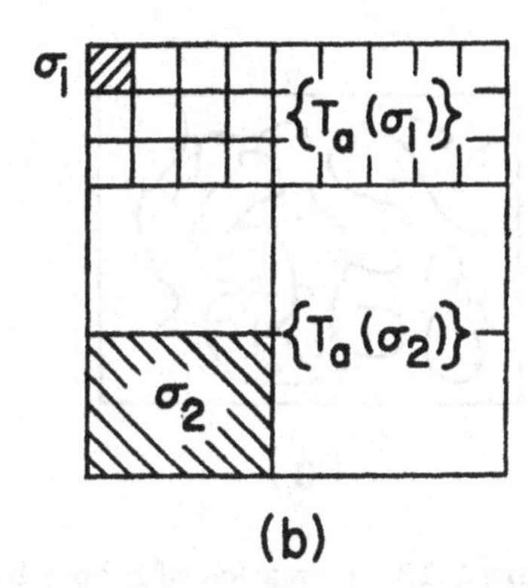

Figure 2.12 Ergodicity breaking accompanying spontaneous symmetry breaking. (a) Symmetry-related ergodic regions exhaust phase space. (b) Symmetry-related regions do NOT exhaust phase space. This situation is thought to occur in some disordered systems, where it is known as "replica symmetry breaking".

with their symmetry-related counterparts, which are required to exhaust phase space, then we conclude that there are two ways for the solid state to exist, one corresponding to the microstates in σ_1, the other to the microstates in σ_2.

This situation is illustrated schematically in figure (2.12). This latter possibility does in fact occur in some systems, most notably in some disordered systems, such as **spin glasses**[15] and **rubber**,[16] where it is referred to, for historical reasons, as **replica symmetry breaking**.[17]

---

[15] See the review by K. Binder and A.P. Young, *Rev. Mod. Phys.* **58**, 801 (1986).

[16] See P. Goldbart and N. Goldenfeld, *Phys. Rev. Lett.* **58**, 2676 (1987); *Phys. Rev.* **39**, 1402 (1989); *ibid.* **39**, 1412 (1989), where a detailed discussion is given of the replica method and the statistical mechanics of cross-linked polymeric materials.

[17] This seems to have been proposed first in the context of spin glasses by A. Blandin, *J. Phys. (Paris) Colloq.* **C6-39**, 1499 (1978). The key developments are given by G. Parisi, *Phys. Rev. Lett.* **43**, 1754 (1979); *J. Phys. A* **13**, L155 (1980); *ibid.* **13**, 1101 (1980); *ibid.* **13**, 1887 (1980); *Phys. Rev. Lett.* **50**, 1946 (1983).

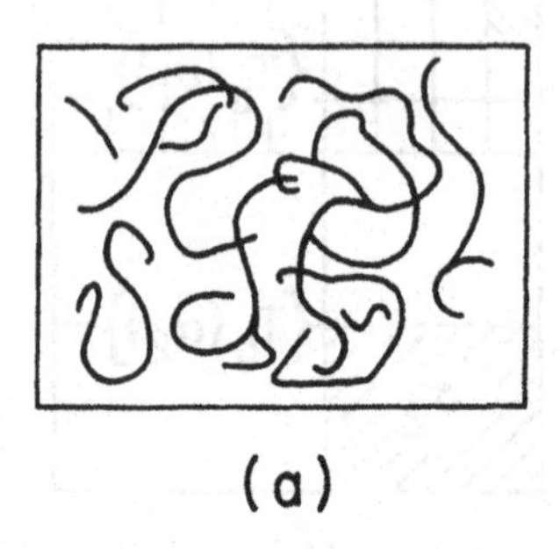

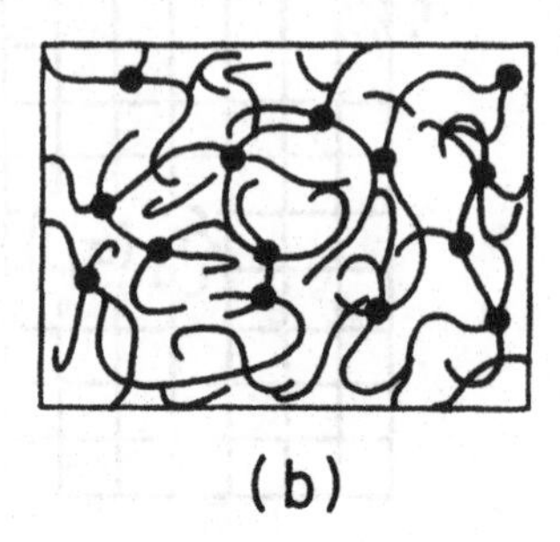

Figure 2.13 Formation of rubber by vulcanisation. (a) A melt of polymer molecules. (b) A cross-linked melt of polymer molecules.

2.10.3 Example of Replica Symmetry Breaking: Rubber

In this section, we will describe a physical example of replica symmetry breaking — the vulcanisation of rubber to form an **equilibrium amorphous solid**. The use of the term "equilibrium" is significant here: common glass is an amorphous solid too, but is almost certainly not in equilibrium. We will postpone a discussion on the origin of the term "replica symmetry breaking" until section 2.10.5.

Consider the situation shown schematically in figure (2.13a), where a melt of polymer molecules is confined within a box. This corresponds to how rubber starts out in life — as liquid latex. Rubber is formed by adding sulphur to the system, and baking; as depicted in figure (2.13b), this causes strong chemical bonds to be formed between the sulphur atoms and two different repeat units (monomers), which may be on the same polymer or on different polymers. When a certain fraction of monomers have been cross-linked, the system is found to become a **solid**.

Here, we will not present the detailed theory of this phase transition, but in the spirit of this chapter, show how, in principle, it may be described using the concept of ergodicity breaking. To do this, we will adopt a simple **minimal model** of a polymer, a model that captures the essential features and incorporates the remaining details into phenomenological parameters. For polymers, such a model has the following characteristics:

(1) A polymer is a featureless line object. Details of the chemical structure, side-group and back-bone composition are ignored. This is sometimes referred to as a *long wavelength* description: when viewed with poor resolution, this description of a polymer would seem to be accurate.

(2) The lines representing the polymers interact with a potential U. This models the complex forces between the actual chemical units

making up the chain. The details of the potential are of course determined by the factors omitted under (1). These find their expression in the precise form of the potential. The interactions are of two sorts: interactions between monomers on the *same* polymer and interactions between monomers on *different* polymers. In both cases, the interaction must have a hard core, representing the fact that atoms cannot occupy the same point in space, and a tail representing the actual forces between monomers. If the polymers are immersed in a solvent, then interactions with the solvent molecules also must be taken into account.

(3) The interactions U in (2) seem formidable. Yet certain observables, such as the spatial extent of a given polymer, turn out to be independent of the finer details of the interactions, for long chains, and crude modelling of the interactions is sufficient. This is reminiscent of the notion of **universality**, and indeed, RG methods have played an essential role in studying the statistical physics of polymers.[18]

(4) Dynamics: polymer molecules may not pass through each other.

(5) Cross-links simply staple together two polymer strands, at some given point along their arclengths. Note that a cross-link does not occupy a fixed position in space, although its position is fixed along the arclength of each of the participating polymer molecules. The cross-links are supposed to have formed at random along the arclength of the chains during the vulcanisation process.

With these ingredients, let us now consider the phase space of a cross-linked polymer system. Let Γ denote the phase space of the system of *cross-linked polymers, but with the dynamics modified so that the polymers may pass through one another.* This phase space is the starting point for the detailed theory of the statistical mechanics of cross-linked systems. This is shown schematically in figure (2.14a). Now imagine that at time $t = 0$, the physical dynamics is restored: polymers may not pass through one another (figure (2.14b)). In the absence of cross-links, the system would still be ergodic within the whole of Γ. However, the combination of the cross-links and the physical dynamics (4) means that there will be configurations of the chains which are *topologically inequivalent.* A simple example is given in figure (2.14c): with the modified dynamics (when $t < 0$) the two configurations may evolve into one another, whereas with the physical dynamics (when $t > 0$) it is impossible to go from one configuration to the other. This observation implies that if the system is at point A in Γ at $t = 0$, then for $t > 0$, only those configurations topologically equivalent to that of A may be sampled. Thus we conclude that Γ is

---

[18] A review of some aspects of polymer science, from the renormalisation group point of view, is given by Y. Oono, *Adv. Chem. Phys.* **61**, 301 (1985).

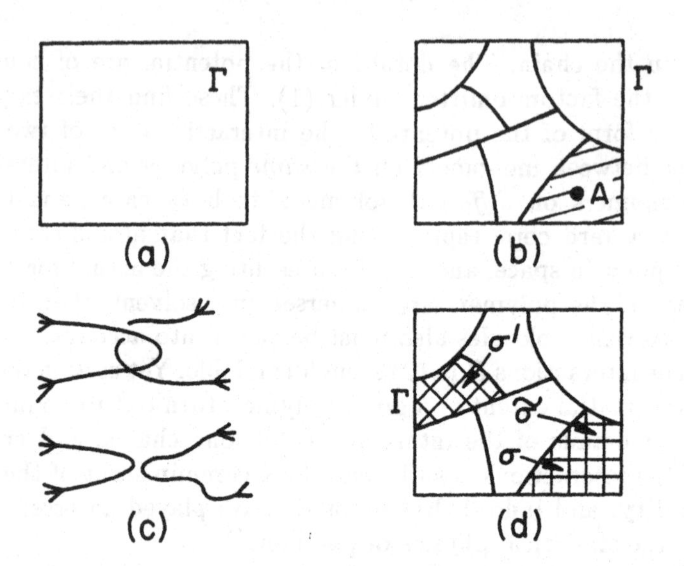

Figure 2.14 Phase space Γ of a system of cross-linked polymers. (a) Polymers allowed to pass through one another. (b) Polymers not allowed to pass through one another after time $t = 0$. If the representative point of the system was at A at time $t = 0$, then the system is later trapped within the shaded ergodic region. (c) Two configurations which are unable to evolve into one another under the physical dynamics. (d) If the system has become a solid, then translational invariance is spontaneously broken. Ergodic regions labelled by σ and σ' are not related by translational symmetry, whereas $\tilde{\sigma}$ is related to σ by translational symmetry.

broken up into disjoint ergodic regions, which bear no particular relation to one another. They simply reflect the fact that, with a given disposition of cross-links along the arclength of the polymers in the network, there are *many* possible topologies of the network.

So far, we have not mentioned spontaneous symmetry breaking. Let us now suppose that beyond a particular number of cross-links, the system becomes a solid. From the considerations outlined in the preceding sections, we know that solidification will entail a breaking of phase space into disjoint ergodic regions, related by translational symmetry. This presumably occurs within each of the ergodic regions created by the topological constraint. Thus the *original* phase space Γ now contains *two* categories of ergodic region: those related by translational symmetry (σ and $\tilde{\sigma}$ in figure (2.14d)), and those which are not related by translational symmetry (σ and σ' in figure (2.14d)). Thus, the simplest possibility for the structure of phase space in this example is that Γ is fragmented into ergodic regions corresponding to the topological constraint, which are themselves further fragmented into translational symmetry-related ergodic regions only. A more complicated scenario is also possible, namely one in which each ergodic region corresponding to the topological constraint is *not* filled completely by one ergodic region and its symmetry-related counterparts; this would imply that with a specified set of cross-links there is more than one way for the system to solidify. Questions of this sort can only be addressed by detailed calculation. Both scenarios, nevertheless, are good examples of replica symmetry breaking.

2.10.4 Order Parameters and Overlaps in a Classical Spin Glass

The notion of order parameter was useful when we were discussing spontaneous symmetry breaking of the simplest kind, where there is only one way for the system to order. In this section, we will extend the notion of order parameter to situations where there is replica symmetry breaking, *i.e.* where there is more than one way for the system to order.

There are two principal complications that arise when replica symmetry breaking occurs. First, it is usually very hard to specify any given ergodic region σ; this, in turn, makes it impossible to perform the computation of statistical expectation values in this ergodic region. Secondly, each family of non-symmetry-related ergodic regions may yield a different value for the expectation value of a given quantity, since they correspond to different physical orderings of the system.

To obviate these difficulties, it has been found useful to employ the concept of the **overlap**. The basic idea is to form a comparison between the microstates in one ergodic region and those in another. The idea is easy to grasp from the simple example of a three dimensional Ising spin glass, which we shall describe here in a qualitative way only.

An Ising spin glass is nothing more than an Ising model, with the exchange constant J being a *random variable*, distributed with some variance about a mean value.[19] This can arise in certain magnetic alloys, where the magnetic moments are randomly distributed in the material. In such a system, it is not surprising that the ground state (at zero temperature) is not simply all spins up or down; instead the configuration which minimises the energy will be some sequence of up and down spins, dictated by the actual spatial variation of J in the particular sample being studied. Thus, the time average magnetic moment m_i at site i is not necessarily the same as that at another site j, but at zero temperature, it is certainly non-zero.

---

[19] J is random in space, but is constant in time. The spatial variation of J reflects the random positions of the magnetic moments and varies from sample to sample.

Now consider the system at temperature slightly above zero. The flipping of each spin does not destroy the low temperature phase, although the average value of the local moment is reduced from the zero temperature value: $0 < m_i(T) < m_i(0)$, as in a ferromagnet. On the other hand, at very high temperature, where entropic considerations outweigh energetics, the system is a paramagnet, and $m_i(T) = 0$. In the low temperature phase, known as the spin glass phase, the net magnetisation ($\sum_i m_i$) of the system is zero if J is distributed symmetrically above and below zero. Consequently, the net magnetisation cannot distinguish a low temperature ordered state of the spins from the high temperature paramagnetic phase, and so it is not a good order parameter here.

One way to circumvent this problem is to use the overlap, constructed as follows. Suppose that the phase space of the system is, at some value of the temperature, external fields and other coupling constants, broken into ergodic regions labelled by σ, σ', *etc.* Then let the overlap $q^{\sigma\sigma'}$ between ergodic regions σ and σ' be defined by

$$q^{\sigma\sigma'} \equiv \lim_{N(\Omega)\to\infty} \left| \frac{1}{N(\Omega)} \sum_{i=1}^{N(\Omega)} m_i^\sigma m_i^{\sigma'} \right|, \tag{2.109}$$

where

$$m_i^\sigma = \frac{1}{Z_\sigma} \mathrm{Tr}_\sigma S_i e^{-\beta H_\Omega} \tag{2.110}$$

is the average moment at site i when the system is in ergodic region σ.

The overlap has several desirable features. It is zero in the paramagnetic phase, but may be non-zero for appropriate choices of σ and σ' in the spin glass phase. In particular, $q^{\sigma\sigma} \neq 0$. If ergodic region σ_1' is the same as ergodic region σ_1, but all the spins are flipped (*i.e.* σ_1 and σ_1' are related by time-reversal symmetry), then $q^{\sigma\sigma_1} = q^{\sigma\sigma_1'}$, showing that the overlap does not favour any particular ergodic region over its symmetry-related counterparts.

If there is only one ergodic region (plus its time-reversal symmetry-related counterpart), then $q^{\sigma\sigma'}$ has only one possible value. On the other hand, if there is replica symmetry breaking, then $q^{\sigma\sigma'}$ can take on a range of values, depending upon which ergodic regions are being compared. Thus, $q^{\sigma\sigma'}$, if it could be calculated for all pairs of ergodic regions, could act not only as an order parameter, but could also indicate whether or not replica symmetry breaking occurs.

Unfortunately, $q^{\sigma\sigma'}$ cannot be calculated directly, in general. Instead, we consider the *probability distribution* for $q^{\sigma\sigma'}$:

$$P(q) \equiv \sum_{\sigma\sigma'} w^\sigma w^{\sigma'} \delta\left(q - q^{\sigma\sigma'}\right) \tag{2.111}$$

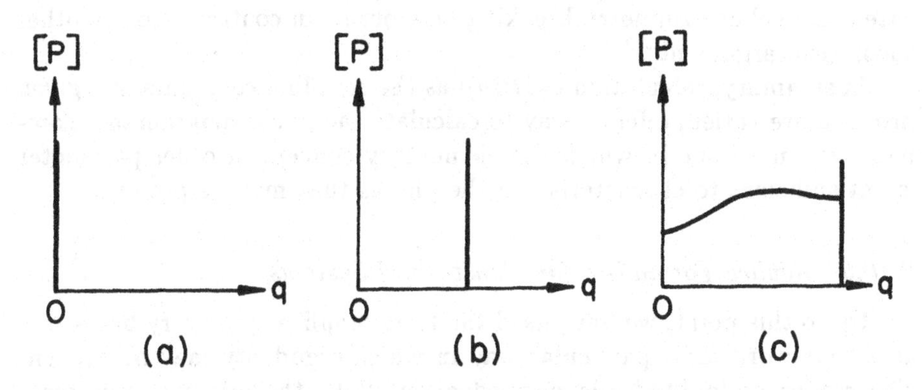

Figure 2.15 Possible forms for $[P]$. (a) No order (*e.g.* paramagnetic phase). (b) Ordering has occured, but there is only one way for the system to order (*e.g.* ferromagnetic phase). (c) Multiple ways for the system to order: replica symmetry breaking (*e.g.* putative spin glass phase).

where w^σ is the normalised Boltzmann weight for the ergodic region σ:

$$w^\sigma \equiv \frac{Z_\sigma}{\sum_\tau Z_\tau}. \tag{2.112}$$

The Boltzmann weights are included in the definition to ensure that P is normalised. The probability distribution, as defined here, varies from sample to sample as the actual values of J vary from sample to sample. So in practice, it is the probability distribution, averaged over the probability distribution for J, that is important and potentially calculable: this we denote by $[P]$. One important virtue of $[P]$ is the fact that, by definition, it is obtained by summing over *all* of phase space! This makes approximate calculations of this quantity feasible.

From the considerations in the preceding paragraph, we expect that $[P]$ has three generic forms, as illustrated in figure (2.15). In the absence of ordering, it is a delta function at the origin, because $m_i^\sigma = 0$. This would occur in the paramagnetic phase. In the presence of ordering, where there is only one way for the system to order (not counting symmetry-related counterparts), as in the ferromagnetic phase, $[P] = \delta(q - \tilde{q})$, where $\tilde{q} \neq 0$ is the unique value of the overlap. Finally, if there are many ways in which ordering can occur (*i.e.* replica symmetry breaking), as is thought to be the case in the spin glass phase then $[P(q)]$ is some non-zero function of q which is not a delta function at one value of q. At the time of writing, the nature of the low temperature phase of experimentally realisable spin glasses is a matter of debate. Certain model calculations[20], computer

[20] A. Georges, M. Mézard and J.S. Yedidia, *Phys. Rev. Lett.* **64**, 2937 (1990).

calculations on finite systems[21] and experiments on small samples [22] indicate that replica symmetry breaking may occur, in contradiction to other theoretical arguments.[23]

In summary, calculation of $[P(q)]$ as the coupling constants in a given problem are varied, offers a way to calculate the phase diagram of a complex system — one in which the elementary concept of order parameter is not sufficient to characterise all the phases that may be present.[24]

2.10.5 Replica Formalism for Constrained Systems

Up to this point, we have used the term "replica symmetry breaking" as a code-word for a particular way in which ergodicity may be broken. The reader undoubtedly is puzzled about both the origin of this term and how to implement the notions of ergodicity that we have described heuristically. In this section, we indicate the answer to these questions, at least in principle.

We begin by considering a system with **quenched disorder**. By this, we mean one whose Hamiltonian can be written in the form $H(\{S_i\}; X)$, where $\{S_i\}$ is the set of degrees of freedom (for convenience associated here with lattice sites $i = 1 \ldots N(\Omega)$) and X denotes some parameter or set of parameters in the Hamiltonian which is distributed at random according to its probability distribution P. In the example of the Ising spin glass discussed earlier,

$$H = -\sum_{ij} J_{ij} S_i S_j \tag{2.113}$$

with X representing the random variables J_{ij}, and the probability of J_{ij} taking on the value J being given by some probability distribution $P(J)$ such as

$$P(J) = \frac{1}{\sqrt{2\pi(\delta J)^2}} e^{-(J-J_0)^2/2(\delta J)^2}. \tag{2.114}$$

In the **infinite-range Ising spin glass** or as it is sometimes called, the **Sherrington-Kirkpatrick model**, the sum over i and j in eqn. (2.113) is

---

[21] J. Reger, R. Bhatt and A.P. Young, *Phys. Rev. Lett.* **64**, 1859 (1990).

[22] N.E. Israeloff, G.B. Alers and M.B. Weissman, *Phys. Rev.* **44**, 12613 (1991).

[23] W. McMillan, *J. Phys. C* **17**, 3179 (1984); A. Bray and M. Moore, in *Glassy Dynamics and Optimization*, edited by J. van Hemmen and I. Morgenstern (Springer, Berlin, 1986); D. Fisher and D. Huse, *Phys. Rev. Lett.* **56**, 1601 (1986).

[24] An explicit example where replica symmetry breaking occurs in a system *without* frustration is given by S.A. Janowsky, *Phys. Lett. A* **125**, 305 (1987).

not restricted to nearest neighbours; since a given spin may then interact with any other spin, no matter how far away in real space, this model Hamiltonian is sometimes considered to represent the physical system in an infinite dimensional space. Alternatively, each spin may be regarded as reacting to the average or mean field of all the other spins in the system; hence, the statement that **mean field theory** is exact for an infinite-range model. Most of the analytic work on spin glasses, with which the replica theory was developed, was for the infinite-range model. The variables X are referred to as *quenched*, because they are considered not to be able to equilibrate with the remainder of the system, as represented by $\{S_i\}$. For example, they might represent the magnetic moments of random impurity atoms in a solid. If the impurity atoms are able to diffuse around in the solid, and thus relax their distribution into an equilibrium state, they would be referred to as **annealed disorder**. On the other hand, quenched impurities are supposed to be bound to particular lattice sites on the timescale of a given experiment, and so cannot equilibrate.

The profound difference is clearly seen in the thermodynamics. For a system with annealed disorder, the free energy F_{ad} satisfies

$$\exp(-\beta F_{ad}) = \mathrm{Tr}_{\{S\},X} \exp[-\beta H(\{S_i\}; X)] \tag{2.115}$$

whereas for a system with quenched disorder, the free energy F_{qd} satisfies

$$\exp(-\beta F_{qd}(X)) = \mathrm{Tr}_{\{S\}} \exp[-\beta H(\{S_i\}; X)]. \tag{2.116}$$

In the expression for F_{ad}, the trace includes X, whereas in that for F_{qd}, the resultant free energy depends on X. In an experimental situation that is well-described by F_{qd}, how do we calculate statistical mechanical observables? Usually it is assumed that the experimental sample is sufficiently large that it may be considered to be composed of a large number of sub-systems, each of which is itself macroscopic and may be considered to be a realisation of the system with a particular choice for the quenched variables X. Thus a measurement of any observable in such a system corresponds to an average over all the sub-systems, *i.e.* an average over the ensemble of all realisations of X. Systems for which this assumption is valid are said to be **self-averaging**.

Thermodynamic quantities may be calculated from the free energy. From the above argument, the experimentally observed free energy F_e is given by

$$\begin{aligned} F_e = [F_{qd}(X)] &\equiv \int P(X) F_{qd}(X)\, dX \\ &= -k_B T \int P(X) \log Z(X)\, dX. \end{aligned} \tag{2.117}$$

The logarithm in this equation creates a technical problem, which may be circumvented using the identity

$$\log Z = \lim_{n\to 0} \frac{Z^n - 1}{n}. \tag{2.118}$$

Thus we must now calculate

$$\begin{aligned} [Z^n] &= \left[\prod_{\alpha=1}^{n} Z_\alpha\right] \\ &= \left[\mathrm{Tr}_{S^1} \ldots \mathrm{Tr}_{S^n} e^{-\beta \sum_{\alpha=1}^{n} H(\{S^\alpha\}, X)}\right] \\ &= \mathrm{Tr}_{S^1} \ldots \mathrm{Tr}_{S^n} \left[e^{-\beta \sum_{\alpha=1}^{n} H(\{S^\alpha\}, X)}\right]. \end{aligned} \tag{2.119}$$

This may be interpreted as the statistical mechanical trace of a system composed of n replicas of the original system (each with the same realisation of the quenched disorder) with $nN(\Omega)$ degrees of freedom $\{S_i^1\}$, $\{S_i^2\}$, ..., $\{S_i^n\}$. Here we label spin i in replica α by S_i^α. The effective Hamiltonian H_n of the replica system is given by

$$e^{-\beta H_n(\{S^\alpha\})} = \left[e^{-\beta \sum_{\alpha=1}^{n} H(\{S^\alpha\}, X)}\right] \tag{2.120}$$

which does not depend upon X because of the disorder average that has already been performed. The Hamiltonian for the n replicas of the system is symmetric under interchange or permutation of the replicas, but is *not* expressible as the sum of n Hamiltonians, one for each replica: the disorder average has *coupled* the replicas! This new statistical mechanical problem has the unusual feature that n is to be taken as tending towards zero. As long as n is an integer, the permutation symmetry of the Hamiltonian H_n is manifest. However, in the limit $n \to 0$, this symmetry can be spontaneously broken!

To appreciate this fact and its significance, we need to make contact between replica theory and ergodicity breaking. The central result of the detailed replica theory[25] is that the disorder average of the overlap probability distribution $[P(q)]$ can be calculated approximately using replicas. In the case of the infinite-range spin glass, the result is

$$[P(q)] = \lim_{n\to 0} \frac{1}{n(n-1)} \sum_{\alpha \neq \beta} \delta\left(q - q^{\alpha\beta}\right), \tag{2.121}$$

---

[25] See the articles by Parisi *op. cit*; see also those by Goldbart and Goldenfeld *op. cit.*

where

$$q^{\alpha\beta} = \left\langle S_i^\alpha S_i^\beta \right\rangle_n . \tag{2.122}$$

The expectation value in eqn. (2.122) is taken with respect to H_n. Since H_n is symmetric under permutation of the replicas (in accord with the rather artificial nature of their introduction into the problem) it might be thought that $q^{\alpha\beta}$ must be independent of which replicas α and β are chosen, *i.e.* that $q^{\alpha\beta}$ is equal to some number $\hat{q}$, dependent upon the temperature, external field, mean and standard deviation of the exchange constant, *etc.* However, it turns out that this is not correct: below the zero field spin glass transition temperature, $q^{\alpha\beta}$ is *not* independent of α and β — replica permutation symmetry is spontaneously broken! The question of how to give meaning to an order parameter which is a $n \times n$ matrix, in the limit that $n \to 0$, is a delicate one, and beyond the scope of these notes; the interested reader is invited to consult the seminal papers by Parisi, to which we have already referred.

The implication of the replica symmetry breaking may be seen by referring to eqn. (2.121) and the discussion in section 2.10.4. If $q^{\alpha\beta} = \hat{q}$, then $[P(q)] = \delta(q - \hat{q})$. This corresponds to simple spontaneous symmetry breaking, with ordering at a single value corresponding to $\hat{q}$. When replica symmetry is broken, however, then $[P(q)]$ is not a single delta function at a unique value of q. This means that the ergodicity has been broken, with *more than one* family of ergodic regions not related by symmetry.

We have now come full circle in our description of how ergodicity can be broken at a phase transition. The replica method may seem to be nothing more than a convenient trick to perform a disorder average. However, the interpretation of replica symmetry breaking in terms of ergodicity breaking indicates that it actually represents a complex encoding of phase space. The replica method has been of great value not only in studies of disordered systems, but in neural network models, graph theory and optimisation.[26] There can be little doubt that this strange and fascinating technique will become better understood and further used in the future.

2.11 FLUIDS

In this section, we will review the statistical mechanical formalism for fluids, and demonstrate a connection between the description of magnets and the description of fluids.

---

[26] For a survey, see M. Mézard, G. Parisi and M.A. Virasoro, *Spin Glass Theory and Beyond* (World Scientific, Singapore, 1987).

We consider a continuum fluid, which may be classical or quantum, in a system Ω. The system may have hard walls (*i.e.* be in a box), specified by the potential $U_i(\mathbf{r})$. Suppose there are N particles, labelled by $i = 1, 2, \ldots N$.

The kinematics is specified by the co-ordinates $\mathbf{r}_i$ and momenta $\mathbf{p}_i$, and the Hamiltonian is

$$H_\Omega = \sum_{i=1}^{N} \left[\frac{p_i{}^2}{2m} + U_1(\mathbf{r}_i) \right] + \frac{1}{2!} \sum_{i \neq j} U_2(\mathbf{r}_i, \mathbf{r}_j) + \frac{1}{3!} \sum_{i \neq j \neq k} U_3(\mathbf{r}_i, \mathbf{r}_j, \mathbf{r}_k) + \ldots \tag{2.123}$$

$U_1, U_2, U_3, \ldots$ are the one-body, two-body, three-body potentials *etc.* In general, the potentials will be *translationally invariant*, and depend only on *differences* between the co-ordinates of the particles. The potential U_1 represents an external field acting on each particle, such as gravity; usually, we will set it equal to zero.

Normally, we work in the grand canonical ensemble, with the grand partition function

$$\Xi_\Omega = \mathrm{Tr}\,_\Omega e^{-\beta(H_\Omega - \mu N)} \tag{2.124}$$

where μ is the chemical potential,

$$\mathrm{Tr}\,_\Omega \equiv \sum_{N=0}^{\infty} \frac{1}{N!} \int \prod_{i=1}^{N} \frac{(d^d\mathbf{p}_i)(d^d\mathbf{r}_i)}{(h)^{dN}} \tag{2.125}$$

in *classical* statistical mechanics and h is Planck's constant, included to provide continuity with the formulae from quantum statistical mechanics. Dimensional analysis requires the presence of a constant with the dimensions of action, so it is conventional to choose it to be h.

It is conventional to separate out the contributions from kinetic energy and potential energy as follows. We substitute eqn. (2.123) into eqn. (2.124), and re-order the integrals to give

$$\Xi_\Omega = \sum_{N=0}^{\infty} \frac{1}{N!} \prod_{i=1}^{N} \left\{ \int_{-\infty}^{\infty} \frac{d^d\mathbf{p}_i}{h^d} e^{-\beta p_i^2/2m} \right\} \int \prod_{i=1}^{N} d^d\mathbf{r}_i \, e^{-\beta(U\{\mathbf{r}_i\} - \mu N)}, \tag{2.126}$$

where $U\{\mathbf{r}_i\}$ represents the potential terms in H_Ω. Now

$$\int_{-\infty}^{\infty} \frac{d^d\mathbf{p}}{h^d} e^{-\beta p^2/2m} = \frac{1}{\Lambda_T^d} \tag{2.127}$$

where the thermal wavelength is defined to be

$$\Lambda_T \equiv \frac{h}{\sqrt{2\pi m k_B T}}. \tag{2.128}$$

Then eqn. (2.126) becomes

$$\Xi_\Omega = \sum_{N=0}^{\infty} \frac{1}{N!} \left(\frac{e^{\beta\mu}}{\Lambda_T^d} \right)^N Q_N \tag{2.129}$$

where the **configurational sum** is

$$Q_N \equiv \int \prod_{i=1}^{N} d^d \mathbf{r}_i e^{-\beta U\{\mathbf{r}_i\}} \tag{2.130}$$

The variable $e^{\beta\mu}$ is sometimes referred to as the **fugacity**.

The **grand free energy** is given by

$$F_\Omega(T, \mu, U_1, U_2, \ldots) = -k_B T \log \Xi_\Omega. \tag{2.131}$$

Sometimes, the grand free energy is referred to as the **grand potential**, and is denoted by the symbol Ω. F_Ω is not singular, despite the infinite sum in eqn. (2.129). The reason is that, in practice, the potentials U_2, $U_3 \ldots$ have a hard-core, and so there is a maximum number of particles N_{cp} that may be packed into the finite volume $V(\Omega)$:

$$N \leq N_{cp} \tag{2.132}$$

and the $\sum_{N=0}^{\infty}$ does in fact have a finite upper limit. Then, the proofs for analyticity and the existence of the thermodynamic limit go through pretty much as in the magnetic case, with similar qualifications (range of potential, shape of region Ω). The grand free energy density is given by

$$f_b(T, \mu, U_1, U_2, \ldots) = \lim_{V(\Omega)\to\infty} \frac{F_\Omega}{V(\Omega)}, \tag{2.133}$$

with differential

$$df_b(T, \mu) = -\sigma dT - \rho d\mu \tag{2.134}$$

where σ is the entropy per unit volume and the number density is

$$\rho = \lim_{V(\Omega)\to\infty} \frac{\langle N \rangle_\Omega}{V(\Omega)}. \tag{2.135}$$

The grand free energy density is minus the pressure p, a result that follows from thermodynamics.[27]

[27] See, for example, L.D. Landau and E.M. Lifshitz *Statistical Physics Part 1 (Third edition)* (Pergamon, New York, 1980) §24.

As in the magnetic case, a variety of analytic properties may be proven, many of which derive from the convexity of f_b:

(a) $p(T, \mu) \geq 0$.
(b) $p(T, \mu)$ is continuous.
(c) $\partial p/\partial T$, $\partial p/\partial \mu$ exist almost everywhere.
(d) $\partial p/\partial \mu = \rho \geq 0$.
(e) $\partial p/\partial T = \sigma \geq 0$ (True in quantum statistical mechanics only, but false for classical systems).
(f) $\partial p/\partial T$ is monotonic non-decreasing, which implies that the heat capacity $C_\mu \geq 0$.
(g) $\partial p/\partial \mu$ is monotonic non-decreasing, which implies that the isothermal compressibility K_T is non-negative.

2.12 LATTICE GASES

One of the reasons for the importance of the Ising model is that a variety of other statistical mechanical systems can be simulated by it. This is the topic of **equivalence** between models, or more precisely **exact equivalence** or **mapping.** Now we discuss a simple model for the statistical mechanics of a fluid — the **lattice gas**, due to Lee and Yang. The basic idea is to relate the local density of particles in a fluid to the local up-spin density of a magnet: We will demonstrate the equivalence in two steps, and in so doing, we will, incidentally, expose the advantage of the grand canonical ensemble over the canonical ensemble.

As a preliminary step, recall that the potential terms in the Hamiltonian (2.123) may be re-written in terms of the microscopic density of the fluid

$$\rho(\mathbf{r}) \equiv \sum_{i=1}^{N} \delta(\mathbf{r} - \mathbf{r}_i). \tag{2.136}$$

This expression is the microscopic density, because

$$\int_V \rho(\mathbf{r}) d^3\mathbf{r} = N(V) \tag{2.137}$$

where $N(V)$ is the number of particles in the arbitrary volume V. Then, using the property of the delta function that

$$\int f(x)\delta(x-a)\,dx = f(a), \tag{2.138}$$

we write

$$\sum_i U_1(\mathbf{r}_i) = \sum_i \int_\Omega U_1(\mathbf{r})\delta(\mathbf{r} - \mathbf{r}_i)\,d^d\mathbf{r} = \int_\Omega d^d\mathbf{r}\, U_1(\mathbf{r})\rho(\mathbf{r}) \tag{2.139}$$

and

$$\sum_{i\neq j} U_2(\mathbf{r}_i - \mathbf{r}_j) = \sum_{i\neq j} \int_\Omega U_2(\mathbf{r}_i - \mathbf{r}')\delta(\mathbf{r}' - \mathbf{r}_j) d^d\mathbf{r}'$$
$$= \sum_{i\neq j} \int_\Omega \int_\Omega U_2(\mathbf{r} - \mathbf{r}')\delta(\mathbf{r} - \mathbf{r}_i)\delta(\mathbf{r}' - \mathbf{r}_j)\, d^d\mathbf{r}\, d^d\mathbf{r}'$$
$$= \int_\Omega \int_\Omega U_2(\mathbf{r} - \mathbf{r}')\rho(\mathbf{r})\rho(\mathbf{r}')\, d^d\mathbf{r}\, d^d\mathbf{r}'. \tag{2.140}$$

These expressions are not directly useful as they stand, because the degrees of freedom in the grand canonical trace operation are the co-ordinates, not the microscopic density. We will shortly see how we can effectively make a change of variables to the density co-ordinates; but first, we need to discuss how to represent a fluid system by a spin system.

2.12.1 Lattice Gas Thermodynamics from the Ising Model

Consider a d-dimensional lattice, with co-ordination number z. Each site can be occupied by a single molecule or not at all. The occupation number of the ith site, n_i, takes the values 0 and 1 only. The total number of particles in the system is

$$N = \sum_{i=1}^{N(\Omega)} n_i. \tag{2.141}$$

The occupation number n_i is rather like the microscopic density $\rho(\mathbf{r})$ in the continuum fluid. Thus we might guess a suitable Hamiltonian for the lattice gas of the form

$$H_\Omega = \sum_{i=1}^{N(\Omega)} U_1(i)n_i + \frac{1}{2}\sum_{i,j=1}^{N(\Omega)} U_2(i,j)n_i n_j + O(n_i n_j n_k), \tag{2.142}$$

so that, in the grand canonical ensemble,

$$H_\Omega - \mu N = \sum_i (U_i - \mu)n_i + \frac{1}{2}\sum_{ij} U_2(i,j)n_i n_j + \ldots \tag{2.143}$$

The factor of 1/2 in the above equations avoids double counting the contribution to the energy from the interaction between two particles. The Hamiltonian (2.142) only represents the potential energy of the gas; but this does not matter, since as we have seen, the kinetic energy only contributes to the fugacity. In the following, we will actually model the configurational sum Q_N, rather than the full partition function.

To make contact with the Ising model, define

$$n_i = \frac{1}{2}(1+S_i), \qquad S_i = \pm 1 \tag{2.144}$$

$$n_i = 0 \Longleftrightarrow S_i = -1 \tag{2.145}$$

$$n_i = 1 \Longleftrightarrow S_i = +1. \tag{2.146}$$

Substitution into eqn. (2.143) gives

$$\sum_i (U_1(i)-\mu)\frac{1}{2}(1+S_i) = \frac{1}{2}\sum_i (U_1(i)-\mu) + \frac{1}{2}\sum_i (U_1(i)-\mu)S_i \tag{2.147}$$

and

$$\begin{aligned}\frac{1}{2}\sum_{ij} U_2(i,j)n_i n_j =& \frac{1}{2}\cdot\frac{1}{4}\sum_{ij} U_2(i,j)(1+S_i)(1+S_j) \\ =& \frac{1}{2}\cdot\frac{1}{4}\sum_{ij} U_2(i,j) + \frac{1}{2}\cdot\frac{1}{4}\sum_{ij} U_2(i,j)S_i\cdot 2 \\ &+ \frac{1}{2}\cdot\frac{1}{4}\sum_{ij} U_2(i,j)S_i S_j\end{aligned} \tag{2.148}$$

If the forces between fluid particles are short-ranged and the density is sufficiently low, then we can ignore three and higher body potentials, and model the two-body potential by

$$U_2(i,j) = \begin{cases} U_2 & i \text{ and } j \text{ nearest neighbors;} \\ 0 & \text{otherwise.} \end{cases} \tag{2.149}$$

Then the right hand side of eqn. (2.148) becomes

$$\frac{1}{4}U_2\cdot N(\Omega)z\cdot\frac{1}{2} + \frac{2}{2}\cdot\frac{1}{4}U_2\cdot z\cdot\sum_i S_i + \frac{U_2}{4}\cdot\sum_{<ij>} S_i S_j. \tag{2.150}$$

Setting $U_1 = 0$, we find that

$$H_\Omega - \mu N = E_0 - \sum_i S_i H_i - J\sum_{<ij>} S_i S_j \tag{2.151}$$

with

$$E_0 = -\frac{1}{2}N(\Omega)\mu + U_2N(\Omega)z/8 \tag{2.152}$$

$$-H = -\frac{1}{2}\mu + \frac{1}{4}U_2z \tag{2.153}$$

$$-J = \frac{U_2}{4}. \tag{2.154}$$

Thus

$$\begin{aligned}\Xi_{\text{lattice gas}} &= \text{Tr}\, e^{-\beta(H_\Omega - \mu N)} \\ &= \left(\prod_{i=1}^{N(\Omega)} \sum_{n_i=0,1}\right) e^{-\beta(H_\Omega-\mu N)} \\ &= e^{-\beta E_0} Z_{\text{Ising}}(H, J, N(\Omega))\end{aligned} \tag{2.155}$$

This is our desired result — the thermodynamic properties of the lattice gas may be obtained from the thermodynamics of the Ising model. The reader is invited to investigate this in a subsequent example, which shows that the ideal gas law and corrections to it, the equation of state *etc.* may all be derived from the lattice gas model.

2.12.2 Derivation of Lattice Gas Model from the Configurational Sum

In the previous section, we showed that the lattice gas model is related to a spin system. In this section, we derive the lattice gas model directly from the configurational sum for a fluid. These two results together serve to illustrate an equivalence between fluid and magnetic systems.

We approximate Q_N by dividing space Ω up into cells of linear dimension a, such that probability of finding more than one molecule per cell is negligible: *i.e.* $a \leq$ hard core radius. Then, the measure in Q_N may be replaced heuristically by

$$\int_\Omega \prod_{i=1}^{N} d^d\mathbf{r}_i \approx a^{dN(\Omega)} \sum_{\alpha=1}^{N(\Omega)}, \tag{2.156}$$

where α labels the cells and

$$1 \leq \alpha \leq N(\Omega) \equiv \frac{V(\Omega)}{a^d}. \tag{2.157}$$

Note the distinction between N, the number of particles, and $N(\Omega)$, the number of cells. Next, we replace the interaction $U\{\mathbf{r}_i\}$ between the *particles* by the interaction energy between *occupied cells*:

$$U_2(\mathbf{r}_i, \mathbf{r}_j) = U_2(\alpha, \beta) \quad \text{if } \mathbf{r}_i \in \text{cell } \alpha \text{ and } \mathbf{r}_j \in \text{ cell } \beta. \tag{2.158}$$

As before, we will assume that the forces are short-ranged, and that $U_2(\alpha, \beta) = 0$ unless cells α and β are neighbours, in which case U_2 has a constant value independent of α and β. Then

$$\begin{aligned} U\{\mathbf{r}_i\} \to U\{n_\alpha\} &= \sum_{\alpha\beta} U_2(\alpha\beta) n_\alpha n_\beta + \ldots \\ &= U_2 \sum_{<\alpha\beta>} n_\alpha n_\beta + \ldots \end{aligned} \tag{2.159}$$

Note that n_α depends upon all the particle co-ordinates: $n_\alpha = n_\alpha\{\mathbf{r}_i\}$, and that

$$\sum_\alpha n_\alpha = N = \int_\Omega d^d\mathbf{r} \sum_{i=1}^{N} \delta(\mathbf{r} - \mathbf{r}_i). \tag{2.160}$$

For each configuration specified by $\{n_\alpha\}$, there correspond $N!$ configurations specified by $\{\mathbf{r}_i\}$, because we can freely interchange the particles. Thus

$$Q_N = N!\, a^{dN} \sum_{\{n_\alpha = 0,1\}}{}' e^{-\beta U\{n_\alpha\}} \tag{2.161}$$

and

$$\Xi_\Omega = \sum_{N=0}^{\infty} \left[e^{\beta\mu} \left(\frac{a}{\Lambda_T} \right)^d \right]^N \sum_{\{n_\alpha\}}{}' e^{-\beta U\{n_\alpha\}} \tag{2.162}$$

where $\sum'$ signifies a sum over the occupation numbers subject to the constraint that the total number of particles is fixed:

$$\sum_\alpha n_\alpha = N. \tag{2.163}$$

This constraint makes calculation of the configurational sum very difficult, *but* the grand partition function is relatively simple, as we now explain.

Consider the constrained sum below, where f is some functional:

$$\begin{aligned} \sum_{N=0}^{\infty} \sum_{\{n_\alpha\}}{}' f(\{n_\alpha\}) &= \underbrace{\sum_{\{n_\alpha\}} f(n_\alpha)}_{\sum_\alpha n_\alpha = 0} + \underbrace{\sum_{\{n_\alpha\}} f(n_\alpha)}_{\sum_\alpha n_\alpha = 1} + \ldots + \underbrace{\sum_{\{n_\alpha\}} f(n_\alpha)}_{\sum_\alpha n_\alpha = \infty} \\ &= \underbrace{\sum_{\{n_\alpha\}} f(n_\alpha)}_{\text{unrestricted}} \end{aligned} \tag{2.164}$$

In the final line, all $2^{N(\Omega)}$ possible states of the variables $\{n_\alpha\}$ are included in the sum, and not just those that satisfy the constraint (2.163). This is the advantage of the grand canonical ensemble. Thus

$$\Xi_\Omega = \sum_{\{n_\alpha\}} \exp -\beta \left\{ \left[-\mu - \frac{d}{\beta} \log(a/\Lambda_T) \right] \sum_\alpha n_\alpha + U_2 \sum_{<\alpha\beta>} n_\alpha n_\beta + \dots \right\}, \tag{2.165}$$

and so we see that

$$\Xi_\Omega = \mathrm{Tr}\ e^{-\beta(H_\Omega - \bar{\mu} N)} = \Xi_{\text{lattice gas}} \tag{2.166}$$

where

$$\bar{\mu} \equiv \mu_{\text{lattice gas}} = \mu_{\text{physical}} + dk_B T \log(a/\Lambda_T). \tag{2.167}$$

Thus, we have shown, albeit heuristically, that the grand partition function of the lattice gas follows from that of the continuum fluid.

2.13 EQUIVALENCE IN STATISTICAL MECHANICS

The preceding analysis has illustrated an **equivalence** between two physical systems. This term is used in two different ways in statistical mechanics, so it is appropriate to mention these usages here.

(A) *Exact equivalence.* This refers to an exact mapping of one model into another, meaning that there is an exact relation between the partition functions of the two models. An example is the Ising model $\longleftrightarrow$ lattice gas equivalence.

(B) *Approximate equivalence.* This refers to models whose partition functions are *not* equal or related by an exact mapping, but which, nevertheless, behave identically near a critical point. Such models are said to be in the same **universality class.** An example, which we shall meet later, is the approximate equivalence between the long wavelength effective Hamiltonian used in the Landau theory of phase transitions and the Ising model. In the former, the free energy density of the system is, crudely speaking, expanded as a power series in the magnetisation M, to order M^4, whereas the free energy density of the Ising model, when written in these terms, actually involves the quantity $\log\cosh M$.

The phase diagram, thermodynamics, correlation functions *etc.* of a system depend on

(a) Specific values of coupling constants K_i.

(b) Symmetry of model: discrete, continuous, actual symmetry group.

(c) Type of lattice (if present).
(d) Dimensionality d.
(e) Everything else you can think of!

whereas the *critical behaviour* turns out only to depend on

(a) Symmetry of order parameter.
(b) Dimensionality.
(c) Nature of critical point.

and *not* on specific values of coupling constants, lattice type, or the precise form of the model Hamiltonian.[28] Clearly, we expect models related by exact equivalence relations to have identical critical behaviour. We would *not* expect *a priori* that systems, which are not related by exact equivalence, would have the same critical behaviour. Nevertheless, experiment and theory both provide many examples where this does in fact occur! This is the phenomenon of universality.

2.14 MISCELLANEOUS REMARKS

2.14.1 History of the Thermodynamic Limit

The importance of the thermodynamic limit was first realized by Hendrik Kramers.[29] He came to the realization that the limiting process $\lim_{N\to\infty}(1/N)\log Z$ can give distinct functions, not necessarily joined analytically. As late as 1937, most physicists were unaware of this work, and even Sommerfeld believed that the partition function described only one phase (and could not describe liquid-gas coexistence, for example)! At a congress in Amsterdam to commemorate the birth of Van der Waals, there was confusion as to whether the partition function could give a sharp phase transition. Kramers, as chairman, put the issue to a vote, but the outcome was "inconclusive"!

In 1941, Kramers and Wannier used the transfer matrix method to find the critical temperature of the $d = 2$ Ising model, assuming the existence of a single transition in the thermodynamic limit. Later, in 1944,

---

[28] A notable exception to the above statement is the Eight-vertex model, where one of the critical exponents *does* depend continuously on the coupling constants. See R.J. Baxter *Exactly Solved Models in Statistical Mechanics* (Academic, New York, 1989). This behaviour is well-understood within the framework of renormalisation group theory, and only occurs for certain models with marginal operators.

[29] H. Kramers, *Commun. Kamerlingh Onnes Lab.* **22**, *Suppl.* 83, 1 (1936)). An interesting account of the history of the thermodynamic limit and Kramers' contributions to statistical mechanics may be found in the article by M. Dresden, *Physics Today*, Sept. 1988, pp. 26–33.

Lars Onsager solved the $d = 2$ Ising model *exactly* in one of the greatest *tour de forces* of mathematical physics.[30] This demonstrated beyond any doubt:

(1) The relevance of the thermodynamic limit;
(2) That critical behaviour can disagree with mean field theory and critical exponents need not be those of the Weiss ferromagnet or the Van der Waals gas (see later).

2.14.2 Do Quantum Effects Matter?

The properties of matter, particularly condensed matter (meaning systems with strong interactions between the constituent parts), are usually attributed to quantum mechanics. Whilst this assertion is indisputable in a literal sense, it is by no means obvious that all the gross or qualitative characteristics of condensed matter are determined principally by quantum mechanical considerations. For example, in section 1.2.4, we mentioned the self-avoiding walk model of a polymer in solution: although quantum mechanics certainly accounts for the existence of the polymer, and the details of the bond energies which result in the flexibility of the chain, the average end-to-end distance and certain other statistical properties on long length scales are independent of the microscopic details.[31]

Is the behaviour at phase transitions sensitive to quantum mechanical effects? For first-order transitions, it is not ruled out, but the author is not aware of any examples. For second-order transitions ($T > 0$), the answer is no! As $\xi \to \infty$, the thermodynamic behaviour is determined by regions whose size is much greater than length scales at which quantum effects are important. Thus intrinsically quantum systems, such as ferromagnets or superconductors, can be successfully modelled by classical systems, such as the Ising model. For $T = 0$ quantum effects *are* important.[32] But a *quantum* system at $T = 0$ in d dimensions is related to a *classical* system in $d + 1$ dimensions: thus even at zero temperature, classical statistical mechanics can be used.[33]

---

[30] L. Onsager, *Phys. Rev.* **65**, 117 (1944).

[31] A complete discussion of the levels of description of condensed matter physics, and the rôle of quantum effects in particular is given by M.E. Fisher, in *Proc. Bohr Symp.*, H. Feshbach (ed.) (Gordon and Breach, New York, 1988), pp 65-115.

[32] Quantum critical phenomena are discussed by J.A. Hertz, *Phys. Rev. B* **14**, 1165 (1976).

[33] The cleanest way to show this is with path integrals; see (*e.g.*) A.M. Polyakov, *Gauge Fields and Strings* (Harwood, New York, 1987), pp. 1-4.

EXERCISES

Exercise 2–1

This question concerns the convexity of the free energy for the nearest neighbour Ising model:

$$-H_\Omega\{S\} = H\sum_i S_i + J\sum_{<ij>} S_i S_j$$

(a) Show that $g(\beta) \equiv \lim_{N(\Omega)\to\infty} \log Z_\Omega(\beta)/N(\Omega)$ is convex down in β.
(b) By considering the second derivative of $g(\beta)$ or otherwise, show that the free energy per unit volume $f(T) = -(1/\beta)g(\beta)$ is convex up in T. Try to be as careful as you can.
(c) The Gibbs free energy is defined as

$$\Gamma_\Omega(M) = F_\Omega(H(M)) + N(\Omega)MH(M)$$

where $H(M)$ is given implicitly by $M(H) = -\partial f/\partial H$. Show that $\Gamma(M)$ is convex down in M.
(d) Sketch the form of $\Gamma(M)$, $f(H)$, $M(H)$, $M(T)$, $H(M)$, for the cases $T > T_c$ and $T < T_c$. Quantities without the subscript Ω are taken in the thermodynamic limit. What is the form of the quantities above for a finite system Ω?

Exercise 2–2

This question concerns the infinite-range Ising model, where the coupling constant $J_{ij} = J$ for all (sometimes abbreviated by the symbol $\forall$) i, j (*i.e.* no restriction to nearest neighbour interactions).

$$-H_\Omega\{S\} = H\sum_i S_i + \frac{J_0}{2}\sum_{ij} S_i S_j$$

You will solve this model using a method referred to as the **Hubbard-Stratonovich transformation, or auxiliary fields.** Although it is nothing more than completing the square, this technique is one of the most useful tricks in the physicist's arsenal. The virtue of the present example is that you will be able to calculate a partition function in an essentially exact way, and see precisely how it is that the thermodynamic limit or the zero temperature limit are required in order for there to be a phase transition.

(a) Explain why this model only makes sense if $J_0 = J/N$, where N is the number of spins in the system.

(b) Prove that

$$\exp\left\{\frac{a}{2N}x^2\right\} = \int_{-\infty}^{\infty} \frac{dy}{\sqrt{2\pi/Na}} e^{-\frac{Na}{2}y^2+axy} \qquad \text{Re } a > 0.$$

(c) Hence show that

$$Z_\Omega = \int_{-\infty}^{\infty} \frac{dy}{\sqrt{2\pi/N\beta J}} e^{-N\beta L}$$

where

$$L = \frac{J}{2}y^2 - \frac{1}{\beta}\ln\{2\cosh(\beta(H+Jy))\}.$$

When can this expression become non-analytic?

(d) In the thermodynamic limit, this integral can be evaluated exactly by the method of steepest descents. Show that

$$Z(\beta, H, J) = \sum_i e^{-\beta NL(H,J,\beta,y_i)}$$

and find the equation satisfied by y_i. What is the probability of the system being in the state specified by y_i? Hence show that the magnetisation is given by

$$M \equiv \lim_{N(\Omega)\to\infty} \frac{1}{\beta N(\Omega)} \frac{\partial \ln Z_\Omega}{\partial H} = y_0$$

where y_0 is the position of the global minimum of L.

(e) Now consider the case $H = 0$. By considering how to solve the equation for y_i graphically, show that there is a phase transition and find the transition temperature T_c. Discuss the acceptability of all the solutions of the equation for y_i both above and below T_c.

(f) Calculate the isothermal susceptibility

$$\chi_T \equiv \frac{\partial M}{\partial H}.$$

For $H = 0$, show that χ_T diverges to infinity both above and below T_c, and find the leading and next to leading behaviour of χ_T in terms of the reduced temperature $t = (T - T_c)/T_c$.

CHAPTER 3

How Phase Transitions Occur in Practice

We have explored in some detail the basic notions of phase transitions, and the manner in which non-analytic thermodynamic behaviour may occur in principle. Now it is time to see these ideas in action. This chapter presents some concrete calculations of phase transition behaviour in magnetic systems. Not only will we see explicitly the occurence of non-analytic behaviour, but we shall also make precise concepts related to spatial correlations. Most of this chapter will be concerned with the **transfer matrix** method, first introduced by Kramers, which reduces the problem of calculating the partition function to the problem of finding the eigenvalues of a certain matrix. We also describe the **low temperature expansion** and Weiss' **mean field theory** for magnets.

3.1 AD HOC SOLUTION METHODS

We start with the Ising model Hamiltonian with nearest neighbor interactions in one dimension:

$$-H_{\Omega}\{S\} = H\sum_{i} S_i + J\sum_{<ij>} S_i S_j, \quad J > 0. \tag{3.1}$$

Define

$$\begin{aligned} h &\equiv \beta H \\ K &\equiv \beta J. \end{aligned} \tag{3.2}$$

Then the partition function for N spins is

$$Z_\Omega(h,K,N) = \sum_{\{S\}} \exp\left[h\sum_i S_i + K\sum_i S_iS_{i+1}\right]. \tag{3.3}$$

3.1.1 Free Boundary Conditions and $H = 0$

First we solve the case with no constraints on the boundary spins, and with the external field $H = 0$. Then

$$Z_\Omega = \sum_{S_1}\cdots\sum_{S_N} \exp\left(K\sum_{i=1}^{N-1} S_iS_{i+1}\right). \tag{3.4}$$

Define a new variable

$$\eta_i = S_iS_{i+1} \quad \text{where } i = 1\ldots N-1 \tag{3.5}$$

so that

$$\eta_i = \begin{cases} +1 & S_i = S_{i+1}; \\ -1 & S_i = -S_{i+1}. \end{cases} \tag{3.6}$$

To specify completely the state of the system, we need to provide the set of numbers $\{S_1, \eta_1, \ldots, \eta_{N-1}\}$. Hence we can write the partition function as

$$\begin{aligned} Z_\Omega &= \sum_{S_1}\sum_{\eta_1}\cdots\sum_{\eta_{N-1}} e^{K(\eta_1+\eta_2+\cdots+\eta_{N-1})} \\ &= 2\,(2\cosh K)^{N-1} \end{aligned} \tag{3.7}$$

The first factor of 2 comes from the sum over S_1 and the $N-1$ factors of $2\cosh K$ arise from factorising the sums over the η variables.

3.1.2 Periodic Boundary Conditions and $H = 0$

Now we solve the same problem, but with periodic boundary conditions:

$$S_{N+1} = S_1. \tag{3.8}$$

The partition function is now

$$Z_\Omega = \sum_{S_1} \cdots \sum_{S_N} \exp\left(K \sum_{i=1}^{N-1} S_i S_{i+1} + K S_N S_1\right). \tag{3.9}$$

In this case, the state of the system is still specified by the variables $\eta_0 \equiv S_1, \eta_1, \ldots, \eta_{N-1}$. In terms of these variables, we can write

$$S_N S_1 = \eta_1 \eta_2 \ldots \eta_{N-1}, \tag{3.10}$$

where we have used the fact that $S_i^2 = 1$. The technical difficulty arises from implementing the periodic boundary conditions properly. One way to proceed is to use eqn. (3.10) in the expression for Z_Ω:

$$Z_\Omega = \sum_{\eta_0} \cdots \sum_{\eta_{N-1}} e^{K(\eta_1 + \cdots + \eta_{N-1}) + K\eta_1\eta_2 \ldots \eta_{N-1}} \tag{3.11}$$

$$= 2\sum_{\eta_1} \cdots \sum_{\eta_{N-1}} e^{K(\eta_1 + \cdots + \eta_{N-1})} \sum_{\alpha=0}^{\infty} \frac{(K\eta_1 \ldots \eta_{N-1})^\alpha}{\alpha!} \tag{3.12}$$

$$= 2\sum_{\alpha=0}^{\infty} \frac{K^\alpha}{\alpha!} \left[\sum_{\eta} \eta^\alpha e^{K\eta}\right]^{N-1} \tag{3.13}$$

$$= 2\sum_{\alpha=0}^{\infty} \frac{K^\alpha}{\alpha!} \left[e^K + (-1)^\alpha e^{-K}\right]^{N-1} \tag{3.14}$$

$$= (2\cosh K)^N + (2\sinh K)^N \tag{3.15}$$

3.1.3 Recursion Method for $H = 0$

Another method for $h = 0$ is to use recursion. The partition function for a chain of N spins is

$$Z(N) = \sum_{S_1} \cdots \sum_{S_N} e^{KS_1S_2 + KS_2S_3 + \cdots KS_{N-1}S_N}. \tag{3.16}$$

where we have used free, not periodic, boundary conditions. From this we work out the partition function for a chain with $N + 1$ spins:

$$Z(N+1) = \sum_{S_1} \cdots \sum_{S_N} \sum_{S_{N+1}} e^{K(S_1S_2 + \cdots + S_{N-1}S_N)} e^{KS_NS_{N+1}}. \tag{3.17}$$

The sum $\sum_{S_{N+1}} \to e^{KS_N} + e^{-KS_N}$. Now do the remaining sums to obtain

$$Z(N+1) = Z(N) 2\cosh K. \tag{3.18}$$

Iterating this recursion relation gives

$$Z(N+1) = Z(1)(2\cosh K)^N, \tag{3.19}$$

i.e., $Z(N) = Z(1)(2\cosh K)^{N-1}$. But

$$Z(1) = \sum_{\{S_1\}} 1 = 2 \tag{3.20}$$

Thus,

$$Z(N) = 2(2\cosh K)^{N-1} \tag{3.21}$$

in agreement with eqn. (3.7).

3.1.4 Effect of Boundary Conditions

Now that we have calculated the partition function in zero field using two different boundary conditions, let us see how they affect the thermodynamics. The free energy, calculated for the case with free boundary conditions is

$$F_\Omega(N, 0, K) = -Nk_BT\left[\log 2 + \frac{N-1}{N}\log(\cosh K)\right]. \tag{3.22}$$

As $N \to \infty$,

$$F_\Omega = -k_BTN\left[\log(2\cosh K) + O\left(\frac{1}{N}\right)\right], \tag{3.23}$$

which is what we would obtain from using periodic boundary conditions. As expected, the difference between boundary conditions becomes negligible as the system size grows.

3.2 THE TRANSFER MATRIX

The *ad hoc* methods of the previous section are difficult to apply in the case when $H \neq 0$. The **transfer matrix** method generalises the ad hoc methods, and can be used for $h \neq 0$, to compute correlations *etc.*

We start with the case of periodic boundary conditions: $S_{N+1} = S_1$.

$$Z_N(h,K) = \text{Tr}\ e^{h\sum_i S_i + K\sum_i S_i S_{i+1}}. \tag{3.24}$$

Now we factorise the summand as follows:

$$Z_N(h,K) = \sum_{S_1}\cdots\sum_{S_N}\left[e^{\frac{h}{2}(S_1+S_2)+KS_1S_2}\right]\cdot\left[e^{\frac{h}{2}(S_2+S_3)+KS_2S_3}\right]\cdots$$
$$\cdots\left[e^{\frac{h}{2}(S_N+S_1)+KS_NS_1}\right]. \tag{3.25}$$

We can think of each term as being like a matrix element of a matrix $\mathbf{T}$:

$$T_{S_1S_2} = e^{\frac{h}{2}(S_1+S_2)+KS_1S_2}. \tag{3.26}$$

Observe that S_1 and S_2 are the *labels* of the matrix elements:

$$\mathbf{T} = \begin{pmatrix} T_{11} & T_{1-1} \\ T_{-11} & T_{-1-1} \end{pmatrix} = \begin{pmatrix} e^{h+K} & e^{-K} \\ e^{-K} & e^{-h+K} \end{pmatrix}. \tag{3.27}$$

Then

$$Z_N(h,K) = \sum_{S_1}\cdots\sum_{S_N} T_{S_1S_2}T_{S_2S_3}T_{S_3S_4}\cdots T_{S_NS_1}. \tag{3.28}$$

To see what this means, recall that the rule of matrix multiplication is that the matrix elements are given by

$$\mathbf{A} = \mathbf{B}\cdot\mathbf{C} \longleftrightarrow A_{ij} = \sum_k B_{ik}C_{kj}. \tag{3.29}$$

Also, the **trace** of a matrix is defined as

$$\text{Tr}\ (\mathbf{A}) = \sum_i A_{ii}. \tag{3.30}$$

Then we see that performing the sums $\sum_{S_2}\cdots\sum_{S_N}$ in equation (3.28) is just matrix multiplication. Thus

$$Z_N(h,K) = \sum_{S_1} T^N_{S_1S_1} = \text{Tr}\ (\mathbf{T}^N). \tag{3.31}$$

How do we compute Tr $(\mathbf{T}^N)$? The factor $\mathbf{T}^N$ is at first sight rather forbidding, but we can readily diagonalise $\mathbf{T}$, by multiplying and post-multiplying with a matrix $\mathbf{S}$ whose rows and columns are eigenvectors of

$\mathbf{T}$, *i.e.* perform a similarity transformation. Since $\mathbf{T}$ is real and symmetric $\mathbf{S}^T = \mathbf{S}^{-1}$ and the diagonalised form of $\mathbf{T}$ is

$$\mathbf{T}' = \mathbf{S}^{-1}\mathbf{T}\mathbf{S}. \tag{3.32}$$

This gives

$$\mathbf{T}' = \begin{pmatrix} \lambda_1 & 0 \\ 0 & \lambda_2 \end{pmatrix} \tag{3.33}$$

where λ_1 and λ_2 are the eigenvalues of $\mathbf{T}$. Now use the cyclic property of the trace operation, which implies that

$$\mathrm{Tr}\,(\mathbf{T}) = \mathrm{Tr}\,(\mathbf{T}') \tag{3.34}$$

to give

$$\mathrm{Tr}\,(\mathbf{T}^N) = \lambda_1^N + \lambda_2^N. \tag{3.35}$$

Now consider the case that $\lambda_1 \neq \lambda_2$; the degenerate case will be discussed in the next section. Assuming that $\lambda_1 > \lambda_2$, we have

$$Z_N(h,K) = \lambda_1^N \left(1 + \left[\frac{\lambda_2}{\lambda_1}\right]^N\right) \tag{3.36}$$

and in the thermodynamic limit $N \to \infty$

$$Z_N(h,K) \simeq \lambda_1^N \left(1 + O(e^{-\alpha N})\right), \tag{3.37}$$

where $\alpha \equiv \log(\lambda_1/\lambda_2)$ is a positive constant. So only the largest eigenvalue of the transfer matrix is important in the thermodynamic limit. The free energy is then given by

$$\lim_{N\to\infty} \frac{F_N(h,K,T)}{N} = -k_B T \log \lambda_1. \tag{3.38}$$

We can easily compute λ_1. The eigenvalues of $\mathbf{T}$ are given by

$$\det \begin{vmatrix} e^{h+K} - \lambda & e^{-K} \\ e^{-K} & e^{-h+K} - \lambda \end{vmatrix} = 0. \tag{3.39}$$

Solving, we obtain

$$\lambda_{1,2} = e^K \left[\cosh h \pm \sqrt{\sinh^2 h + e^{-4K}}\right]. \tag{3.40}$$

and

$$\begin{aligned} \frac{F_N(h,K)}{N} &= -k_B T \log \left\{ e^K \left[\cosh h + \sqrt{\sinh^2 h + e^{-4K}}\right]\right\} \\ &= -J - k_B T \log \left[\cosh h + \sqrt{\sinh^2 h + e^{-4K}}\right] \end{aligned} \tag{3.41}$$

since $K \equiv \beta J$. This is the general result for the free energy of the one dimensional Ising model in an external magnetic field.

3.3 PHASE TRANSITIONS

How does a phase transition arise? For $T > 0$, the argument of the square root in eqn. (3.41) for the free energy is positive, for real h and K. This expression for f is manifestly analytic for $T > 0$, which agrees with our earlier heuristic argument that there can be no phase transition for $T > 0$ in the one-dimensional case.

More systematically, the only possibilities for the occurence of a phase transition are: that λ_1 is a non-analytic function of K and h, that the square root vanishes, in which case the eigenvalues become degenerate: $\lambda_1 = \lambda_2$, or that $\lambda_1 = 0$. For $T > 0$ none of these things can happen. We see this here by inspection, but it also follows from **Perron's theorem**[1]

Theorem: For an $N \times N$ matrix ($N < \infty$) **A** with $A_{ij} > 0$ for all i, j, the eigenvalue of largest magnitude is:

(a) real and positive,
(b) non-degenerate,
(c) an analytic function of A_{ij}.

In one dimension, the transfer matrix for finite-ranged interactions satisfies the requirements of Perron's theorem. Thus, we immediately find that

$$(a) \Rightarrow \lambda_1 \neq 0$$
$$(b) \Rightarrow \lambda_1 \neq \lambda_2$$
$$(c) \Rightarrow \lambda_1 \text{ is analytic}$$

and hence that there is no phase transition for $T > 0$. Perron's theorem can be used to show the absence of phase transitions at non-zero temperature in one dimension for systems with sufficiently short-ranged interactions. In terms of the mathematics, what is special about one dimension: why can the theorem not be applied in *higher* dimensions to show that phase transitions *never* occur for $T > 0$? In one dimension the transfer matrix for the Ising model is a 2×2 matrix. A little reflection (see the examples at the end of the chapter) reveals that in two and higher dimensions, the transfer matrix is a $\infty \times \infty$ matrix in the thermodynamic limit, and so Perron's theorem does not apply.

Now let us see what happens at $T = 0$ or equivalently $K \to \infty$. The largest eigenvalue of the transfer matrix becomes

$$\lambda_1 = e^K \left[\cosh h + \sqrt{\sinh^2 h \left(1 + O(e^{-4K})\right)}\right]. \qquad (3.42)$$

---

[1] See F.R. Gantmacher, *Applications of the Theory of Matrices* (Interscience, New York, 1959), p. 64 *et seq.*

Recalling that

$$\sqrt{x^2} = |x|, \tag{3.43}$$

we obtain

$$\lambda_1 = e^K \left[\cosh h + |\sinh h| \left(1 + O(e^{-4K})\right)\right]. \tag{3.44}$$

Now

$$\cosh h + |\sinh h| = \begin{cases} \frac{1}{2}\left(e^h + e^{-h} + e^h - e^{-h}\right) & h > 0; \\ \frac{1}{2}\left(e^h + e^{-h} - e^h + e^{-h}\right) & h < 0. \end{cases} \tag{3.45}$$

In other words,

$$\cosh h + |\sinh h| = e^{|h|} \tag{3.46}$$

and

$$\lambda_1 = e^{K+|h|}. \tag{3.47}$$

Thus,

$$F = -Nk_BT\,(K + |h|) + O(T^2) \tag{3.48}$$

so that at $T = 0$,

$$F = -N(J + |H|). \tag{3.49}$$

As mentioned earlier, we do not need to take the thermodynamic limit to obtain non-analytic behaviour at $T = 0$. The magnetisation is

$$M = -\frac{1}{N}\frac{\partial F}{\partial H} = \begin{cases} 1 & H > 0; \\ -1 & H < 0, \end{cases} \tag{3.50}$$

as expected. Note that the non-analytic behaviour came from a term $\sqrt{h^2} = |h|$. For $T > 0$, this term is of the form (see eqn. (3.40)) $\sqrt{h^2 + \epsilon^2}$ which is analytic at $h = 0$ as long as the constant $\epsilon \neq 0$.

3.4 THERMODYNAMIC PROPERTIES

We start with the case $h = 0$. Then

$$\lambda_1 = e^K(1 + e^{-2K}) = 2\cosh K \tag{3.51}$$

and $Z = (2\cosh K)^N$ as $N \to \infty$. This is not in contradiction with eqn. (3.15), because in the thermodynamic limit, we have dropped the contribution of λ_2 to eqn. (3.51), which in this case is $(2\sinh K)^N$.

The free energy is then

$$F = -k_BT\,N\left[K + \log(1 + e^{-2K})\right] \tag{3.52}$$

with low and high temperature limits

$$f_N \equiv F/N = \begin{cases} -J & T \to 0 \ (K \to \infty); \\ -k_B T \log 2 & T \to \infty \ (K \to 0). \end{cases} \tag{3.53}$$

As anticipated, the high temperature limit corresponds to the entropy dominating the free energy, whereas the low temperature behaviour is determined mainly by the energy.

The specific heat is simply obtained from the internal energy E:

$$E = -\frac{\partial}{\partial \beta} \log Z \tag{3.54}$$

$$= -N \frac{\partial}{\partial \beta} \log(2 \cosh \beta J) \tag{3.55}$$

$$= -NJ \tanh \beta J. \tag{3.56}$$

Therefore

$$C = \frac{dE}{dT} = -\frac{1}{k_B T^2} \frac{dE}{d\beta}$$

$$= \frac{NJ^2}{k_B T^2} \mathrm{sech}^2(J/k_B T). \tag{3.57}$$

The heat capacity does not exhibit any singularity, but note the presence of a peak near $J \sim k_B T$, which is sometimes known as a **Schottky anomaly**.

To calculate the magnetisation, write the free energy per spin in the form

$$\frac{F}{N} = -J - k_B T \log \left[\cosh h + \sqrt{\sinh^2 + w^2} \right], \tag{3.58}$$

where $w^2 \equiv e^{-4K}$ is the relative probability of the two configurations below, which differ by a single spin flip:

$$\uparrow\uparrow\uparrow\uparrow\uparrow \qquad \uparrow\uparrow\downarrow\uparrow\uparrow \, .$$

Thus

$$M = -\frac{1}{N} \frac{\partial F}{\partial H} = -\frac{1}{N k_B T} \frac{\partial F}{\partial h}$$

$$= \frac{\partial}{\partial h} \log \left[\cosh h + \sqrt{\sinh^2 h + w^2} \right]$$

$$= \frac{\sinh h}{\sqrt{\sinh^2 h + w^2}}. \tag{3.59}$$

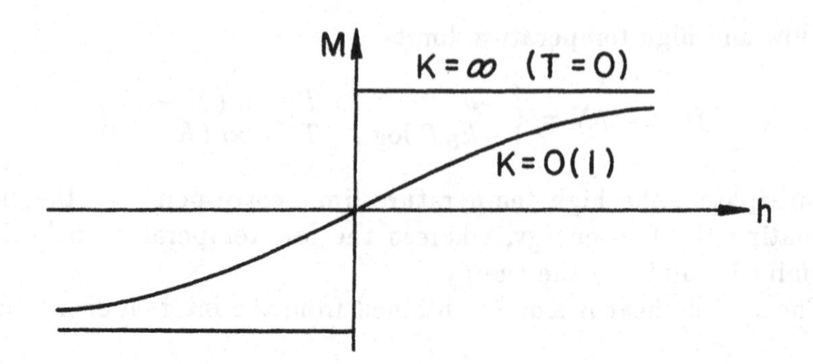

Figure 3.1 Magnetisation as a function of external field at zero temperature ($K = \infty$) and at non-zero temperature ($K = O(1)$).

The isothermal magnetic susceptibility χ_T describes how the magnetisation changes in response to an external field:

$$\chi_T \equiv \frac{\partial M}{\partial H}. \tag{3.60}$$

What happens for small fields $h \to 0$? Using $\sinh h \sim h$ and

$$M \simeq \frac{h}{w} = he^{2K} \tag{3.61}$$

$$= e^{2K}\frac{H}{k_B T} \tag{3.62}$$

we find

$$\chi_T = \frac{e^{2K}}{k_B T} = \frac{e^{2J/k_B T}}{k_B T}. \tag{3.63}$$

The low and high temperature behaviours are

$$\chi_T \sim \begin{cases} 1/k_B T, & \text{as } T \to \infty \text{ (Curie's Law);} \\ e^{(2J/k_B T)}/k_B T, & \text{as } T \to 0. \end{cases} \tag{3.64}$$

Note the exponential divergence at zero temperature.

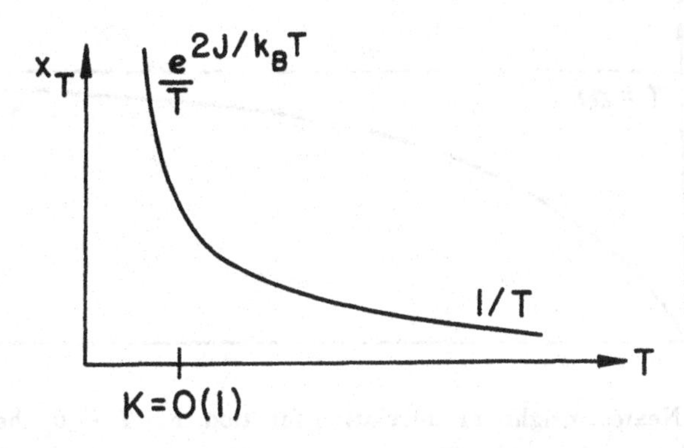

Figure 3.2 Isothermal susceptibility as a function of temperature at vanishing field.

3.5 SPATIAL CORRELATIONS

We can use both the transfer matrix and an *ad hoc* approach to calculate correlation functions.

3.5.1 Zero Field: Ad Hoc Method

For $T > 0$, we have seen that

$$\langle S_i \rangle = 0. \tag{3.65}$$

The two-point **correlation function** is defined as

$$G(i,j) = \langle S_i S_j \rangle - \langle S_i \rangle \langle S_j \rangle = \langle S_i S_j \rangle \tag{3.66}$$

for $h = 0$ and $i < j$. The two-point correlation function may be written in the slightly more transparent form:

$$G(i,j) = \langle (S_i - \langle S_i \rangle)(S_j - \langle S_j \rangle) \rangle \tag{3.67}$$

showing that G actually measures the correlation in the *fluctuation* of the spins at different sites. $G(i,j)$ is also related to the probability P_{ij} that spins i and j have the same value:

$$P_{ij} \equiv \langle \delta_{S_i S_j} \rangle \tag{3.68}$$

$$= \left\langle \frac{1}{2}(1 + S_i S_j) \right\rangle \tag{3.69}$$

$$= \frac{1}{2} + \frac{1}{2}\left[G(i,j) + \langle S_i \rangle \langle S_j \rangle\right]. \tag{3.70}$$

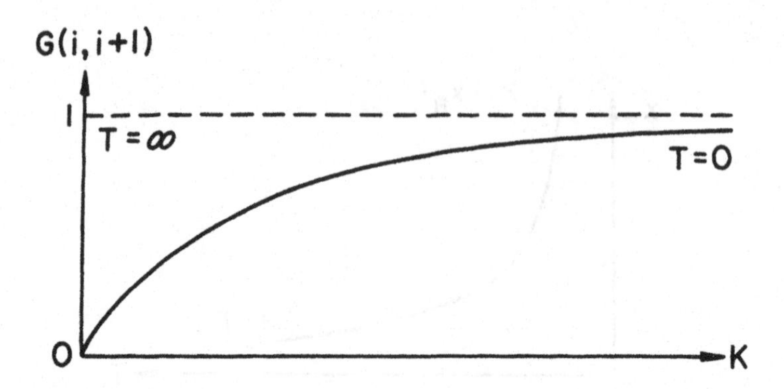

Figure 3.3 **Nearest neighbour correlation function. For $T \to 0$, the neighbours are strongly correlated, whereas for $T \to \infty$, the neighbours are not correlated at all.**

Proceeding to the calculation, we first replace the exchange between spins i and j by J_{ij}, a value that, in principle, varies from bond to bond on the chain. Then

$$G(i,j) = \frac{1}{Z_N\{K_i\}} \sum_{S_1} \cdots \sum_{S_N} S_i S_j e^{K_1 S_1 S_2 + K_2 S_2 S_3 + \ldots K_{N-1} S_{N-1} S_N}. \quad (3.71)$$

Now we will derive the form of $G(i,j)$ by successive differentiation. There are two steps to the argument. First, we let $j = i+1$, and calculate $G(i,i+1)$. Then we show how we can obtain $G(i,i+j)$ from this.

Step 1: $j = i+1$. Write eqn. (3.71) as

$$\begin{aligned} \langle S_i S_{i+1} \rangle &= \frac{1}{Z_N} \frac{\partial}{\partial K_i} \sum_{S_i} \cdots \sum_{S_N} e^{K_1 S_1 S_2 + \ldots + K_{N-1} S_{N-1} S_N} \\ &= \frac{1}{Z_N} \frac{\partial}{\partial K_i} Z_N = \frac{\partial}{\partial K_i} \log Z_N\{K_i\}. \end{aligned} \quad (3.72)$$

We can find $Z_N\{K_i\}$ from our previous argument when all the $K_i = K$ — we simply get factors of $\cosh K_i$ instead of $\cosh K$ at each recursion. Thus

$$Z_N\{K_i\} = 2^N \prod_{i=1}^{N-1} \cosh K_i \quad (3.73)$$

and

$$\langle S_i S_{i+1} \rangle = \tanh K_i = \tanh \beta J_i. \quad (3.74)$$

Putting $K_i = K$, and inspecting the sketch, shown in figure (3.3), we see that neighbours are strongly correlated at low temperature ($K \to \infty$), whereas at high temperature ($K \to 0$), the correlations diminish.

We could also have got the result (3.73) by calculating the energy.

$$E = \langle H_\Omega \rangle = -J \sum_i \langle S_i S_{i+1} \rangle \tag{3.75}$$

$$= -NJ \tanh \beta J \quad \text{(from eqn. (3.56))}. \tag{3.76}$$

But $\langle S_i S_{i+1} \rangle$ does not depend on i in a translationally invariant state, so

$$-J \sum_i \langle S_i S_{i+1} \rangle = -NJ \langle S_i S_{i+1} \rangle = -NJ \tanh \beta J. \tag{3.77}$$

Thus

$$\langle S_i S_{i+1} \rangle = \tanh \beta J \tag{3.78}$$

in agreement with eqn. (3.74).

Step 2: How do we go beyond $\langle S_i S_{i+1} \rangle$? Notice that

$$\frac{1}{Z_N} \frac{\partial}{\partial K_i} \frac{\partial}{\partial K_{i+1}} Z_N = \langle S_i S_{i+1} S_{i+1} S_{i+2} \rangle$$

$$= \left\langle S_i S_{i+2} (S_{i+1})^2 \right\rangle \tag{3.79}$$

$$= \langle S_i S_{i+2} \rangle . \tag{3.80}$$

where we have used the fact that $S_i^2 = 1$. Therefore

$$G(i, i+2) = \tanh K_i \cdot \tanh K_{i+1}. \tag{3.81}$$

Proceeding by induction,

$$G(i, i+j) = \langle S_i S_{i+j} \rangle$$

$$= \frac{1}{Z_N} \frac{\partial}{\partial K_i} \frac{\partial}{\partial K_{i+1}} \cdots \frac{\partial}{\partial K_{i+j-1}} Z_N\{K_i\} \tag{3.82}$$

$$= \tanh K_i \tanh K_{i+1} \ldots \tanh K_{i+j-1}. \tag{3.83}$$

Now set all the $K_i = K$:

$$G(i, i+j) = (\tanh K)^j . \tag{3.84}$$

The result is translationally invariant as expected when K_i does not depend on i. $G(i, i+j)$ only depends on j. In a continuous system such as a fluid, the analogous statement is that the two-point correlation function satisfies $G(\mathbf{r}, \mathbf{r}') = G(\mathbf{r} - \mathbf{r}')$. This applies even for a finite system, as long as i, j are not near the boundaries, and for a system with periodic boundary conditions.

3.5.2 Existence of long-range order

What do we learn from the two-point correlation function? As expected, there are two possible cases:

For $T = 0$ ($K = \infty$), we have $\tanh K = 1$ and

$$G(i, i + j) = 1 \quad \text{for all } j. \tag{3.85}$$

Thus, at zero temperature, the probability that $S_i = S_j$ is unity for all j. This is a perfectly correlated state, and one that exhibits **long range order.** The two states of the system consistent with this are

$$\begin{aligned} &\uparrow\uparrow\uparrow\uparrow\uparrow \cdots \uparrow \\ \text{and } &\downarrow\downarrow\downarrow\downarrow\downarrow \cdots \downarrow \end{aligned} \tag{3.86}$$

For $T \neq 0$ ($K \neq \infty$) we have that $\tanh K < 1$ and $\coth K > 1 \Rightarrow \log \coth K > 0$. Therefore, for $j > 0$, we can write

$$G(i, i + j) = e^{-j \log(\coth K)}, \tag{3.87}$$

showing that correlations decay *exponentially* for $T > 0$. The **correlation length** ξ is defined by

$$G(i, i + j) = e^{-j/\xi}, \tag{3.88}$$

where ξ is measured in units of the lattice spacing a. From eqn. (3.87) we can read off

$$\xi = \frac{1}{\log(\coth K)}. \tag{3.89}$$

From eqn. (3.88), we see that ξ is a measure of the length over which spins are correlated with probability ~ 1. As $K \to \infty$ *i.e.* as $T \to 0$, the correlation length diverges to infinity, whereas at high temperature, $\xi \to 0$. How does ξ approach infinity as $T \to 0$? For $K \gg 1$

$$\coth K = \frac{e^K + e^{-K}}{e^K - e^{-K}} \simeq 1 + 2e^{-2K} + O(e^{-4K}). \tag{3.90}$$

Therefore

$$\xi = \frac{e^{2K}}{2} = \frac{1}{2} e^{J/k_B T} \tag{3.91}$$

as $T \to 0$. There is an essential singularity in the correlation length as the temperature approaches zero, which is the transition temperature in this model: ξ diverges exponentially fast. In fact, we shall generally find that near a continuous transition, the correlation length diverges with some exponent ν, but the correlation length varies according to $\xi(T) \sim (T - T_c)^{-\nu}$ as $T \to T_c$, and not $\xi \sim \exp[J/k_B(T - T_c)]$, as we have in this one dimensional example.

3.5.3 Transfer Matrix Method

Correlation functions can also be computed using the transfer matrix formalism. The basic trick is as follows:

$$\langle S_i \rangle = \frac{1}{Z} \sum_{S_1} \cdots \sum_{S_N} e^{-\beta H_\Omega} S_i \tag{3.92}$$

$$= \frac{1}{Z} \sum_{S_1} \cdots \sum_{S_N} \left[T_{S_1 S_2} T_{S_2 S_3} \cdots T_{S_{i-1} S_i} S_i T_{S_i S_{i+1}} \cdots \right] \tag{3.93}$$

Focus on the section of the string of transfer matrices near S_i:

$$\cdots \sum_{S_i} T_{S_{i-1} S_i} S_i T_{S_i S_{i+1}} \cdots \tag{3.94}$$

The result of this string is itself a matrix, **A**, whose *ab* element is

$$A_{ab} = \sum_{S_i} T_{a S_i} T_{S_i b} S_i. \tag{3.95}$$

But this is the same as

$$\mathbf{A} = \mathbf{T} \begin{pmatrix} 1 & 0 \\ 0 & -1 \end{pmatrix} \mathbf{T}. \tag{3.96}$$

The matrix sandwiched between the transfer matrices is one of the **Pauli matrices**, usually denoted by σ_z. Finally, using the cyclic property of the trace operation,

$$\langle S_i \rangle = \frac{1}{Z} \mathrm{Tr}\ (\sigma_Z T^N). \tag{3.97}$$

We can evaluate this expression by using the similarity transformation eqn. (3.32) and the cyclic property of the trace:

$$\langle S_i \rangle = \frac{\mathrm{Tr}\ \left[\mathbf{S}^{-1} \sigma_Z \mathbf{S}\, (\mathbf{T}')^N \right]}{\mathrm{Tr}\ (\mathbf{T}')^N}. \tag{3.98}$$

We can explicitly compute **S**: let

$$X^{(1)} = \begin{pmatrix} a \\ b \end{pmatrix} \tag{3.99}$$

$$X^{(2)} = \begin{pmatrix} c \\ d \end{pmatrix} \tag{3.100}$$

be normalised eigenvectors of $\mathbf{T}$ with eigenvalues $\lambda_{1,2}$. Then

$$\mathbf{S} = \begin{pmatrix} a & c \\ b & d \end{pmatrix}. \tag{3.101}$$

Write

$$\mathbf{S}^{-1}\sigma_Z\mathbf{S} = \begin{pmatrix} e & g \\ f & k \end{pmatrix}, \tag{3.102}$$

so that

$$\langle S_i \rangle = \frac{e\lambda_1^N + k\lambda_2^N}{\lambda_1^N + \lambda_2^N}, \tag{3.103}$$

using the result that

$$(\mathbf{T}')^N = \begin{pmatrix} \lambda_1^N & 0 \\ 0 & \lambda_2^N \end{pmatrix}. \tag{3.104}$$

Note that in the above equations, e, f, g, k *etc.* are all functions of h, K, which can be explicitly evaluated (see the exercises at the end of this chapter!). In the limit that $N \to \infty$

$$\langle S_i \rangle \sim e(T, H, J). \tag{3.105}$$

Similarly, the two-point correlation function is

$$\langle S_i S_{i+j} \rangle = \frac{1}{Z} \mathrm{Tr}\ \{(\mathbf{S}^{-1}\sigma_Z\mathbf{S})(\mathbf{T}')^j(\mathbf{S}^{-1}\sigma_Z\mathbf{S})(\mathbf{T}')^{N-j}\} \tag{3.106}$$

In the limit $N \to \infty$

$$\langle S_i S_{i+j} \rangle \simeq e^2 + gf\left(\frac{\lambda_2}{\lambda_1}\right)^j, \tag{3.107}$$

and the correlation function becomes

$$\begin{aligned} G(i, i+j) &= \langle S_i S_{i+j} \rangle - \langle S_i \rangle \langle S_j \rangle \\ &= gf\left(\frac{\lambda_2}{\lambda_1}\right)^j \\ &= gf \exp\left[-j \log(\lambda_1/\lambda_2)\right]. \end{aligned} \tag{3.108}$$

We can read off the correlation length

$$\xi = \frac{1}{\log(\lambda_1/\lambda_2)}, \tag{3.109}$$

and since $\lambda_1 > \lambda_2$ we see again that correlations decay exponentially to zero for $T > 0$.

We can also check our previous results for $h = 0$: in that case

$$\begin{aligned} \lambda_1 &= 2\cosh K \\ \lambda_2 &= 2\sinh K \end{aligned} \tag{3.110}$$

and

$$\xi = \frac{1}{\log \coth K} \tag{3.111}$$

which agrees with our earlier result (3.89).

Note:

(1) ξ cannot diverge unless $\lambda_1 = \lambda_2$. There cannot be a phase transition unless the largest eigenvalue of the transfer matrix becomes degenerate. This is a general result.

(2) For $h \neq 0, \lambda_1 > \lambda_2$, so there cannot be a phase transition for $h \neq 0$.

(3) For $h = 0, \lambda_1 = \lambda_2$ when $K = \infty$, so that a phase transition can occur at $T = 0$, as we have already seen.

3.6 LOW TEMPERATURE EXPANSION

Before we leave the topic of the $d = 1$ Ising model, let us briefly introduce another method for studying the free energy systematically. This is the **low temperature expansion**. The basic idea is to start at $T = 0$, and then raise T slightly. If the ground state is stable with respect to thermal fluctuations, then it should be possible to do perturbation theory about it, in a variable which corresponds to the number of flipped spins. Here, we will perform this expansion in d dimensions on a hypercubic lattice with coordination number $z = 2d$. At zero temperature, all the spins are aligned. Choose the ground state $\uparrow$, and recall the Hamiltonian in zero field

$$\mathcal{H}_\Omega\{S\} = -J \sum_{<ij>} S_i S_j . \tag{3.112}$$

For $T > 0$ it will be possible for $1, 2, \ldots k, \ldots$ spins to be flipped. We will portray the spin configurations with flipped spins in a schematic way, showing the portion of the lattice with the flipped spins only. The probability of a large number of spin flips is small, so we can perturb as follows, only going up to the case of two flipped spins here. Without loss of generality, we will consider them to be in adjacent rows. Let g denote the **degeneracy** of each configuration, and E its energy.

Ground state:

$$\begin{matrix}\uparrow\uparrow\uparrow\uparrow\uparrow\uparrow\uparrow \\ \uparrow\uparrow\uparrow\uparrow\uparrow\uparrow\uparrow\end{matrix} \quad \begin{matrix} E = E_0 = -\dfrac{JNz}{2} \\ g_0 = 1. \end{matrix} \tag{3.113}$$

One spin down:

$$\begin{matrix}\uparrow\uparrow\uparrow\downarrow\uparrow\uparrow\uparrow \\ \uparrow\uparrow\uparrow\uparrow\uparrow\uparrow\uparrow\end{matrix} \quad \begin{matrix} E = E_0 + 2Jz \\ g_1 = N. \end{matrix} \tag{3.114}$$

Two spins down:
(a) Spins are nearest neighbours

$$\begin{matrix}\uparrow\uparrow\uparrow\downarrow\downarrow\uparrow\uparrow \\ \uparrow\uparrow\uparrow\uparrow\uparrow\uparrow\uparrow\end{matrix} \quad \begin{matrix} E = E_0 + 2J(2z-2) \\ g_{2a} = Nz/2! \end{matrix} . \tag{3.115}$$

(b) Spins are not nearest neighbours

$$\begin{matrix}\uparrow\uparrow\downarrow\uparrow\uparrow\downarrow\uparrow \\ \uparrow\uparrow\uparrow\uparrow\uparrow\uparrow\uparrow\end{matrix} \quad \begin{matrix} E = E_0 + 2J(2z) \\ g_{2b} = \begin{pmatrix} N \\ 2 \end{pmatrix} - dN = \dfrac{N(N-1-2d)}{2}. \end{matrix} \tag{3.116}$$

This process is clearly rather tedious! Now we can write the partition function as

$$Z = g_0 e^{-\beta E_0} + g_1 e^{-\beta E_1} + g_2 e^{-\beta E_2} + \ldots \tag{3.117}$$

$$= e^{NJd\beta}\left(1 + Ne^{-4\beta Jd} + dNe^{-4\beta J(2d-1)} + \frac{N(N-1-2d)}{2}e^{-8\beta Jd} + \ldots\right). \tag{3.118}$$

Recalling the variable $w^2 = e^{-4J\beta}$, we write the free energy as

$$\begin{aligned} F &= -k_B T \log Z \\ &= -dJN - k_B T \log\left[1 + N(w^2)^d + dN(w^2)^{2d-1} + (w^2)^{2d}\frac{N(N-1-2d)}{2} + \ldots\right]. \end{aligned} \tag{3.119}$$

It is not at all obvious that the expansion in the log will give a free energy per unit volume (per site) independent of N. Let us check that it does. Write

$$F = -dJN - k_B T \log\left(1 + \sum_k B_k\right) \tag{3.120}$$

where B_k are the terms from k flipped spins:

$$B_1 = N(w^2)^d \tag{3.121}$$

$$B_2 = dN(w^2)^{2d-1} + \frac{N(N-1-2d)}{2}(w^2)^{2d} \tag{3.122}$$

and consider separately the two cases $d > 1$ and $d = 1$.

3.6.1 $d > 1$

For $d > 1$, $B_k \ll B_{k-1}$. Now expand

$$\log(1+\epsilon) = \epsilon - \frac{\epsilon^2}{2} + \ldots \tag{3.123}$$

considering B_k to be of order ϵ^k. To second order we find that:

$$\log\left(1 + \sum_k B_k\right) \approx B_1 + B_2 - \frac{1}{2}B_1^2. \tag{3.124}$$

Note that for $d > 1$, $B_2 = O(B_1^2)$. This is not true in $d = 1$, and we will say more about this shortly. Evaluating

$$\begin{aligned} B_1 + B_2 - \frac{B_1^2}{2} =& N(w^2)^d + dN(w^2)^{2d-1} + \frac{N^2}{2}(w^2)^{2d} - \frac{N}{2}(w^2)^{2d} - \\ & dN(w^2)^{2d} - \frac{N^2}{2}(w^2)^{2d} \\ =& N\left[w^{2d} + d(w^2)^{2d-1} - (d+\frac{1}{2})(w^2)^{2d}\right] \end{aligned} \tag{3.125}$$

The N^2 terms do cancel! Therefore

$$f = \frac{F}{N} = -dJ - \frac{1}{\beta}\left(w^{2d} + d(w^2)^{2d-1} - (d+\frac{1}{2})(w^2)^{2d}\right). \tag{3.126}$$

3.6.2 d=1

What happens in $d = 1$? From the transfer matrix calculation after the thermodynamic limit has been taken,

$$\begin{aligned} f &= -k_BT\log(2\cosh K) = -k_BT\log\left(e^K + e^{-K}\right) \\ &= -k_BT\log e^K\left(1 + e^{-2K}\right) = -J - k_BT\log\left(1 + e^{-2K}\right) \\ &= -J - k_BTe^{-2J/k_BT} + \ldots \end{aligned} \tag{3.127}$$

The factor e^{-2J/k_BT} is in apparent disagreement with the result from the low temperature expansion, eqn. (3.126). What has gone wrong? One way to state the problem is the non-commutability of the limits $T \to 0$ and $N \to \infty$. To see that this is the case, consider the transfer matrix result at *finite* N. From the partition function

$$\begin{aligned} Z_N &= (2\cosh K)^N + (2\sinh K)^N \\ &= e^{NK}\left[\left(1+e^{-2K}\right)^N + \left(1-e^{-2K}\right)^N\right] \end{aligned} \tag{3.128}$$

we obtain

$$F_N = -NJ - k_BTN^2e^{-4K} + O\left((N^2e^{-4K})^2\right) \tag{3.129}$$

i.e.

$$f = \frac{F}{N} = -J - k_BTNe^{-4K} + \ldots \tag{3.130}$$

In the thermodynamic limit, this expression does not reduce to eqn. (3.127). Furthermore, as $N \to \infty$, the free energy becomes unbounded from below, indicating that the ground state with energy $E/N = -J$ is not stable for $T > 0$. This non-commutability of limits is another reflection of the fact that long range order is destroyed in $d = 1$ for $T > 0$.

We can now see that the low temperature expansion, attempted in eqn. (3.118) is only valid for $d > 1$, where the requirement that B_k decreases with increasing k is satisfied. In $d = 1$, the terms coming from clusters of down-turned spins, which are next to each other, violate this requirement. The "problem term" in the low temperature expansion is that of eqn. (3.115): two spins down being nearest neighbors contribute the term $d(w^2)^{2d-1}$ to eqn. (3.127). In $d = 1$, this term, from a cluster of two spins is of the same order as the term $(w^2)^d$ coming from a single spin reversal. This pattern is repeated as we consider three spin reversals *etc.* Thus, in $d = 1$, *blocks* of adjacent reversed spins are the dangerous thermal fluctuations which are excited for any $T > 0$, and which destroy the low temperature, or zero temperature ground state. This calculation makes concrete the physical argument presented in section 2.8.2

3.7 MEAN FIELD THEORY

We end this chapter with a brief look at an important topic – **mean field theory**. Mean field theory is the simplest treatment of an interacting statistical mechanical system, apart from the approximation of ignoring the interactions all together. Often it is almost trivial to perform the mean field theory calculations, although not always: in complex situations, where spontaneous symmetry breaking is accompanied by replica symmetry breaking, even the mean field theory is exceedingly difficult.

The term "mean field theory" conveys an impression of uniqueness, but this is false: there are many ways to generate mean field theories. All share in common the nature of the scaling near the critical point (which is erroneous usually, for low enough dimension), but other non-universal features may be better calculated in one form of mean field theory than another. In the next section, we will present a simple, non-systematic derivation of mean field theory for the Ising model. This mean field theory can be easily improved by applying it to clusters of spins rather than individual spins.[2] The reader is invited to work through various systematic developments of mean field theory, presented in the exercises to this and the previous chapter. These systematic techniques provide a good starting point for the renormalisation group theory of critical phenomena.

3.7.1 Weiss' Mean Field Theory

As usual, we start with the nearest neighbour Ising model Hamiltonian in d dimensions:

$$H_\Omega\{S\} = -J\sum_{\langle ij\rangle} S_i S_j - H\sum_i S_i. \tag{3.131}$$

Suppose for the moment that the spins are independent: $J = 0$. Then

$$Z_\Omega\{0, H\} = \prod_{i=1}^{N}\left(e^{\beta H} + e^{-\beta H}\right) = [2\cosh(H/k_BT)]^N. \tag{3.132}$$

The magnetisation is given by

$$\begin{aligned} M &= -\frac{1}{N}\frac{\partial F}{\partial H} \\ &= \tanh(H/k_BT). \end{aligned} \tag{3.133}$$

So far, with $J = 0$, we have been describing a paramagnet. Weiss tried to understand what happens when $J \neq 0$ by postulating that each spin experiences the presence of an effective field H_{eff} due to the magnetic moment of all the other spins. This magnetic moment would be proportional to the magnetic moment M of all the other spins, which is itself

---

[2] For example, the cluster method of R. Kikuchi, *J. Chem. Phys.* **53**, 2713 (1970).

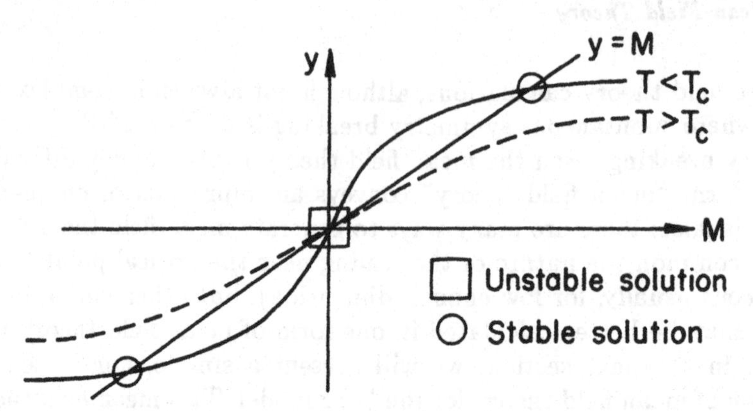

Figure 3.4 Graphical solution of the mean field equation for the spontaneous magnetisation.

unknown *ab initio*. Thus, a given spin experiences both the externally applied field and the effective field due to the other spins. The combination of the two fields then determine the response of the given spin *i.e.* its average magnetic moment. But there is nothing special about the spin which we have chosen; thus, its moment must be the average magnetic moment M. In this way, we are able to calculate M self-consistently.

To see this explicitly, we observe that we can write the Hamiltonian in the form appropriate to a paramagnetic spin in a site-dependent effective field H_i:

$$\mathcal{H}_\Omega\{S\} = -\sum_i S_i H_i \tag{3.134}$$

where

$$H_i = H + \sum_j J_{ij} \langle S_j \rangle + \sum_j J_{ij} (S_j - \langle S_j \rangle). \tag{3.135}$$

Here, the first term is the external field, the second is the mean field, and the final term is the fluctuation, which we will now ignore. If the spins reside on the vertices of a d-dimensional hypercubic lattice, then the co-ordination number z of each site is $2d$, and

$$H_i = H + 2dJM. \tag{3.136}$$

From eqn. (3.133), we find

$$M = \tanh\left(\frac{H + 2dJM}{k_B T}\right). \tag{3.137}$$

Even in the absence of an external field H, we can apply the same idea to find the **spontaneous magnetisation**. Setting $H = 0$, we obtain

$$M = \tanh(2dJM/k_B T). \tag{3.138}$$

This is best visualised graphically, as shown in figure (3.4). For $T > T_c \equiv 2dJ/k_B$, the tanh curve lies *below* $y = M$, and the only intersection is at $M = 0$. For $T < T_c$ it intersects at $\pm M_s(T)$. Non-analyticity arises because of the way that the solution to eqn. (3.138) changes as T varies. Thus, in mean field theory, we find that the critical temperature is

$$T_c = \frac{2dJ}{k_B}. \tag{3.139}$$

We can study the critical behaviour afforded by this description, by expanding the equation of state (3.137) in the vincinity of T_c. Let $\tau = T_c/T$. First, we invert eqn. (3.137) to obtain the equation of state:

$$M = \tanh\left(H/k_BT + M\tau\right) = \frac{\tanh H/k_BT + \tanh M\tau}{1 + \tanh H/k_BT \tanh M\tau}. \tag{3.140}$$

Thus

$$\tanh H/k_BT = \frac{M - \tanh M\tau}{1 - M\tanh M\tau}. \tag{3.141}$$

For small H and M, we can expand in powers of M to find

$$\frac{H}{k_BT} \approx M(1-\tau) + M^3\left(\tau - \tau^2 + \frac{\tau^3}{3} + \ldots\right) + \ldots \tag{3.142}$$

Now we can extract the critical exponents for the ferromagnetic transition, as calculated in mean field theory: for $H = 0$ and $T \to T_c^-$, eqn. (3.142) implies that

$$M^2 \approx 3\frac{(T_c - T)}{T_c} + \ldots \tag{3.143}$$

where the dots indicate corrections to this leading order formula. We can read off the **critical exponent** β: $\beta = 1/2$. The **critical isotherm** is the curve in the H–M plane corresponding to $T = T_c$. Its shape near the critical point is described by the critical exponent δ:

$$H \sim M^\delta. \tag{3.144}$$

Setting $\tau = 1$ in the equation of state (3.142), we find

$$\frac{H}{k_BT} \sim M^3, \tag{3.145}$$

showing that the mean field value of δ is 3. The isothermal magnetic susceptibility χ_T also diverges near T_c:

$$\chi_T \equiv \left.\frac{\partial M}{\partial H}\right|_T. \tag{3.146}$$

Differentiating the equation of state (3.142), gives

$$\frac{1}{k_B T} = \chi_T(1-\tau) + 3M^2\chi_T(\tau - \tau^2 + \frac{1}{3}\tau^3). \tag{3.147}$$

For $T > T_c$, $M = 0$ and

$$\chi_T = \frac{1}{k_B}\frac{1}{T - T_c} + \ldots. \tag{3.148}$$

Comparing with the definition of the critical exponent γ:

$$\chi_T \sim |T - T_c|^{-\gamma}, \tag{3.149}$$

we conclude that $\gamma = 1$. For $T < T_c$,

$$M = \sqrt{3}\left(\frac{T_c - T}{T}\right)^{1/2} + \ldots \tag{3.150}$$

Substituting into eqn. (3.147) gives

$$\chi_T = \frac{1}{2k_B}\frac{1}{T - T_c} + \ldots, \tag{3.151}$$

which shows that the divergence of the susceptibility below the transition temperature is governed by the critical exponent $\gamma' = \gamma = 1$. The calculation of the critical exponent α, which governs the divergence of the heat capacity near the transition, is straightforward. The result is that the heat capacity exhibits a **discontinuity** at $T = T_c$, in mean field theory.

3.7.2 Spatial Correlations

The mean field theory presented here is not at the level of sophistication that we can extract complete details of the spatial correlations. One might have thought that mean field theory and spatial correlations were antithetical: how can there be interesting spatial correlations when the fluctuations have been averaged out over the entire system? The answer is that the correlation length not only governs the spatial extent of fluctuations of the order parameter, but also governs the way in which the order parameter varies in space in response to an *inhomogeneous external field.* Given such a perturbation on the system, one can still construct a self-consistent solution for the order parameter in the spirit of mean field theory, and we will see this in detail when we discuss the general framework of mean field theory, namely Landau theory.

For now, we can proceed using thermodynamic arguments and an interesting identity, known as the **static susceptibility sum rule.** This is an important relation between a thermodynamic quantity – the isothermal susceptibility – and the two-point correlation function. The derivation is as follows.

First, we define the two-point function:

$$G(\mathbf{r}_i - \mathbf{r}_j) = \langle S_i S_j \rangle - \langle S_i \rangle \langle S_j \rangle , \tag{3.152}$$

where $\mathbf{r}_i$ is the spatial position of the spin S_i. For future reference, the expectation values above may be obtained by differentiating the partition function with respect to the external field:

$$Z_\Omega = \mathrm{Tr}\ \exp\left[\beta J \sum_{<ij>} S_i S_j + H\beta \sum_i S_i\right] \tag{3.153}$$

and thus

$$\begin{aligned} \sum_i \langle S_i \rangle &= \frac{1}{Z_\Omega} \mathrm{Tr} \sum S_i e^{-\beta H_\Omega} \\ &= \frac{1}{\beta Z_\Omega} \frac{\partial Z_\Omega(H,J)}{\partial H} \end{aligned} \tag{3.154}$$

and

$$\sum_{ij} \langle S_i S_j \rangle = \frac{1}{\beta^2 Z_\Omega} \frac{\partial^2 Z_\Omega}{\partial H^2}. \tag{3.155}$$

Now let us construct a general expression for the isothermal susceptibility:

$$\chi_T = \frac{\partial M}{\partial H} = \frac{1}{N\beta} \frac{\partial^2 \log Z_\Omega}{\partial H^2} \tag{3.156}$$

$$= \frac{1}{N} k_B T \left[\frac{1}{Z_\Omega} \frac{\partial^2 Z_\Omega}{\partial H^2} - \frac{1}{Z_\Omega^2} \left(\frac{\partial Z_\Omega}{\partial H} \right)^2 \right] \tag{3.157}$$

$$= \frac{1}{N} (k_B T)^{-1} \left[\sum_{ij} \langle S_i S_j \rangle - \left(\sum_i \langle S_i \rangle \right)^2 \right] \tag{3.158}$$

$$= \frac{1}{N} (k_B T)^{-1} \sum_{ij} G(\mathbf{r}_i - \mathbf{r}_j) \tag{3.159}$$

$$= (k_B T)^{-1} \sum_i G(\mathbf{x}_i) \tag{3.160}$$

$$= (a^d k_B T)^{-1} \int_\Omega d^d\mathbf{r}\, G(\mathbf{r}). \tag{3.161}$$

This is an important result, because it connects the divergence in χ_T with the two-point correlation function G: apparently, G must reflect the divergence of χ_T. We will see later that for $|\mathbf{r}| \gg \xi$,

$$G(\mathbf{r}) \sim \frac{e^{-|\mathbf{r}|/\xi}}{|\mathbf{r}|^{(d-1)/2}\xi^{(d-3)/2}}, \tag{3.162}$$

where ξ is the **correlation length.** Since G decays so rapidly at infinity, the integral in eqn. (3.161) should be convergent. Yet, we know that as $T \to T_c$, χ_T diverges, and therefore, so must the integral. The resolution of this apparent conundrum is that the *correlation length* also diverges, as we saw explicitly in the transfer matrix calculations for $d = 1$ and $T \to 0$.

We can calculate how the correlation length diverges, using the result that $\gamma = 1$. From eqns. (3.161) and (3.162), we have

$$\begin{aligned} \left(\frac{T-T_c}{T_c}\right)^{-1} &\sim \int \frac{r^{d-1}e^{-r/\xi}}{r^{(d-1)/2}\xi^{(d-3)/2}}\,dr \\ &\sim \left(\int z^{(d-1)/2}e^{-z}\,dz\right)\xi^2, \end{aligned} \tag{3.163}$$

where we made the substitution $z = r/\xi$ (*i.e.* the scaling trick). The integral is a well-defined constant, and thus we find that the correlation length does indeed diverge as

$$\xi \sim \left(\frac{T-T_c}{T_c}\right)^{-\nu} \tag{3.164}$$

with

$$\nu = 1/2. \tag{3.165}$$

Finally, we mention the critical exponent η. It describes how the two-point correlation function behaves at large distances *exactly at the critical point.* We will later see that the same mean field theory calculation which yields eqn. (3.163) for the long distance behaviour *near* the critical point also predicts that

$$G(r) \sim r^{-(d-2+\eta)} \tag{3.166}$$

with $\eta = 0$. In principle η can be non-zero.

Table 3.1 CRITICAL EXPONENTS FOR THE ISING UNIVERSALITY CLASS

Exponent	Mean Field	Experiment	Ising ($d = 2$)	Ising ($d = 3$)
α	0 (disc.)	0.110 – 0.116	0 (log)	0.110(5)
β	1/2	0.316 – 0.327	1/8	0.325±0.0015
γ	1	1.23 – 1.25	7/4	1.2405±0.0015
δ	3	4.6 – 4.9	15	4.82(4)
ν	1/2	0.625±0.010	1	0.630(2)
η	0	0.016 – 0.06	1/4	0.032±0.003

3.7.3 How Good is Mean Field Theory?

Table 3.1 compares critical exponents calculated in mean field theory with those measured in experiment or deduced from theory for the Ising model in two and three dimensions. The table is essentially illustrative: the values given are not necessarily the most accurate known at the time of writing. In addition the experimental values are just given approximately, with a range reflecting inevitable experimental uncertainty. The values for ν and δ are not independent from the other values, obtained using **scaling laws.** The experimental values quoted are actually obtained from experiments on fluid systems.[3] Our discussion of the lattice gas model implies that these fluid systems should be in the universality class of the Ising model. Indeed, this expectation is borne out by the comparison of the experimental values and those from the three dimensional Ising model. The latter values are in some cases given with the number in brackets representing the uncertainty in the last digit quoted.[4]

The numerical values of the critical exponents calculated from mean field theory are in reasonable agreement with those given by experiment and the Ising model in three dimensions, although there are clearly systematic differences. First of all, the mean field theory exponents here do not depend on dimension, whereas it is clear that the exact critical exponents do. It is possible for mean field theory to exhibit exponents with a value dependent upon dimension. An example is the mean field theory

[3] J.V. Sengers in *Phase Transitions*, Proceedings of the Cargèse Summer School 1980 (Plenum, New York, 1982).

[4] J.C. Le Guillou and J. Zinn-Justin, *Phys. Rev. B* **21**, 3976 (1980); numerical values and details of the calculational techniques used to obtain these estimates are given by J. Zinn-Justin, *Quantum Field Theory and Critical Phenomena* (Clarendon, Oxford, 1989), Chapter 25.

(MFT) for the size R of a self-avoiding random walk of N steps:

$$R \sim N^{\nu}, \quad \nu_{\mathrm{MFT}} = \frac{3}{2+d}. \tag{3.167}$$

In this case, however, the dimension dependence is quantitatively inaccurate. Secondly, the values are slightly but definitely in error.

Despite these observations, it is worth dwelling on the success of mean field theory. In the next chapter, we will describe mean field theory for fluid systems, and we shall find that the critical exponents have the same values there. So mean field theory does exhibit universality, a feature that emerges from the general framework of Landau theory. Furthermore, mean field theory has predicted the correct phase diagram in this case, and provided an expression for the critical temperature. These predictions are not always reliable, but always provide a good first step to a more complete theory. Unfortunately, we will need to work very hard to improve mean field theory.

EXERCISES

Exercise 3–1

This question is an exercise in the use of the transfer matrix method. In the first parts of the question, we deal with the $d = 1$ Ising model with periodic boundary conditions.

(a) Construct the matrix $\mathbf{S}$ which diagonalises the transfer matrix $\mathbf{T}$: that is, $\mathbf{T}' = \mathbf{S}^{-1}\mathbf{T}\mathbf{S}$ is diagonal. You will find it helpful to write down the matrix elements in terms of the variable ϕ given by

$$\cot(2\phi) = e^{2K}\sinh(h).$$

(b) Check that you understand why

$$\langle S_i \rangle = \frac{\mathrm{Tr}\,(\mathbf{S}^{-1}\sigma_z\mathbf{S}(\mathbf{T}')^N)}{Z_N}$$

and use your answer to part (a) to show that $\langle S_i \rangle = \cos(2\phi)$ as $N \to \infty$. In a similar fashion calculate $\langle S_i S_j \rangle$ and hence show that in the thermodynamic limit

$$G(i, i+j) \equiv \langle S_i S_j \rangle - \langle S_i \rangle \langle S_j \rangle = \sin^2(2\phi)\left(\frac{\lambda_2}{\lambda_1}\right)^j.$$

(c) Calculate the isothermal susceptibility χ_T from the formula given for $M(H)$ in the text. Verify explicitly that $\chi_T = \sum_j G(i, i+j)/k_B T$. (Caution: In the thermodynamic limit, the sum runs over $-\infty$ to $+\infty$.)

(d) Now we will examine what happens when the system has boundaries. Consider the partition function with free boundary conditions:

$$Z_N(h, K) = \sum_{S_1} \cdots \sum_{S_N} e^{h(S_1 + \cdots + S_N) + K(S_1 S_2 + \cdots + S_{N-1} S_N)}.$$

In this case, the partition function is not simply Tr $(\mathrm{T}')^N$. Work out what the correct expression is (you will need to introduce a new matrix in addition to T), and show that the free energy F_N is given by

$$F_N = N f_b(h, K) + f_s(h, K) + F_{fs}(N, h, K)$$

where f_b is the bulk free energy, f_s is the surface free energy due to the boundaries, and $F_{fs}(N, h, K)$ is an intrinsically finite size contribution which depends on the system size as $e^{-C(h,K)N}$, where C is a function of h and K.

(e) Check that in the case $h = 0$ and $N \to \infty$ your result for the surface free energy agrees with that obtained from $\lim_{N\to\infty} F_N^{free} - F_N^{periodic}$ (notation should be obvious).

Exercise 3-2

This question invites you to generalise the transfer matrix formalism to the two dimensional Ising model on a square lattice. Suppose that there are N rows parallel to the x axis and M rows parallel to the y axis. We will require that $N \to \infty$ whilst we will calculate the transfer matrix for $M = 1$ and $M = 2$. Periodic boundary conditions apply in both directions, so that our system has the topology of a torus. The Hamiltonian H_Ω is given by

$$-\beta H_\Omega = K \sum_{n=1}^{N} \sum_{m=1}^{M} S_{mn} S_{m+1n} + S_{mn} S_{mn+1}$$

(a) For the case $M = 1$ show that the transfer matrix is a 2×2 matrix, and show that its eigenvalues are

$$\lambda_1 = 1 + x^2 \qquad \lambda_2 = x^2 - 1$$

where

$$x \equiv e^K.$$

(b) Now consider the case $M = 2$. We need to extend the transfer matrix formalism. Consider the vector

$$\mathbf{v}_n = (S_{1n} S_{2n} \dots S_{mn}).$$

This vector gives the configuration of a row n. Show that

$$H_\Omega = \sum_{n=1}^{N} E_1(\mathbf{v}_n, \mathbf{v}_{n+1}) + E_2(\mathbf{v}_n)$$

where E_1 is the energy of interaction between neighbouring rows and E_2 is the energy of a single row. Hence show that

$$Z = \sum_{\mathbf{v}_1 \cdots \mathbf{v}_N} T_{\mathbf{v}_1 \mathbf{v}_2} T_{\mathbf{v}_2 \mathbf{v}_3} \cdots T_{\mathbf{v}_N \mathbf{v}_1}$$

where T is a transfer matrix of dimensions $2^M \times 2^M$, whose form you should give.

(c) Calculate T for the case $M = 2$.

(d) Show that the two largest eigenvalues are

$$\lambda_1 = \left(x^4 + 2 + x^{-4} + \sqrt{x^8 + x^{-8} + 14}\right)/2$$

and

$$\lambda_2 = x^4 - 1.$$

Exercise 3-3

In the previous chapter, you studied the infinite range Ising model, and showed how a phase transition can occur in the thermodynamic limit. Your results were very similar to those of the Weiss model of ferromagnetism, because in a mean field description such as the Weiss model, every spin is interacting with every other spin. In this question, we will look at the nearest neighbour Ising model, and systematically derive mean field theory by the method of steepest descents. This approach, based on the Hubbard-Stratonovich transformation is probably the most general method for turning statistical mechanics problems into field theories. We will see that mean field theory just comes from taking the maximum term in the partition function. The Hamiltonian is

$$H_\Omega\{S\} = -\frac{1}{2}\sum_{i \neq j} J_{ij} S_i S_j - \sum_i H_i S_i$$

where $J_{ij} = J > 0$ if i and j are nearest neighbours and $J_{ij} = 0$ otherwise. In this question, we will use Einstein's summation convention: repeated indices are to be summed over, *i.e.* $A_{ij}x_j \equiv \sum_j A_{ij}x_j$.

(a) Prove the identity

$$\int_{-\infty}^{\infty} \prod_{i=1}^{N} \left(\frac{dx_i}{\sqrt{2\pi}}\right) \exp\left(-\frac{1}{2}x_i A_{ij} x_j + x_i B_i\right) = \frac{1}{\sqrt{\det \mathrm{A}}} e^{\frac{1}{2}B_i (A^{-1})_{ij} B_j}$$

where A is a real symmetric positive matrix, and **B** is an arbitrary vector. [*Hint: Make change of variables* $y_i = x_i - (A^{-1})_{ij}B_j$.]

(b) We want to use the above identity to make the term in the Hamiltonian with S_iS_j linear in S_i, just as you did for the infinte range model. Why can you not do that straight away? Show that this technical point is easily dealt with by a trivial redefinition of the zero of energy.

(c) Apply the identity of part (a), making the identification $A_{ij}^{-1} = J_{ij}$ and $B_i = S_i$. Show that

$$Z = \int_{-\infty}^{\infty} \prod_{i=1}^{N} d\psi_i e^{-\beta S(\{\psi_i\},\{H_i\},\{J_{ij}\})}$$

where

$$S = \frac{1}{2}(\psi_i - H_i)J_{ij}^{-1}(\psi_j - H_j) - \frac{1}{\beta}\sum_i \log(2\cosh\beta\psi_i).$$

This form for the partition function is really what is meant by the term **functional integral**. The dummy variable ψ_i is like a function $\psi(\mathbf{r})$ in the limit that the lattice spacing $a \to 0$ and $N \to \infty$.

(d) Assume that we can approximate Z by the maximum term in the functional integral: $Z \approx \exp -S(\bar{\psi}_i)$ where $\bar{\psi}_i$ is the value of the field ψ_i which minimises S. Find the equation satisfied by $\bar{\psi}_i$, and show that the magnetisation at site i,

$$m_i \equiv \langle S_i \rangle = -\partial F/\partial H_i \approx -\partial S/\partial H_i$$

is given by $m_i = \tanh\beta\bar{\psi}_i$. Hence find $H_i(\{m_j\})$.

(e) Let $\bar{S}$ be the value of S at $\psi_i = \bar{\psi}_i$. The mean field approximation is that the free energy $F \approx \bar{S}$. Show that

$$\bar{S}(\{m_i\}) = \frac{1}{2}J_{ij}m_im_j - \frac{1}{\beta}\sum_i \log\left(\frac{2}{\sqrt{1-m_i^2}}\right).$$

Hence calculate the mean field approximation to the Gibbs free energy from the Legendre transform

$$\Gamma\{m_i\} = \bar{S} + \sum_i H_i(\{m_j\})m_i.$$

Verify that the equation of state is correctly given by $H_i = \partial\Gamma/\partial m_i$.

Exercise 3–4

Now we will use our mean field solution of the Ising model to solve the lattice gas model. Set $U_1 = 0$, and check that you understand the correspondence between the lattice gas variables and the Ising variables. In particular, write down the relation between the pressure and the free energy of the Ising model. Also, write down the relation between the mean density ρ of the lattice gas and the mean value of the magnetisation of the Ising model.

(a) Express E_0 in terms of H and J. Using the result of 3–2, rewrite this in terms of H and T_c. Write down the relation between the pressure p, $H(m)$ and $S(m)$, using the results from 3–2 for the uniform magnetisation case. Hence show that the equation of state in the mean field approximation is

$$p = k_B T \log\left(\frac{1}{1-\rho}\right) - 2k_B T_c \rho^2$$

(b) Show that at the critical point for the fluid (p^*, ρ^*, T^*),

$$p^* = k_B T_c(\log 2 - 1/2),$$

and $T^* = T_c$, $\rho^* = 1/2$. This corresponds to the critical point $H = 0$, $T = T_c$ in the Ising model.

CHAPTER 4

Critical Phenomena in Fluids

In this chapter, we shall discuss the critical phenomena at the liquid-gas transition, with particular emphasis on the description given by the Van der Waals equation. The Van der Waals equation gives a qualitatively accurate account of the phase diagram of fluids, exhibits properties such as the law of corresponding states and gives a mean-field theory of the liquid gas critical point. Apart from its intrinsic interest, the mean field theory of the liquid-gas critical point is presented here to emphasise the similarities with the mean field theory of magnets, discussed in the previous chapter. These structural similarities form the basis of the general theory of phase transitions at the mean field level — the Landau theory.

4.1 THERMODYNAMICS

4.1.1 Thermodynamic Potentials

Here we give a rapid review of the fundamental thermodynamics of fluids and phase equilibria. Working in the grand canonical ensemble, we consider the thermodynamic potentials energy E, Helmholtz free energy F and Gibbs free energy G, which are functions of combinations of entropy S, volume V, particle number N, chemical potential μ, pressure p and temperature T. The energy and its differential satisfy

$$E = E(S, V, N) \tag{4.1}$$
$$dE = TdS - pdV + \mu dN. \tag{4.2}$$

The Helmholtz free energy and its differential satisfy

$$F = F(T, V, N) = E - TS \tag{4.3}$$
$$dF = -SdT - pdV + \mu dN. \tag{4.4}$$

The Gibbs free energy and its differential satisfy

$$G = G(T, p, N) = F + pV = \mu N \tag{4.5}$$
$$dG = -SdT + Vdp + \mu dN. \tag{4.6}$$

These relations just follow from the First Law of Thermodynamics and ordinary calculus. From them we can read off thermodynamic identities, such as

$$S = -\left.\frac{\partial G}{\partial T}\right|_{p,N} = -\left.\frac{\partial F}{\partial T}\right|_{V,N}. \tag{4.7}$$

4.1.2 Phase diagram

If we compute G subject to the constraint that the system is in a particular phase, then the realised phase is the one with the lowest G. Thus, the coexistence line between two phases, I and II, corresponds to the locus on the phase diagram where $G_I = G_{II}$, *i.e.* $\mu_I = \mu_{II}$. From our perspective, this is somewhat simplistic, since we have already remarked that the partition function describes the entire phase diagram. Nevertheless, we proceed with an important consequence: the **Clausius-Clapeyron relation.**

Consider the coexistence line between liquid and solid in the $p - T$ plane:

$$G_l(T, p, N) = G_s(T, p, N). \tag{4.8}$$

Now move along the line: $T \rightarrow T + \delta T$, $p \rightarrow p + \delta p$. This implies that

$$G_l(T + \delta T,\ p + \delta p, N) = G_s(T + \delta T,\ p + \delta p, N). \tag{4.9}$$

Expand to first order, and use the thermodynamic identities

$$\frac{\partial G}{\partial T} = -S \quad \frac{\partial G}{\partial p} = V \tag{4.10}$$

to find the desired relation:

$$p\frac{dp}{dT}\bigg|_{\text{transition}} = \frac{S_l - S_s}{V_l - V_s}. \tag{4.11}$$

Typically $S_l > S_s$ and $V_l > V_s$ for fixed particle number, so that $\partial p/\partial T > 0$. Exceptions include water, where $V_l < V_s$ (*i.e.* ice floats!) and ^{3}He, for which $V_l > V_s$ but $S_l < S_s$ below about 0.5 K, due principally to the spin disorder, present in the solid phase, but not in the liquid phase.

The fact that $S_l \neq S_s$ implies that **latent heat** is released at this first order transition. Moving along a coexistence line, we may encounter a **critical point**, where (in the example of the liquid-gas transition) $S_l \to S_g$ and $V_g \to V_l$; hence the latent heat becomes vanishingly small.

4.1.3 Landau's Symmetry Principle

In general, two phases of matter with different symmetry must be separated by a line of transitions. This reflects the fact that one cannot continuously change symmetry; that is, a symmetry is either present or absent. Thus a liquid and a solid are believed to be always separated by a line of transitions. On the other hand, the liquid- gas transition can end at the critical point, because there is no symmetry difference between a liquid and a gas. Note also that the symmetry principle does **not** tell you how the transition occurs between the two phases. The line between liquid and solid could be either first order or second order, or both (*i.e.* one or the other depending on the pressure). In rubber, the transition to the solid state at fixed temperature, as the number of cross-links increases, is thought to be a continuous transition.

Although a symmetry is either present or absent, it can *emerge* either continuously or discontinuously *as the coupling constants are changed.*

4.2 TWO-PHASE COEXISTENCE

There are several ways of understanding the phase diagram shown in figure (4.1).

4.2.1 Fluid at Constant Pressure

Consider a liquid maintained at constant *pressure*. Apply heat. At first, the liquid expands, and then begins to boil. As more matter becomes gaseous, the *temperature* remains constant in addition to the pressure. Correspondingly, the volume increases. Eventually, all the liquid has

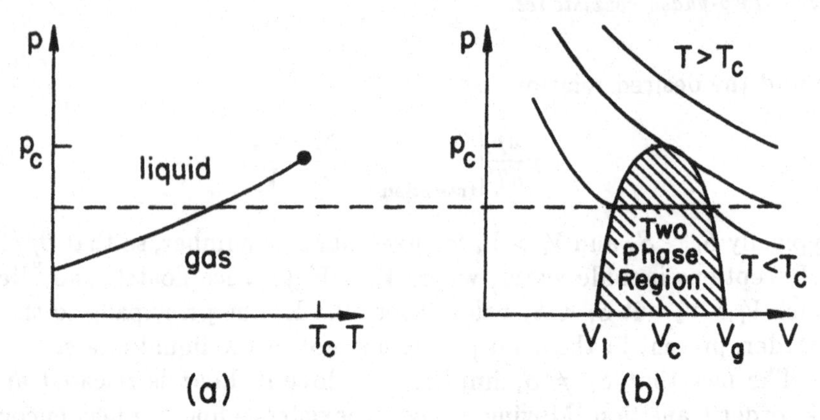

Figure 4.1 (a) Coexistence curve in the $p-T$ plane. (b) Isotherms above and below the critical temperature in the $p-V$ plane.

become gas, and subsequent heating causes the gas to expand. This situation is shown in figure (4.1a). To the left of the coexistence line, there is a unique phase – liquid – and the system moves horizontally at constant pressure until meeting the coexistence line. When it meets the coexistence line, gas begins to appear in the system. During the entire time that liquid is being converted into gas, the system remains at the coexistence line, at the point corresponding to the initial pressure. This point represents the system at all stages of the coexistence, as the volume fraction of gas varies from zero to unity. This region is shown in figure (4.1b). Once all the fluid has become gas, the system begins to move again horizontally off the coexistence line.

4.2.2 *Fluid at Constant Temperature*

Another way to think of the coexistence region is to consider how the system behaves as it is moved along an *isotherm*. Start on the right in figure (4.1b), and consider the gas as it is compressed. Initially, $\partial p/\partial V < 0$, which is equivalent to the fact that the **compressibility**

$$\kappa_T \equiv -\frac{1}{V}\frac{\partial V}{\partial p}\bigg|_T > 0. \tag{4.12}$$

i.e. the system is mechanically stable and exerts a restoring force when it is compressed.

As we move to the left, increasing p, decreasing V and hence increasing $\rho \equiv N/V$, we eventually reach the equilibrium vapor pressure of the gas and the associated **equilibrium volume** $V_g(T)$. Under further compression, the gas condenses, forming droplets of liquid, at *constant pressure.*

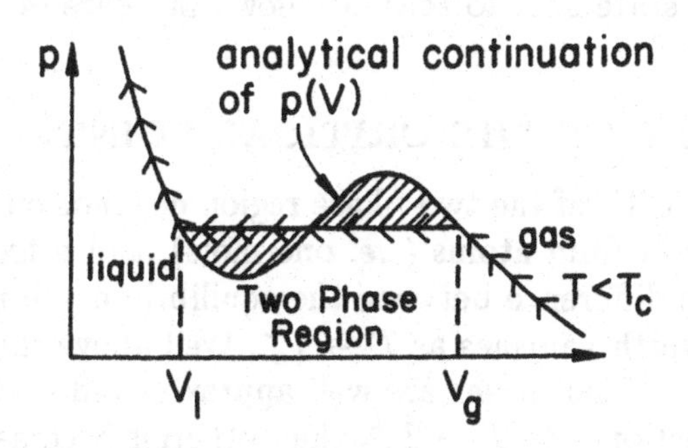

Figure 4.2 Maxwell's equal area construction.

Eventually, the system is all liquified, and the pressure begins to rise again as the liquid is compressed.

4.2.3 *Maxwell's Equal Area Rule*

As we have seen, the isotherms in the $p-V$ plane exhibit non-analytic behaviour at the boundaries of the two-phase region. However, model equations of state, such as that due to Van der Waals, are *analytic* all through the two-phase region. This artifact of these equations reflects the fact that the approximation leading to the model equation of state has not ensured that the equilibrium state of the system globally minimises the Gibbs free energy. As a result, the (*e.g.*) liquid phase is artificially continued into a region of the phase diagram where it is not stable, but only metastable or even unstable.

We can take into account the global minimisation, by using the equivalent statement that at coexistence, the chemical potentials of the liquid and gas phases are equal: $\mu_l = \mu_g$. In addition, we require not only thermal equilibrium, but also **mechanical equilibrium**: this implies that at coexistence, the pressure in the gas phase equals that in the liquid phase. Hence, the isotherm in the two-phase region of the $p-V$ plane must be horizontal.

We start from eqn. (4.5): differentiating and subtracting eqn. (4.6), we obtain

$$d\mu = -\frac{S}{N}dT + \frac{V}{N}dp. \tag{4.13}$$

Along an isotherm, $dT = 0$. Then

$$\mu_l - \mu_g = \int_{\text{gas}}^{\text{liq}} d\mu = \int_{\text{gas}}^{\text{liq}} \frac{V}{N}dp = 0. \tag{4.14}$$

Geometrically, this corresponds to the statement that the horizontal isotherm in the two-phase region must be drawn so that areas bounded by the equation of state sum to zero, as shown in figure (4.2).

4.3 VICINITY OF THE CRITICAL POINT

The width, in V, of the two-phase region depends on T of course. For a given number of fluid atoms (*i.e.* one mole), and a fixed temperature, the width is the difference between the equilibrium volumes of the liquid and gas. This width vanishes as $T \to T_c^-$. Well above the transition temperature, the $p - V$ isotherms are well-approximated by the ideal gas law plus small corrections. As $T \to T_c^+$, the isotherms become flatter, exhibiting a horizontal tangent at the critical point p_c, V_c, T_c. Consequently, the compressibility κ_T diverges as the critical point is approached. We will see below that the compressibility is analogous to the isothermal susceptibility in magnetic systems.

Let us characterise the phenomena associated with the critical point, as we did for magnetic systems. The isothermal compressibility κ_T diverges as $T \to T_c^{\pm}$ with an exponent γ (+) or γ'(−):

$$\kappa_T = -\frac{1}{V}\frac{\partial V}{\partial p}\bigg|_T \sim |T - T_c|^{-\gamma}. \tag{4.15}$$

As we go through the critical point, the coexistence curve has equilibrium volumes $V_l(T)$, $V_g(T)$, which are not equal. In terms of the volume per particle in the liquid (gas) phase, $v_{l(g)} \equiv V_{l(g)}/N$,

$$v_g - v_l \sim |T_c - T|^{\beta}. \tag{4.16}$$

The width of the coexistence curve is analogous to the spontaneous magnetisation in a magnetic system. The specific heat at constant volume C_V diverges as $T \to T_c^{\pm}$ with an exponent α (+) or α'(−):

$$C_V \sim |T - T_c|^{-\alpha}. \tag{4.17}$$

On the critical isotherm, $T = T_c$ and

$$|p - p_c| \sim |V - V_c|^{\delta}. \tag{4.18}$$

4.4 VAN DER WAALS EQUATION

In terms of the volume per particle $v = V/N$ in a gas or a liquid, the equation of state of an ideal gas is

$$pv = k_B T. \tag{4.19}$$

Van der Waals proposed an equation of state to take into account the hard core potential of the atoms (*i.e.* the **excluded volume** due to the non-zero radius of the atoms) and the attractive interactions between the atoms. The former reduces the volume by a certain amount denoted by b. The latter reduces the *energy per particle on average* by an amount proportional to the density: since $p = -\partial F/\partial V$, this in turn reduces the pressure by an amount proportional to V^{-2}. Hence we arrive at the Van der Waals equation

$$p = \frac{k_B T}{v - b} - \frac{a}{v^2}. \tag{4.20}$$

Here, $b \approx$ volume of hardcore of the fluid particles and a is a measure of the **attraction** between particles ($a > 0$). The motivation for the equation given above shows that it has the status of mean field theory. The parameters a and b can be determined by fitting the proposed equation of state to experimental data at high temperatures well above the critical point.

4.4.1 Determination of the Critical Point

As $T \to T_c$, the equation $p = p(v)$ has an inflexion point:

$$\frac{\partial p}{\partial v} = \frac{\partial^2 p}{\partial v^2} = 0 \tag{4.21}$$

at $T = T_c$. We can readily find p_c, v_c, T_c by noting that $p(v)$ is a cubic. Thus the equation $p(v)$ =constant should have 3 solutions. For $T > T_c$, there are 1 real and 2 imaginary solutions, whereas for $T < T_c$ there are 3 real solutions. Hence, at the critical point, the three solutions all merge. Write Van der Waals equation as

$$v^3 - \left(b + \frac{k_B T}{p}\right) v^2 + \frac{a}{p} v - \frac{ab}{p} = 0. \tag{4.22}$$

and note that at p_c, v_c, T_c all three roots should be equal, *i.e.* eqn. (4.22) must be of the form

$$(v - v_c)^3 = 0. \tag{4.23}$$

Table 4.1 **CRITICAL POINT OF ARGON**

Parameter	Experiment	Prediction of eqn. (4.25)
$T_c/^\circ$C	-122	-119.7
p_c/atm	48	48
v_c/cm^3/mol	75	97

Hence, equating coefficients of powers of v in eqns. (4.22) and (4.23), we read off

$$3v_c = b + \frac{k_B T_c}{p_c}; \quad 3v_c^2 = \frac{a}{p_c}; \quad v_c^3 = \frac{ab}{p_c}, \tag{4.24}$$

and thus find

$$v_c = 3b; \quad p_c = a/27b^2; \quad k_B T_c = 8a/27b^2. \tag{4.25}$$

This is a remarkable result: the high temperature fit of a and b to experiment *predicts* T_c, v_c and p_c. How well does it work? For argon, $a = 1.354$ l^2 atm/mol and $b = 0.0322$ l/mol $= 53.4$ Å/molecule. The results are summarised in table (4.1). The agreement is quite good, considering the modest theoretical effort.

The theory also predicts that

$$\frac{p_c v_c}{k_B T_c} = \frac{3}{8} = 0.375, \tag{4.26}$$

a *universal number*, independent of a or b, and thus the same for *all* fluids. Experimentally this ratio is 0.292 for argon, 0.23 for water and 0.31 for ^{4}He. The agreement is reasonable – the number is clearly not varying by orders of magnitude, or even by factors of 2.

4.4.2 Law of Corresponding States

The Van der Waals equation may be expressed in dimensionless form by rescaling. Define the reduced pressure, volume and temperature by

$$\pi \equiv p/p_c; \quad \nu \equiv v/v_c; \quad \tau \equiv T/T_c. \tag{4.27}$$

Then the Van der Waals equation becomes, using eqn. (4.25),

$$\left(\pi + \frac{3}{\nu^2}\right)(3\nu - 1) = 8\tau. \tag{4.28}$$

This is also remarkable! When scaled by p_c, v_c, T_c **all** fluids are predicted to have the same equation of state, with no other parameters involved. This is the **law of corresponding states**. Since the equation of state is, after rescaling, predicted to be the same for all fluids, all thermodynamic properties which follow from the equation of state should also be universal.

Note that this is a form of universality, but is quite different from the universality at the critical point. The law of corresponding states is predicted to apply *everywhere* on the phase diagram. In fact, it can be shown that the law of corresponding states is a simple consequence of dimensional analysis, and is of greater generality than the derivation from the Van der Waals equation might suggest; the reader is invited to explore this in the exercises at the end of this chapter.

Experimentally, the law of corresponding states is well-satisfied, *even by fluids which do not obey the Van der Waals equation.* For example, the ratio of pv/k_BT plotted against reduced pressure at fixed reduced temperature should be independent of the particular fluid. The data from a wide variety of fluids do indeed fall on the same curve to a high degree of accuracy.[1]

4.4.3 Critical Behaviour

In this section, we will calculate the critical exponents of the Van der Waals fluid. As a first step, let us calculate the free energy corresponding to the Van der Waals equation of state; this will enable us to compute thermodynamic quantities by differentiation.

We start from the relation $p = -\partial F/\partial V$, where the derivative is at constant T. Integrating eqn. (4.20) with respect to V gives

$$-F(V,T) = Nk_BT \log(V - Nb) + \frac{aN^2}{V} + f(T), \tag{4.29}$$

where $f(T)$, the constant of integration, is some function of T. When $a = b = 0$, we should just recover the result for ideal gas *i.e.*

$$-F_{ideal}(V,T) = Nk_BT \log(V) + f(T). \tag{4.30}$$

Thus

$$F(V,T) - F_{ideal} = -Nk_BT \log\left(\frac{V - Nb}{V}\right) - \frac{aN^2}{V}. \tag{4.31}$$

---

[1] The experimental data are reproduced by H.E. Stanley, *Introduction to Phase Transitions and Critical Phenomena* (Oxford University Press, New York, 1971), p. 73.

Now we compute the thermodynamics near the critical point. We begin with the heat capacity at constant volume C_V. Using $S = -\partial F/\partial T$ and $C_V = T\partial S/\partial T$ we obtain

$$C_V^{VdW} = C_V^{ideal} = \frac{3}{2}Nk_B. \tag{4.32}$$

Thus C_V does not diverge at the critical point: $\alpha = 0$. On the other hand, the heat capacity at constant pressure C_p is given by the thermodynamic identity[2]

$$C_p - C_V = -T\left(\frac{\partial p}{\partial T}\right)_V^2 \frac{\partial V}{\partial p}\bigg|_T. \tag{4.33}$$

Performing the differentiation gives

$$C_p - C_V = \frac{Nk_B}{1-(2aN)(V-Nb)^2/V^3k_BT}. \tag{4.34}$$

At the critical point, or near to it, put $V = V_c$ and let $T \to T_c$, to give

$$\frac{C_p - C_v}{Nk_B} = \frac{T}{T-T_c}. \tag{4.35}$$

Thus $C_p \sim (T - T_c)^{-1}$. Equation (4.33) shows that $C_p \geq C_V$ in equilibrium, because of the fact that the compressibility $\kappa_T = -V^{-1}\partial V/\partial p$ is non-negative, and that in general C_p diverges in the same way as κ_T. Hence $\gamma = 1$. The divergence of κ_T has a direct physical interpretation: κ_T is an example of a **response function**, describing how the volume of the system changes under applied pressure. The divergence of this quantity means that the system becomes extraordinarily sensitive to an applied pressure near the critical point. This is not unreasonable, because for T slightly less than T_c, the system is thermodynamically unstable towards phase separation. The critical point therefore represents a state of marginal stability.

The exponent β is given from the shape of the coexistence curve as $T \to T_c^-$. We can approximate this using the law of corresponding states. Let

$$t \equiv \tau - 1 = \frac{T-T_c}{T_c}; \quad \phi \equiv \nu - 1 = \frac{V-V_c}{V_c}. \tag{4.36}$$

---

[2] See (*e.g.*) L.D. Landau and E.M. Lifshitz, *Statistical Physics Part 1 (Third Edition)* (Pergamon, New York, 1980), p. 53.

Then, in the vicinity of p_c, T_c, V_c, *i.e.* $\pi = \tau = \nu = 1$, we expand the equation of state (4.28):

$$\begin{aligned}\pi =& \frac{8(1+t)}{3(1+\phi)-1} - \frac{3}{(1+\phi)^2} \\ =& 1 + 4t - 6t\phi - \frac{3}{2}\phi^3 + O(t\phi^2, \phi^4)\end{aligned} \tag{4.37}$$

The terms omitted from this expression are justified *post hoc*: in fact, we will see that $\phi \sim t^{1/2}$ so eqn. (4.37) is indeed the lowest non-trivial order approximation to the equation of state near the critical point. To find the coexistence volumes $v_l(p)$, $v_g(p)$, or equivalently the corresponding values of ϕ, namely ϕ_l and ϕ_g, we use the Maxwell construction $\oint v\, dp = 0$. For fixed $t < 0$, the differential

$$dp = p_c \left[-6t\, d\phi - \frac{9}{2}\phi^2\, d\phi \right]$$

giving

$$\int_{\phi_l}^{\phi_g} \phi \left(-6t - \frac{9}{2}\phi^2 \right) d\phi = 0. \tag{4.38}$$

This must be satisfied for all t in the range of validity of approximation (4.37), so we conclude that $\phi_g = -\phi_l$. To find how ϕ_g or ϕ_l depends on t, write eqn. (4.37) in terms of ϕ_g and ϕ_l:

$$\pi = 1 + 4t - 6t\phi_g - \frac{3}{2}\phi_g^3 \quad \phi_g \text{ eqn.} \tag{4.39}$$

$$\pi = 1 + 4t + 6t\phi_g + \frac{3}{2}\phi_g^3 \quad \phi_l \text{ eqn.}, \tag{4.40}$$

where we have used $\phi_g = -\phi_l$ in eqn. (4.40). Subtracting and solving for (*e.g.*) ϕ_g gives

$$\phi_g = 2\sqrt{-t} \sim \left(\frac{T_c - T}{T_c} \right)^{1/2}. \tag{4.41}$$

i.e., the critical exponent $\beta = 1/2$.

We can calculate the shape of the critical isotherm, by setting $t = 0$ in eqn. (4.37); this yields the critical exponent δ:

$$\pi - 1 = \frac{p - p_c}{p_c} = -\frac{3}{2}\left(\frac{V - V_c}{V_c} \right)^3 \sim -\left(\frac{V - V_c}{V_c} \right)^\delta. \tag{4.42}$$

Hence, $\delta = 3$.

The values of the thermodynamic critical exponents α, β, γ and δ are the same as those obtained for the mean field theory of a ferromagnet. See table (3.1) for the comparison to experiment.

4.5 SPATIAL CORRELATIONS

Now we will discuss the spatial correlations in a fluid near the critical point, again within the framework of mean field theory. In the preceding chapter, we were able to relate the two-point correlation function with the isothermal susceptibility through the static susceptibility sum rule. We will now perform the corresponding analysis for a fluid. In a fluid, the two-point correlation function describes the statistical fluctuations in *density*. Thus, we will begin by showing how fluctuations in the *number* of particles in a fixed volume V are related to the compressibility.

4.5.1 Number Fluctuations and Compressibility

We work in the grand canonical ensemble. One way to visualise this is as follows: consider a volume of space V embedded in a much larger system Ω. Then particles of fluid in Ω will wander in and out of the volume V, causing the number N of particles in the volume V to fluctuate. The mean number of particles in the volume V is

$$\langle N \rangle = k_B T \left. \frac{\partial \log \Xi}{\partial \mu} \right|_{T,V} \tag{4.43}$$

where

$$\Xi = \mathrm{Tr}\; e^{-\beta(H-\mu N)}. \tag{4.44}$$

Similarly

$$\begin{aligned} \langle N^2 \rangle \equiv & \frac{\mathrm{Tr}\; N^2 e^{-\beta(H-\mu H)}}{\mathrm{Tr}\; e^{-\beta(H-\mu H)}} = \frac{1}{\beta^2} \frac{1}{\Xi} \frac{\partial^2 \Xi}{\partial \mu^2} \\ = & \frac{1}{\beta^2} \frac{\partial^2 \log \Xi}{\partial \mu^2} + \langle N \rangle^2 . \end{aligned} \tag{4.45}$$

Therefore

$$\langle N^2 \rangle - \langle N \rangle^2 = \frac{1}{\beta^2} \left. \frac{\partial^2 \log \Xi}{\partial \mu^2} \right|_{T,V} = \frac{1}{\beta} \frac{\partial}{\partial \mu} \langle N \rangle \bigg|_{T,V} . \tag{4.46}$$

Thus, the fluctuation in particle number ΔN is given by

$$\Delta N^2 \equiv \langle N^2 \rangle - \langle N \rangle^2 = \frac{k_B T}{\partial \mu / \partial N|_{T,V}} . \tag{4.47}$$

As it stands this is not very useful, but we can use thermodynamic trickery to express this in terms of measurable quantities, using **Jacobians**:[3]

$$\left. \frac{\partial \mu}{\partial N} \right|_{V,T} = \frac{\partial(\mu, V)}{\partial(N, V)} = \frac{\partial(\mu, V)}{\partial(N, p)} \cdot \frac{\partial(N, p)}{\partial(N, V)} = \frac{\partial(\mu, V)}{\partial(N, p)} \Big/ \frac{\partial(N, V)}{\partial(N, p)}$$

[3] The Jacobian $\partial(u, v)/\partial(x, y)$ is defined to be det $\begin{vmatrix} \partial u/\partial x & \partial u/\partial y \\ \partial v/\partial x & \partial v/\partial y \end{vmatrix}$.

$$= \left(\left.\frac{\partial \mu}{\partial N}\right|_{pT} \left.\frac{\partial V}{\partial p}\right|_{NT} - \left.\frac{\partial \mu}{\partial p}\right|_{NT} \left.\frac{\partial V}{\partial N}\right|_{pT} \right) \left(\left.\frac{\partial V}{\partial p}\right|_{NT} \right)^{-1} . \quad (4.48)$$

This can be simplified, because the fact that $G(T, p, N) = \mu(T, p)N$ means that $\partial \mu / \partial N|_p = 0$. Hence, the first term in eqn. (4.48) vanishes. The second term is simplified, because eqn. (4.6) implies the Maxwell relation

$$\left.\frac{\partial \mu}{\partial p}\right|_N = \left.\frac{\partial V}{\partial N}\right|_p . \quad (4.49)$$

Thus, the second term just involves the particle density

$$\rho(T, p) = \left.\frac{\partial N}{\partial V}\right|_p = \left(\left.\frac{\partial \mu}{\partial p}\right|_N \right)^{-1} . \quad (4.50)$$

Finally, using the definition of the compressibility κ_T given in eqn. (4.12), we obtain

$$\langle N^2 \rangle - \langle N \rangle^2 = k_B T \rho^2 V \kappa_T . \quad (4.51)$$

4.5.2 Number Fluctuations and Correlations

Let us define the *dimensionless* two-point correlation function

$$G(\mathbf{r} - \mathbf{r}') = \frac{1}{\rho^2} \left[\langle \rho(\mathbf{r}) \rho(\mathbf{r}') \rangle - \rho^2 \right] . \quad (4.52)$$

We expect two widely separated points to be uncorrelated, so

$$\langle \rho(\mathbf{r}) \rho(\mathbf{r}') \rangle \to \langle \rho(\mathbf{r}) \rangle \langle \rho(\mathbf{r}') \rangle = \rho^2 \quad (4.53)$$

as $|\mathbf{r} - \mathbf{r}'| \to \infty$. Hence $G(\mathbf{r} - \mathbf{r}') \to 0$ as $|\mathbf{r} - \mathbf{r}'| \to \infty$. Note that we can write

$$G(\mathbf{r} - \mathbf{r}') = \frac{1}{\rho^2} \langle (\rho(\mathbf{r}) - \rho)(\rho(\mathbf{r}') - \rho) \rangle , \quad (4.54)$$

showing that G represents fluctuations about the mean density, $\langle \rho(\mathbf{r}) \rangle = \rho$. The total number of particles in the volume V is given by

$$N = \int \rho(\mathbf{r}) d^d \mathbf{r} . \quad (4.55)$$

Hence,

$$\begin{aligned} \int d^d\mathbf{r}\, d^d\mathbf{r}'\, G(\mathbf{r} - \mathbf{r}') &= \frac{1}{\rho^2} \int d^d\mathbf{r}\, d^d\mathbf{r}' \left[\langle \rho(\mathbf{r}) \rho(\mathbf{r}') \rangle - \rho^2 \right] \\ &= \frac{1}{\rho^2} \left[\langle N^2 \rangle - \langle N \rangle^2 \right] . \end{aligned} \quad (4.56)$$

But translational invariance implies that

$$\int d^d\mathbf{r}\, d^d\mathbf{r}'\, G(\mathbf{r}-\mathbf{r}') = V \int d^d\mathbf{r}\, G(\mathbf{r}). \tag{4.57}$$

Finally, collecting together eqns. (4.51), (4.56) and (4.57), we obtain the desired relation between the correlation function and the isothermal compressibility:

$$\int d^d\mathbf{r}\, G(\mathbf{r}) = k_B T\, \kappa_T. \tag{4.58}$$

As mentioned previously, for a fluid in d dimensions, well away from the critical regime, G is of the form

$$G(\mathbf{r}) \sim \frac{e^{-|\mathbf{r}|/\xi}}{|\mathbf{r}|^{(d-1)/2}\xi^{(d-3)/2}}, \tag{4.59}$$

where $\xi(T)$ is the correlation length.

4.5.3 Critical Opalescence

The relation between the two-point correlation function and the isothermal compressibility leads to exactly the same consequences at the critical point that we saw in the previous chapter for the isothermal susceptibility: a diverging correlation length is implied by the divergence of the compressibility, with exponent $\nu = 1/2$.

In a fluid, the two-point correlation function measures the density fluctuations, which are able to scatter light. Thus, if light of wavelength λ is incident on the fluid, the resultant intensity I of the light scattered through an angle θ is proportional to the **structure factor**

$$S(\mathbf{k}) = \rho \int d^d\mathbf{r}\, e^{-i\mathbf{k}\cdot\mathbf{r}}\, G(\mathbf{r}), \tag{4.60}$$

with $|\mathbf{k}| = 4\pi \sin(\theta/2)/\lambda$. Using eqn. (4.59), and performing the Fourier transform gives

$$I \propto S(k) = \frac{k_B T \kappa_T \rho}{1 + k^2\xi^2(T)} \tag{4.61}$$

Measurements of the intensity I as a function of k are an accurate way to determine $\xi(T)$. At the critical point,

$$S(k) \sim k^{-2+\eta}, \tag{4.62}$$

enabling η also to be determined by light scattering.

The divergence of κ_T is reflected in the divergence of

$$\int d^d\mathbf{r}\, G(\mathbf{r}) = \int d^d\mathbf{r}\, e^{-i\mathbf{k}\cdot\mathbf{r}} G(\mathbf{r})\Big|_{\mathbf{k}=0} \propto S(0). \tag{4.63}$$

The divergence of $S(k)$ as $k \to 0$ for $T \to T_c$ is easily observable, because the scattering of light increases dramatically near the critical point. When there are correlated density fluctuations with $\xi(T) \sim \lambda_{\text{light}}$, the light is strongly scattered, multiple scattering becomes important, and the light cannot be transmitted through the medium. It becomes milky or opaque. This phenomenon is known as **critical opalescence**.[4]

4.6 MEASUREMENT OF CRITICAL EXPONENTS

To conclude this chapter, let us make some general comments about the definition and measurement of critical exponents.

4.6.1 Definition of Critical Exponents

Perhaps the most important point to emphasise is that critical exponents are only defined as **limiting power laws** as $T \to T_c$. Thus

$$f(t) \sim t^{\lambda} \tag{4.64}$$

means that

$$\lambda = \lim_{t\to 0} \frac{\log f(t)}{\log t}. \tag{4.65}$$

Sometimes, a critical exponent will be quoted as having the value zero. This can mean either that the corresponding variable has a discontinuity at T_c, or may show a **logarithmic divergence**. The latter corresponds to the exponent vanishing through the identity

$$\log t = \lim_{\alpha\to 0} \left[\frac{1 - e^{-\alpha \log t}}{\alpha}\right] = \lim_{\alpha\to 0} \left[\frac{1 - t^{-\alpha}}{\alpha}\right]. \tag{4.66}$$

In general, a critical exponent describes only the **leading** behavior. There may well be, and usually are, subdominant corrections, known as **corrections to scaling**. For example, the heat capacity (at constant volume) has the following form in general:

$$C_V(t) = A|t|^{-\alpha}\left(1 + B|t|^{\theta} + \ldots\right), \tag{4.67}$$

---

[4] A detailed discussion is given by D. Beysens in *Phase Transitions*, Proceedings of the Cargèse Summer School 1980 (Plenum, New York, 1982), p.25.

where $\theta > 0$ and B is a constant of order unity. The correction to the leading scaling behaviour vanishes as $|t| \to 0$, but may still make a significant contribution to C_V for small but non-zero values of $|t|$. In particular, when $\theta < 1$, the correction to scaling is actually singular, and such a term is often referred to as a **confluent singularity.** Corrections to scaling cannot be neglected in practice if reliable values for the actual critical exponents are required. Confluent singularities turn out to be particularly important near the superfluid transition in ^4He.[5] The origin of corrections to scaling is discussed from the RG point of view in section 9.10.

In addition to the non-analytic critical behavior, there is usually a smoothly varying backround piece which is not singular at T_c. For example, in the Van der Waals gas, the free energy has a part which is that of an ideal gas, and a part coming from the interaction. The former is always smooth, whilst it is the latter which can give rise to critical behavior. For example, the formula for C_p:

$$C_p = C_V + Nk_B \frac{T}{T - T_c}. \tag{4.68}$$

In these notes, we will be mainly concerned with critical exponents, but there are also constants of proportionality that must be considered too. For example,

$$C_V(t) = \begin{cases} At^{-\alpha} & t > 0; \\ A'(-t)^{-\alpha'} & t < 0. \end{cases} \tag{4.69}$$

A and A' are called **critical amplitudes**, whereas α and α' are **critical exponents.** The prime denotes "below T_c". The primed exponents are always equal to the un-primed exponents, and are universal, as we have discussed. The amplitudes are *not* universal, however. Nevertheless, the **amplitude ratio** A/A' is universal.

4.6.2 Determination of Critical Exponents

It is actually notoriously difficult to measure critical exponents. Let us see why. Suppose we want to measure the specific heat exponent α. In a theorist's ideal dream world, the experimentalist measures $C_V(T)$, makes a log-log plot versus $t \equiv (T - T_c)/T_c$, and reads off α from the slope of the line. What happens in practice?

In practice C_v is measured as $\Delta E/\Delta T$ when putting ΔE of energy into the system and measuring the temperature change ΔT. This is a good

[5] See the discussion by G. Ahlers in *Phase Transitions*, Proceedings of the Cargése Summer School 1980 (Plenum, New York, 1982), p.1.

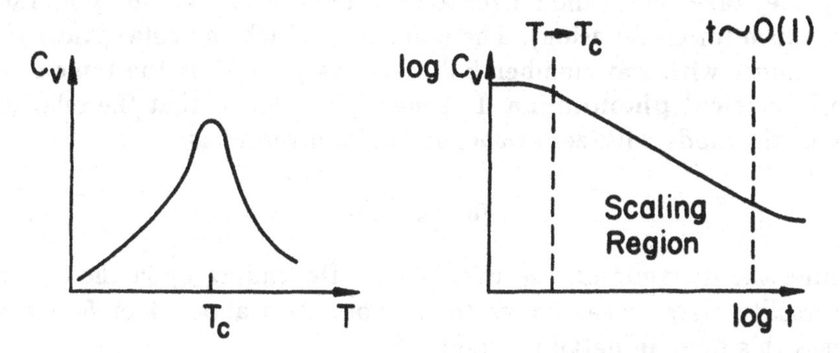

Figure 4.3 Schematic plot of real experimental data on the heat capacity near a continuous phase transition. Near T_c, the data are rounded due to instrumental resolution, impurity effects or the finite size of the system. Away from T_c, corrections to scaling and the background become important.

approximation to the heat capacity as long as $\Delta T \ll T$. However, to look for critical behaviour, we require much finer resolution of temperature differences $\Delta T \ll |T - T_c|$. The limiting sensitivity of the thermometer δT means that $\Delta T > \delta T$, and hence that $|t| \gg \delta T/T_c$. This is a nuisance, because from the discussion above, we really wish to measure the limit $t \to 0$. Thus, very high resolution thermometry is required.

Now suppose that we have managed to take some plausible data. The quantity of interest (C_V) in this example has an analytic background superimposed upon the singular behaviour we are trying to determine. To subtract off the background requires some modelling of it: in other words, curve fitting. To really eliminate background effects requires one to make t small enough that the diverging contribution overwhelms the background.

Even if we have been lucky enough to obtain good data on the singular contribution, when we plot it, we are sure to find that it does not scale! Close to T_c, instrumental resolution, impurity effects or the finite size of the system cause rounding of the divergence. Away from T_c, corrections to scaling and the background may be significant. In practice, one is lucky to obtain two decades of convincing power law behaviour.

Lastly, another difficulty has been swept under the rug. We do not know *a priori* the value of T_c. It has to be treated as an adjustable parameter, along with the background and corrections to scaling.

All of the above assumes that the data can actually be taken. This is not a trivial remark — as $T \to T_c$, it takes longer and longer to *equilibrate* the system. This is a phenomenon known as **critical slowing down**. As the correlation length diverges, the regions of the system, which represent fluctuations about the equilibrium state, get larger and larger; correspondingly, they take longer and longer to relax by whatever is the equilibration mechanism (often diffusion). The manner in which the **relaxation time** τ_k of modes with wavenumber k diverges as $T \to T_c$ is the topic of **dynamic critical phenomena.** In general, it is found that the relaxation time of the mode with zero wavenumber τ_0 diverges as

$$\tau_0 \sim \xi(T)^z, \tag{4.70}$$

defining the dynamic critical exponent z. Depending upon the dynamic universality class, z can range from about 2 to about 4 or 5. We will discuss this topic in detail in chapter 8.

EXERCISES

Exercise 4-1

In this question, you will investigate the microscopic origin of the law of corresponding states. Although this is a scaling law, it is not a result of scaling near the critical point.

(a) Consider a gas of particles in a volume V interacting via a pair potential $U(r)$. Sketch a typical form of $U(r)$. Suppose that for a particular class of substances, $U(r)$ has the form $U(r) = \epsilon u(r/\sigma)$; an example is the class of inert gases, well described by the Lennard-Jones potential. The meaning of this is that the energy scale is set by ϵ and the length scale by σ. For example, different gases might have different hard core radii σ and binding energies ϵ. Working in the canonical ensemble, show that all substances in this class have the same equation of state when expressed in suitably scaled variables. i.e. $p^* = \Pi(v^*, T^*)$, where starred quantities are scaled pressure, volume per particle and temperature.

(b) Show that if there is a critical point for this class of fluids, then $p_c v_c / T_c$ is a constant independent of the particular fluid.

CHAPTER 5

Landau Theory

We have seen, by example, how two apparently different physical systems can be described by **mean field theory**. In the Van der Waals gas, each particle interacts with the average density due to all the other gas particles. In the Weiss model of ferromagnetism, each spin interacts with the average magnetic field due to the magnetisation of the other spins. Such theories are called mean field theories, because each degree of freedom is assumed to couple to the average of the other degrees of freedom. Explicitly, in a magnetic system, the total field experienced by the spin at site i, due to the external field H and the neighbouring spins in a given configuration, may be written as

$$H_{\text{total}}(i) = H + \sum_{j(\neq i)} J_{ij} \langle S_j \rangle + \sum_{j \neq i} J_{ij}(S_j - \langle S_j \rangle). \tag{5.1}$$

In the Weiss model, the **fluctuation** term

$$\sum_{j(\neq i)} J_{ij}(S_j - \langle S_j \rangle) \tag{5.2}$$

is ignored. In fact these fluctuations contain the most important physics near a critical point, as we will shortly see. This chapter begins by addressing the question of why the critical exponents are the same for both the Van der Waals fluid and the Weiss ferromagnet. In answering this question, we will convince ourselves about the universality and the inevitability of these values of the critical exponents; nevertheless, in the following chapter we shall prove that mean field theory cannot possibly be correct near the critical point.

5.1 ORDER PARAMETERS

In chapters 3 and 4, we saw that mean field theory (MFT) gave the so-called *classical* values for the critical exponents by expanding the equation of state near the critical point. The expansion parameter was the spontaneous magnetisation in the case of the ferromagnet and the half-width of the coexistence curve in the case of the fluid. Both quantities exhibited the same temperature dependence, and are examples of the concept of an **order parameter.**

In the examples considered so far, the order parameter was a **scalar.** However, in general this is not necessary. The order parameter can be a vector, a tensor, a pseudo-scalar, or a group element of a symmetry group such as SU(N) *etc.* The order parameter for a given system is not unique; any thermodynamic variable that is zero in the un-ordered phase and non-zero in an adjacent (on the phase diagram), usually ordered phase, is a possible choice for an order parameter. Trivially, we could perfectly well choose M^3 as the order parameter in a ferromagnet. Let us consider some examples of order parameters, with differing numbers of components.

5.1.1 Heisenberg Model

The Hamiltonian is

$$H_\Omega = -J \sum_{\langle ij \rangle} \mathbf{S}_i \cdot \mathbf{S}_j - \mathbf{H} \cdot \sum_i \mathbf{S}_i, \tag{5.3}$$

where each site i on a lattice has a vector spin of constant magnitude (chosen to be 1) but variable direction: $|\mathbf{S}_i|^2 = 1$. When the system orders below T_c in more than two dimensions, the magnetisation is non-zero and points in some arbitary direction $\hat{\mathbf{n}}$. Thus, an order parameter is

$$\mathbf{M} = \frac{1}{N(\Omega)} \sum_i \langle \mathbf{S}_i \rangle = \begin{cases} \mathbf{0} & T > T_c; \\ M\hat{\mathbf{n}} & T < T_c. \end{cases} \tag{5.4}$$

Since $\mathbf{M}$ is a vector in three dimensions, the order parameter has 3 components. The Hamiltonian in zero external field has $O(3)$ symmetry, which means that

$$H_\Omega\{S_i\} = H_\Omega\{R\mathbf{S}_i\}, \quad (\text{for } \mathbf{H} = \mathbf{0}), \tag{5.5}$$

where R is an arbitary **rotation matrix** in three dimensions, acting on *all* the $\{\mathbf{S}_i\}$. The Hamiltonian in zero field is invariant under this arbitrary *global rotation in spin space* because it only depends on the scalar product of $\mathbf{S}_i$ and $\mathbf{S}_j$, *i.e.* the angle between them. Note that R rotates all of the *spins* by the same amount; it does not rotate the *lattice* on which the spins live. In zero field, the $O(3)$ symmetry of the Heisenberg model is spontaneously broken for $T < T_c$.

When $\mathbf{H} \neq \mathbf{0}$, the Hamiltonian is still invariant about rotations of the spins in the plane perpendicular to $\mathbf{H}$. So, now the system is invariant under $O(2)$ rotations.

5.1.2 XY Model

The Hamiltonian of the so-called **XY model** is given by eqn. (5.3), but the spins are unit vectors confined to rotate in a plane. The lattice on which the spins reside is still d dimensional, however, and the order parameter is once again the magnetisation $\mathbf{M} = (M_x, M_y)$. The order parameter has only 2 components, even though the dimensionality of the system may be $d = 1, 2, 3, 4 \ldots$. Spontaneous symmetry breaking can occur at non-zero temperature for $d > 2$. In two dimensions, a phase transition can occur, although it is not described by any local order parameter such as the magnetisation. This transition, the so-called **Kosterlitz-Thouless transition**, will be described separately in chapter 11.

5.1.3 ^{3}He

The order parameter for the superfluid transition in ^{3}He is an **anomalous pairing amplitude,** Ψ, which is a product of quantum mechanical field operators, whose expectation value at $T \neq 0$ is taken. It turns out that Ψ describes a state in which ^{3}He atoms form Cooper pairs in a p-wave state. The order parameter Ψ has 18 different components and is said to form a representation of the group $SO(3) \times SO(3) \times O(2)$. Crudely speaking, this means that with a p-wave state, the orbital angular momentum has three values, and therefore the spin has three possible values also. Finally, the spatial part of the Cooper pair wave function is a complex number, which for a given amplitude, may have an arbitrary phase.

5.2 COMMON FEATURES OF MEAN FIELD THEORIES

Now we will try to understand the reason for the universality that we have seen in mean field theory: *i.e.* the prediction that both a fluid and a ferromagnet near their critical points will exhibit the same critical exponents. We begin with the Van der Waals equation of state

$$\pi = \frac{8\tau}{3\nu - 1} - \frac{3}{\nu^2}, \tag{5.6}$$

$$\pi \equiv p/p_c, \tag{5.7}$$

$$\nu \equiv v/v_c = (N/V)^{-1}/(N/V)_c^{-1} = \frac{\rho_c}{\rho}, \tag{5.8}$$

$$\tau \equiv T/T_c; \quad t \equiv \frac{T - T_c}{T_c} = \tau - 1. \tag{5.9}$$

For later convenience, we also define $\eta \equiv (1/\nu) - 1 = (\rho/\rho_c) - 1$, which will serve as our order parameter. Then

$$\begin{aligned} \phi \equiv \nu - 1 &= \frac{1}{(1+\eta)} - 1 \\ &\approx -\eta. \end{aligned} \tag{5.10}$$

The equation of state (4.37) becomes

$$\pi = 1 + 4t + 6t\eta + \frac{3}{2}\eta^3 + O(\eta^4, \eta^2 t), \tag{5.11}$$

which can be derived from the Gibbs free energy

$$G(p, T, \eta) = G_0(p, T) + \frac{N}{\rho_c^2}\left[-(\pi - 1 - 4t)\,\eta + 3t\eta^2 + \frac{3}{8}\eta^4\right] \tag{5.12}$$

where we regard η as an ***independent*** parameter, whose value is to be determined by ***minimising*** $G(p, T, \eta)$. Note that η is ***not*** now regarded as a function of (p, T). The precise way in which η is determined will be explained shortly. The term (N/ρ_c^2) is unimportant for present purposes. It is present so that once η has its equilibrium value, $G(p, T, \eta)$ is dimensionally correct.

Now we perform the analogous calculation for the equation of state of the Weiss ferromagnet. Start with

$$\frac{H}{k_B T} = M(1 - \tau) + M^3(\tau - \tau^2 + \tau^3/3 + \cdots) \tag{5.13}$$

where, in contrast to eqn. (5.9), $\tau^{-1} \equiv T/T_c$. In terms of

$$t = \frac{T - T_c}{T_c} = \frac{1}{\tau} - 1 \tag{5.14}$$

and the order parameter $\eta \equiv M$, the equation of state becomes

$$\frac{H}{k_B T} = \eta t + \eta^3 + O(t\eta^3), \tag{5.15}$$

which can be derived from the Gibbs free energy

$$\Gamma(\eta, T, H) = \Gamma_0(T, H) - \frac{\eta H}{k_B T} + \frac{t\eta^2}{2} + \frac{1}{4}\eta^4 \tag{5.16}$$

by minimizing with respect to η. The similarity between eqns. (5.12) and (5.16) is why the critical exponents of the two models, obtained by differentiating the thermodynamic potentials[1] are the same. In the Van der Waals case, the role of the external magnetic field is played by $\pi - 1 - 4t$.

5.3 PHENOMENOLOGICAL LANDAU THEORY

In retrospect, it was inevitable that we obtained the mean field exponents that we did. The Taylor expansions of the equation of state in terms of the order parameter η and reduced temperature t involved the most general lowest order term that can be written down, while the absence of terms quadratic in η was guaranteed by the fact that the thermodynamic potential was even in η (for zero external field). These observations motivated Landau to suggest that we can apply similar considerations, at least in spirit if not in detail, to *all* phase transitions. The resulting theory, based upon rather general considerations of symmetry and analyticity, will be referred to sometimes as *phenomenological Landau theory*. Landau theory may also be motivated by a systematic calculation from the microscopic Hamiltonian, as discussed in exercise 3–3, and below; however, this has limited use, as we will see.

---

[1] The distinction between Gibbs and Helmholtz free energies in the present context is a matter of what are the constraints on the system, *e.g.* whether the electromagnetic field is included as part of the system. These distinctions will not turn out to be important for present purposes: the quantity whose expansion is given by eqns. (5.12) and (5.16) is actually neither the Gibbs nor the Helmholtz free energy.

5.3.1 Assumptions

Landau theory *postulates* that we can write down a function L known as the **Landau free energy**, or sometimes **the Landau functional**, which depends on the coupling constants $\{K_i\}$ and the order parameter η. L has the property that the state of the system is specified by the absolute minimum (*i.e.* global) of L with respect to η. L has dimensions of energy, and is related to, but, as we will see, is not identical with the Gibbs free energy of the system. We assume that thermodynamic functions of state may be computed by differentiating L, as if it were indeed the Gibbs free energy. The precise interpretation of L will be discussed in section 5.6. To specify L, it is sufficient to use the following constraints on L:

(1) L has to consistent with the symmetries of the system.

(2) Near T_c, L can be expanded in a power series in η. *i.e.* L is an analytic function of both η, and $[K]$. In a spatially uniform system of volume V, we can express the **Landau free energy density** $\mathcal{L}$ as

$$\mathcal{L} \equiv \frac{L}{V} = \sum_{n=0}^{\infty} a_n([K])\eta^n. \tag{5.17}$$

(3) In an inhomogeneous system with a spatially varying order parameter profile $\eta(\mathbf{r})$, $\mathcal{L}$ is a local function: *i.e.* it depends only on $\eta(\mathbf{r})$ and a *finite* number of derivatives. We will explain later the precise meaning of $\eta(\mathbf{r})$.

(4) In the disordered phase of the system, the order parameter $\eta = 0$, whilst it is small and non-zero in the ordered phase, near to the transition point. Thus, for $T > T_c$, $\eta = 0$ solves the minimum equation for L; for $T < T_c$, $\eta \neq 0$ solves the minimum equation. Thus, for a homogeneous system,

$$\mathcal{L} = \sum_{n=0}^{4} a_n([K], T)\eta^n, \tag{5.18}$$

where we have expanded $\mathcal{L}$ to $O(\eta^4)$ in the expectation that η is small, and all the essential physics *near* T_c appears at this order, as it did for the Van der Waals fluid and the Weiss ferromagnet. Whether or not the truncation of the power series for $\mathcal{L}$ is valid will turn out to depend on both the dimensionality of the system and the codimension of the singular point of interest.

5.3.2 Construction of $\mathcal{L}$

Now we use the assumptions to explicitly construct $\mathcal{L}$. Start with (1):

$$\frac{\partial \mathcal{L}}{\partial \eta} = a_1 + 2a_2\eta + 3a_3\eta^2 + 4a_4\eta^3 = 0. \tag{5.19}$$

For $T > T_c$, $\eta = 0$. Hence $a_1 = 0$.

Now consider the symmetry constraint. This is best seen by example: for the ferromagnet $\eta = M$, and if we have an Ising system with $H = 0$, then the probability distribution P for η is even in a finite system: $P(\eta) = P(-\eta)$. Assuming that $P \sim \exp(-\beta L)$, we require $\mathcal{L}(\eta) = \mathcal{L}(-\eta)$. We shall discuss the precise connection between P and L later. The evenness of $\mathcal{L}$ in this case implies that,

$$a_3 = a_5 = a_7 = \cdots = 0. \tag{5.20}$$

and thus

$$\mathcal{L} = a_0([K],T) + a_2([K],T)\eta^2 + a_4([K],T)\eta^4. \tag{5.21}$$

Note that the requirement that $\mathcal{L}$ be analytic in η precludes terms like $|\eta|$ in eqn. (5.21). Also, the coefficient $a_4([K]) > 0$, otherwise the Landau function can be minimized by $|\eta| \to \infty$, whereas we wish to describe how the order parameter rises from zero and has a *finite* value as the coupling constants are varied through the transition point.

Continuing with our example of the Ising ferromagnet in $H = 0$, we now ask for the form of the coefficients $a_n([K],T)$.

$a_0([K],T)$ is simply the value of $\mathcal{L}$ in the high temperature phase, and we expect it to vary smoothly through T_c. It represents the degrees of freedom in the system which are not described by the order parameter, and so may be thought of as the smooth background, on which the singular behaviour is superimposed. It is sometimes said that $(\mathcal{L} - a_0)$ represents the change in the Gibbs free energy due to the presence of the ordered state, but this statement is erroneous because L is not the Gibbs free energy. We will usually set a_0 to a constant (zero) in what follows.

Now consider a_4. Expanding in temperature near T_c, we obtain

$$a_4 = a_4^0 + (T - T_c)a_4^1 + \ldots \tag{5.22}$$

It will be sufficient to just take a_4 to be a positive constant. The temperature dependence of eqn. (5.22) will turn out not to dominate the leading behavior of the thermodynamics near T_c.

Similarly, we expand a_2:

$$a_2 = a_2^0 + \frac{(T - T_c)}{T_c}a_2^1 + O\left((T - T_c)^2\right). \tag{5.23}$$

We can determine the form of this from (4):

$$\begin{cases} \eta = 0 & T > T_c; \\ \eta \neq 0 & T < T_c. \end{cases}$$

Solving eqn. (5.19) for η gives

$$\eta = 0 \text{ or } \eta = \sqrt{\frac{-a_2(T)}{2a_4}}. \tag{5.24}$$

If η is to be non-zero when $T < T_c$, then

$$a_2^0 = 0 \tag{5.25}$$

and

$$a_2 = a_2^1 \left(\frac{T - T_c}{T_c} \right) + O\left(\left(\frac{T - T_c}{T_c} \right)^2 \right). \tag{5.26}$$

The higher order terms in (5.26) do not contribute to the leading behavior near T_c.

Now let us extend the treatment to the case $H \neq 0$. For the Ising ferromagnet, we know that $\eta = M$ and that the appropriate energy is $-H \sum_i S_i = -HNM$. In conclusion, we obtain our final expression

$$\begin{aligned} L &= V\mathcal{L} = N\mathcal{L} \quad \text{(for lattice systems)}, \\ \mathcal{L} &= at\eta^2 + \frac{1}{2}b\eta^4 - H\eta. \end{aligned} \tag{5.27}$$

The coefficients a and b are phenomenological parameters, which in principle could be obtained from the appropriate microscopic theory. In principle, a term proportional to $H\eta^3$ is also allowed by symmetry in eqn. (5.27), but this is not a leading term near the critical point.

We have only considered the Ising universality class; in general, however, the Landau function is constructed by writing down *all* possible scalar terms which are powers and products of the order parameter components, consistent with the symmetry requirements of the particular system.[2]

---

[2] For a detailed discussion, see L.D. Landau and E.M. Lifshitz, *Statistical Physics Part 1 (Third Edition)* (Pergamon, New York, 1980), §145.

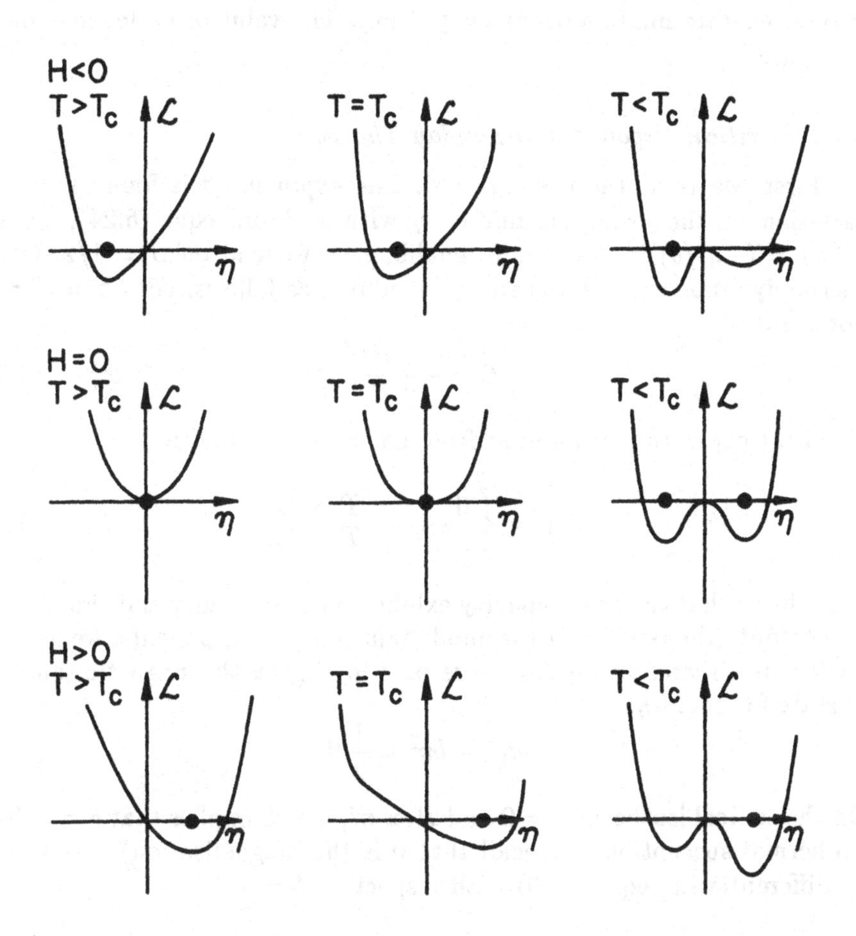

Figure 5.1 The Landau free energy density for various values of T and H. The • indicates the value of η at which $\mathcal{L}$ achieves its global minimum. The right-most column of graphs depicts the first order transition, which occurs for $T < T_c$ as H is varied from a negative to a positive value. The central row depicts the continuous transition, which occurs for $H = 0$ as T is varied from above T_c to below T_c.

5.4 CONTINUOUS PHASE TRANSITIONS

Now we will see how Landau theory accounts for the non-analytic behaviour near and below T_c. It is very helpful to sketch the form of $\mathcal{L}(\eta, H, T)$, as shown in figure (5.1).

Here we discuss the continuous transition, which occurs for $H = 0$, as shown in the central row of graphs in figure (5.1). For $T > T_c$, the minimum of $\mathcal{L}$ is at $\eta = 0$. For $T = T_c$, the Landau function has zero curvature at $\eta = 0$, but $\eta = 0$ is still the global minimum. For $T < T_c$, two degenerate minima occur at $\eta = \pm\eta_s$. The value of η_s depends on T: $\eta_s = \eta_s(T)$.

5.4.1 Critical Exponents in Landau Theory

First, we treat the case $H = 0$. The exponent β is found from the variation of the order parameter η with t. From eqn. (5.24), we get $\eta_s(T) = (-at/b)^{1/2}$, for $t < 0$, enabling us to read off $\beta = 1/2$. Other thermodynamic exponents can be calculated as follows. For $t > 0$, $\mathcal{L} = 0$. For $t < 0$,

$$\mathcal{L} = -\frac{1}{2}\frac{a^2t^2}{b}. \tag{5.28}$$

The heat capacity is computed from $C_V = -T\partial^2\mathcal{L}/\partial T^2$:

$$C_V = \begin{cases} 0 & T > T_c; \\ a^2/bT_c & T < T_c. \end{cases} \tag{5.29}$$

This shows that the heat capacity exhibits a discontinuity and thus $\alpha = 0$. To compute the remaining thermodynamic critical exponents, we need to let $H \neq 0$. Differentiating $\mathcal{L}$ with respect to H gives the magnetic equation of state for small η:

$$at\eta + b\eta^3 = \frac{1}{2}H. \tag{5.30}$$

On the critical isotherm, $t = 0$ and $H \propto \eta^3$, which implies that $\delta = 3$. The isothermal susceptibility (recall that η is the magnetisation) is obtained by differentiating eqn. (5.30) with respect to H:

$$\chi_T(H) \equiv \left.\frac{\partial\eta(H)}{\partial H}\right|_T = \frac{1}{2(at + 3b\eta(H)^2)}, \tag{5.31}$$

where $\eta(H)$ is a solution of eqn. (5.30). We are interested in the response function at zero external field. For $t > 0$, $\eta = 0$ and $\chi_T = (2at)^{-1}$. For $t < 0$, $\eta^2 = -at/b$ and $\chi_T = (-4at)^{-1}$. Hence $\gamma = \gamma' = 1$.

Now we have found all of the mean field thermodynamic exponents; as expected, they agree with those of the Weiss and Van der Waals equations. What about the critical exponents describing the spatial correlations in the system? To calculate these, we have to make an extension of Landau theory to deal with inhomogeneous systems. But first, while we are dealing with uniform systems, it is appropriate to mention the topic of **first order phase transitions.**

5.5 FIRST ORDER TRANSITIONS

Landau theory is predicated on the assumption that the order parameter $\eta(t)$ is arbitrarily small as $t \to 0$. We saw that if $\mathcal{L} = at\eta^2 + \frac{1}{2}b\eta^4 - H\eta$, then for $H = 0$, $t \to 0$, phenomenological Landau theory gives a description of a continuous transition *i.e.* $\eta(t) \to 0$ as $t \to 0$. Now we consider more general expressions for $\mathcal{L}$.

In constructing $\mathcal{L}$, we saw that there cannot be a term linear in η if the symmetric phase corresponds to $\eta = 0$. However, it was only the symmetry of the problem that prevented us from writing down a term cubic in η. Let us now examine the effect of such a term, by considering

$$\mathcal{L} = at\eta^2 + \frac{1}{2}b\eta^4 + C\eta^3 - H\eta \tag{5.32}$$

with a and b positive. Then, for $H = 0$, the equilibrium value of η is obtained by differentiation with respect to η and setting the result equal to zero:

$$\eta = 0, \quad \eta = -c \pm \sqrt{c^2 - at/b}, \tag{5.33}$$

where we have defined $c \equiv 3C/4b$. The solution $\eta \neq 0$ becomes acceptable (*i.e.* real, not complex) for reduced temperatures t satisfying

$$c^2 - \frac{at}{b} > 0 \tag{5.34}$$

i.e. $t < t^* \equiv bc^2/a$. Since t^* is positive, this occurs at a temperature greater than T_c (T_c is the temperature where the coefficient of the term quadratic in η vanishes). In the description of a continuous transition, we found that $\eta \neq 0$ only became acceptable for $t < 0$, *i.e.* $T < T_c$. We can see what is going on by sketching the form of $\mathcal{L}$, as shown in figure (5.2) (without loss of generality, we choose $c < 0$).

For $t < t^*$, a secondary minimum and maximum have developed, in addition to the minimum at $\eta = 0$. As t is reduced further to the value t_1, the value of $\mathcal{L}$ at the secondary minimum is equal to the value at $\eta = 0$. For $t < t_1$, the secondary minimum is now the global minimum, and the value of the order parameter which minimises $\mathcal{L}$ jumps *discontinuously* from $\eta = 0$ to a non-zero value. This is a first order transition.

Note that at the first order transition, $\eta(t_1)$ is not arbitrarily small as $t \to t_1^-$, so Landau theory is *not* valid. Thus, if there is no symmetry reason that forces $C = 0$, then the cubic term will in general cause a first order transition to occur. In more general situations, where the order parameter has more than one component, then $\mathcal{L}$ – a scalar – is constructed out of combinations of the order parameter components which are invariants

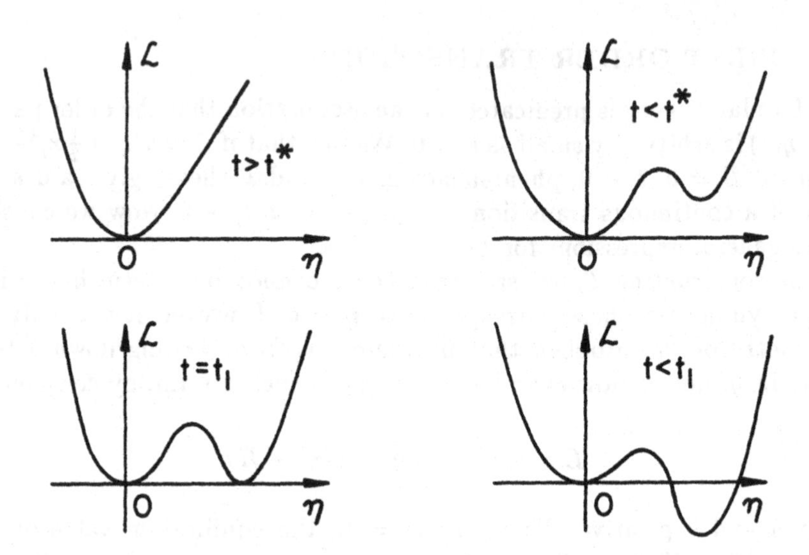

Figure 5.2 $\mathcal{L}$ as a function of η for various temperatures, showing the Landau theory description of a first order transition.

under the symmetries of the physical situation. Then, a sufficient, but not necessary condition for the existence of a continuous transition is that there are no cubic invariants consistent with the symmetry of the problem.

That the condition is not a necessary condition is shown by the following example: in the replica theory of spin glasses and rubber,[3] the order parameter is a $n \times n$ matrix $Q^{\alpha\beta}$, which in the limit that $n \to \infty$, becomes a *non-negative* function $q(x)$: the discrete indices α and β become a continuous argument x. The corresponding Landau free energy is a **functional**[4] of $q(x)$, with a term cubic in q. However, the *sign* of the coefficient of the cubic term is such that the secondary minimum in $\mathcal{L}$, which, in the analysis above causes the first order transition to preempt the continuous transition, only occurs for negative and hence unphysical values of q. Thus the transition in these cases is continuous despite the presence of the cubic term.

---

[3] See the references in section (2.10): G. Parisi, *op. cit*; P. Goldbart and N. Goldenfeld, *op. cit.*

[4] A functional is a map from a function to a number, just as a function is a map from a number to another number. An example of a functional is a definite integral — the result of the integral (a number) is determined by the integrand (a function). We will use the notation $F\{y(x)\}$ to denote the functional F of the function $y(x)$.

One final remark about cubic terms in Landau theory. The Landau criterion is, of course, only a statement made within the context of mean field theory. It is important to realize that the fluctuation effects omitted by mean field theory can change the **order of a transition**. Two examples are:

(a) The three state Potts model[5] in two dimensions. Mean field theory predicts a first order transition, whereas the actual transition is continuous.

(b) Type I superconductor-normal metal transition in three dimensions. Mean field theory predicts a continuous transition, whereas in fact, fluctuations of the electromagnetic field *induce* a first-order phase transition.[6]

5.6 INHOMOGENEOUS SYSTEMS

We now ask what happens if the order parameter η is allowed to vary in space: $\eta = \eta(\mathbf{r})$. In a uniform system, η is a constant in equilibrium. A spatially varying order parameter can arise as a result of an inhomogeneous external field $H(\mathbf{r})$. We will show below that it is conceptually important to consider a spatially varying order parameter, in order to obtain a correct understanding of what the Landau free energy really is, even in the absence of an inhomogeneous external field.

5.6.1 Coarse Graining

We have already seen that the spatial correlations grow as $T \to T_c$, and that the system contains patches of spins, of linear dimension $\approx \xi(T)$, in which the magnetisation is approximately constant. This suggests that we divide the system up into blocks of linear dimension[7] $\Lambda^{-1} \approx \xi(T)$. Within each block, the system is approximately uniform, and as long as there are a large number of spins within each block, then it is sensible to define the **local magnetisation** $M_\Lambda(\mathbf{r})$ within each block centred at $\mathbf{r}$,

---

[5] For a review, see F.Y. Wu, *Rev. Mod. Phys.* **54**, 235 (1982).

[6] B.I. Halperin, T.C. Lubensky and S.-K. Ma, *Phys. Rev. Lett.* **32**, 292 (1974). A similar effect occurs in liquid crystal systems. For a discussion of observations of this effect, see M.A. Anisimov *et al.*, *JETP Lett.* **45**, 425 (1987)[*Pis'ma Zh. Eksp. Teor. Fiz.* **45**, 336 (1987)] and P.E. Cladis *et al.*, *Phys. Rev. Lett.* **62**, 1764 (1989).

[7] We define the size of the blocks to be Λ^{-1} rather than Λ because it will be later convenient to work in Fourier space, where the wavenumber can be restricted to be smaller than Λ.

and to assign a value of the Landau free energy for each block. The local magnetisation is then

$$M_\Lambda(\mathbf{r}) = \frac{1}{N_\Lambda(\mathbf{r})} \sum_{i\in\mathbf{r}} \langle S_i \rangle, \tag{5.35}$$

where $N_\Lambda(\mathbf{r}) = \Lambda^{-d}/a^d$ is the number of spins in the block at $\mathbf{r}$. Note that it is *always* sensible to define $M_\Lambda(\mathbf{r})$ in this way, because near T_c, $\xi \gg a$, and so if we choose $\Lambda^{-1} \leq \xi(T)$, then

$$a \ll \Lambda^{-1} \leq \xi(T) \tag{5.36}$$

and there will be a large number $(O(\xi^d/a^d))$ of spins in the block at $\mathbf{r}$, especially as $T \to T_c$. We must not choose $\Lambda^{-1} > \xi(T)$ or else the magnetisation will not be approximately uniform within each block.

Dividing the system up into blocks like this is often called **coarse graining** and $M_\Lambda(\mathbf{r})$ is often called the **coarse-grained magnetisation.** Note that we have attached the label Λ to $M(\mathbf{r})$, to remind ourselves that $M(\mathbf{r})$ is *not* uniquely defined; one needs to specify Λ in the prescription for defining $M_\Lambda(\mathbf{r})$. By construction, the **coarse-grained order parameter** does not fluctuate wildly on a scale of the lattice spacing a, but varies smoothly in space, with no Fourier components corresponding to wavenumbers greater than Λ. Although we need to specify Λ in defining coarse-grained variables, the procedure will only be useful if the results do not depend on the choice of Λ, at least within the stated range. Thus, in practice, Λ is usually not mentioned explicitly: however, we will see later, when we discuss fluctuation effects in chapter 6, that it is important in principle to recognise the existence of Λ.

What will be the Landau free energy of the system? Clearly, it cannot be

$$L = \sum_{\mathbf{r}} \mathcal{L}(M_\Lambda(\mathbf{r})), \tag{5.37}$$

because the minimisation would just give the equilibrium value for $M_\Lambda(\mathbf{r})$ in each block. But it is obviously energetically unfavourable to have large differences between $M_\Lambda(\mathbf{r})$ in adjacent blocks; domain walls are costly. What is needed is an extra term which acts to penalize differences between the magnetisation in adjacent blocks. The simplest *analytic* expression of this sort is

$$\sum_{\mathbf{r}} \sum_{\delta} \frac{\gamma'}{2} \left(\frac{M_\Lambda(\mathbf{r}) - M_\Lambda(\mathbf{r}+\delta)}{\Lambda^{-2}} \right)^2. \tag{5.38}$$

In this expression, δ is a vector of magnitude Λ^{-1} pointing to the nearest neighbor block to $\mathbf{r}$, and the value of the energy cost is independent of the sign of the magnetisation difference. The positive constant γ' is in principle temperature dependent.

Using the fact that $M_\Lambda(\mathbf{r})$ is slowly varying in space (compared to a), we may write

$$L = \int_\Omega d^d\mathbf{r} \left[\mathcal{L}(M_\Lambda(\mathbf{r})) + \frac{1}{2}\gamma (\nabla M_\Lambda(\mathbf{r}))^2 \right], \tag{5.39}$$

where the function $\mathcal{L}(x)$ is the homogeneous Landau free energy density of eqn. (5.27), and γ is another positive constant, whose temperature dependence is so weak as to be negligible near the critical point. The Landau free energy L is a **functional** of $M_\Lambda(\mathbf{r})$. It depends on the entire function $M_\Lambda(\mathbf{r})$, *i.e.* to compute the value of L, we need to know $M_\Lambda(\mathbf{r})$ at all points in space $\mathbf{r}$. Sometimes, L will be called the **coarse-grained free energy** or the **effective Hamiltonian.**

The reader may be wondering why the gradient term in $\mathcal{L}$ does not also include $M\nabla^2 M$. This term seems perfectly acceptable. It is of the same order in $M_\Lambda(\mathbf{r})$ as the term we have actually used; it is isotropic because of the ∇^2 and is invariant under $M \leftrightarrow -M$, so this term is allowed by symmetry. However, it is not written down because the identity

$$\nabla \cdot (M\nabla M) = M\nabla^2 M + (\nabla M)^2 \tag{5.40}$$

implies that

$$\int_\Omega M\nabla^2 M d^d\mathbf{r} = \int_\Omega M\nabla M \cdot d\mathbf{S} - \int_\Omega (\nabla M)^2 d^d\mathbf{r} \tag{5.41}$$

and the surface term is negligible in the thermodynamic limit. Hence, the term actually written down in eqn. (5.39) is the most general that satisfies the requirements above.

5.6.2 *Interpretation of the Landau Free Energy*

We now need to address the issue of what L actually is. In fact, it is easiest to say what L is not. It clearly is *not* the Gibbs free energy Γ_Ω of the system Ω, defined by

$$\Gamma_\Omega(M) = F_\Omega(H(M)) + NMH(M), \tag{5.42}$$

where F_Ω is the Helmholtz free energy

$$e^{-\beta F_\Omega(H)} = \mathrm{Tr}\, e^{-\beta H_\Omega}, \tag{5.43}$$

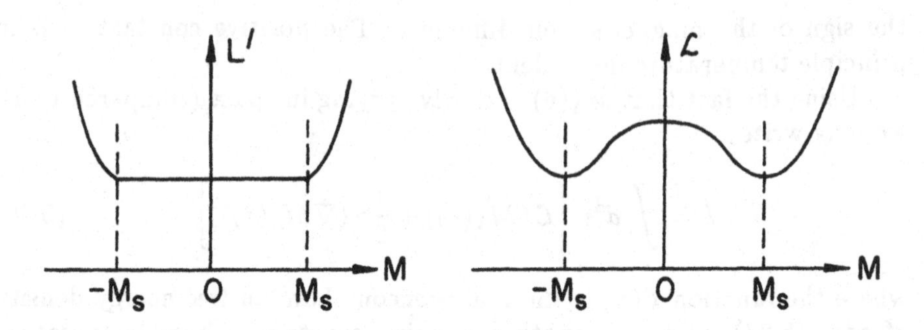

Figure 5.3 Sketch of the functions $L'(M,0) = \Gamma(M,0)$ and the Landau free energy density $\mathcal{L}$ at zero external field.

and $H(M)$ is the function defined implicitly from

$$N(\Omega)M(H) = -\frac{\partial F_\Omega}{\partial H}. \tag{5.44}$$

$H(M)$ is the external magnetic field required for the magnetisation to have the value M. The reasons are two fold:

(1) $\Gamma(M)$ is convex down whilst $L(M)$ is not convex.

(2) $\Gamma(M)$ is a *thermodynamic* function: thus, a spatially varying $M(\mathbf{r})$ has no meaning in this context, since all spatial information has been integrated out in the Tr operation. In other words, $\Gamma(M)$ depends upon Tr $e^{-\beta H}$, and therefore cannot depend on particular microscopic configurations, which have been summed over in Tr .

It is useful to revise (1). Dropping the subscript Ω, we can find the equation of the state as follows:

$$\frac{\partial \Gamma}{\partial M} = \frac{\partial F}{\partial H}\frac{\partial H}{\partial M} + NH + NM\frac{\partial H}{\partial M} = NH, \tag{5.45}$$

i.e. $\partial\Gamma/\partial(NM) = H$. At zero field,

$$\frac{\partial \Gamma}{\partial(NM)} = 0, \tag{5.46}$$

so it seems that minimising Γ with respect to M, the order parameter, specifies the state of the system. This resembles the defining property of the Landau free energy, at least for zero external field, but there is an important difference, as we now discuss.

In non-zero field, the generalisation of the above suggests that we consider as a candidate for the Landau free energy, the function

$$L'(M,\hat{H}) \equiv \Gamma(M) - M\hat{H}N$$
$$= F(H(M)) + NM\left(H(M) - \hat{H}\right). \tag{5.47}$$

Here, $\hat{H}$ is simply a parameter, and $H(M)$ is the inverse of the function $M(H)$ defined by (5.44). This always satisfies

$$\frac{\partial L'}{\partial M}(M,\hat{H}) = 0. \tag{5.48}$$

This identification is incorrect, however, as can be seen when we sketch L and L'. Here we discuss the case $\hat{H} = 0$, leaving the more general case to the exercises at the end of this chapter. For $\hat{H} = 0$, $L' = \Gamma$. The form of Γ can be deduced from inspection of figures (2.9) and (2.10). In particular, for $-M_s < M < M_s$, $H(M) = 0$ (inverting figure (2.10)), and thus, from eqn. (5.46), $\Gamma(M)$ must be a horizontal line, as sketched in figure (5.3). Hence Γ is indeed a convex function, whereas the Landau free energy density is not. In summary, L' is a qualitatively correct identification of the Landau free energy *outside* the region $-M_s < M < M_s$ *i.e.* the coexistence region, and above T_c.

Having disposed of the attempt to give the Landau free energy a strictly thermodynamic interpretation, let us consider the meaning of coarse graining. $L\{M_\Lambda(\mathbf{r})\}$ is the Gibbs free energy for the system, *constrained* to be in the configuration specified by $M_\Lambda(\mathbf{r})$. So L depends on Λ implicitly, and should really be denoted by L_Λ. Clearly, there are *many* configurations of the spin $\{S_i\}$ consistent with a configuration $M_\Lambda(\mathbf{r})$; for example, given a spin configuration consistent with the coarse-grained order parameter profile $M_\Lambda(\mathbf{r}) = \phi(\mathbf{r})$, where ϕ is a specified function, then another microscopic configuration consistent with the coarse-grained order parameter profile may be obtained by flipping spins in pairs (up-down) within each block.

Formally, we can define the Landau free energy as

$$e^{-L\{M_\Lambda(\mathbf{r})\}/k_BT} = \mathrm{Tr}\left\{e^{-\beta H_\Omega\{S_i\}}\delta^{(K)}\left[\sum_{i\in\mathbf{r}} S_i - M_\Lambda(\mathbf{r})N_\Lambda(\mathbf{r})\right]\right\}, \tag{5.49}$$

where the Kronecker delta is defined as

$$\delta^{(K)}(a-b) = \begin{cases} 0 & a \neq b; \\ 1 & a = b. \end{cases} \tag{5.50}$$

Sometimes, the expression (5.50) will be written as

$$e^{-\beta L\{M_\Lambda(\mathbf{r})\}} = \mathrm{Tr}' e^{-\beta H\{S_i\}}. \tag{5.51}$$

The prime on Tr indicates a sum over *microscopic degrees of freedom* S_i, with the constraint that only configurations with the magnetisation $M_\Lambda(\mathbf{r})$ are included. This *partial trace* is clearly *not* equal to the partition function $Z = \mathrm{Tr}\ e^{-\beta H_\Omega}$, where *all* microscopic configurations are included.

So, how is $L\{M_\Lambda(\mathbf{r})\}$ related to Z? To get L, we did a partial trace of the microscopic degrees of freedom. We will denote by $\{\bar{S}_i\}$ the microscopic configurations consistent with a *particular* coarse-grained order parameter profile $\bar{M}_\Lambda(\mathbf{r})$ and by $\{S_i^*\}$ the configurations of the microscopic degrees of freedom which are not consistent with that *same* profile. Since the set $\{S_i^*\}$ is defined with reference to $\bar{M}_\Lambda(\mathbf{r})$, we can write symbolically that $\{S_i^*\} = \{S_i^*\}(\bar{M}_\Lambda(\mathbf{r}))$. Of course, $\bar{M}_\Lambda(\mathbf{r})$ is just some particular function in space, slowly varying on a scale of the lattice spacing a. Each of the microscopic configurations in $\{S_i^*\}$ is consistent with other slowly varying coarse-grained order parameter profiles. The sum over all microstates can be separated into two parts:

$$\sum_{\{S\}} = \sum_{\{S^*\}} \sum_{\{\bar{S}\}}. \tag{5.52}$$

Thus, it follows that

$$\begin{aligned} Z = e^{-\beta F} &= \sum_{\{S\}} e^{-\beta H_\Omega\{S\}} \\ &= \sum_{\{S^*\}} \sum_{\{\bar{S}\}} e^{-\beta H_\Omega} \\ &= \sum_{\{S^*\}} e^{-\beta L\{\bar{M}_\Lambda(\mathbf{r})\}}. \end{aligned} \tag{5.53}$$

The sum $\sum_{\{S^*\}}$ is nothing other than a sum over all configurations of the semi-macroscopic or coarse-grained variable $\bar{M}_\Lambda(\mathbf{r})$. Sometimes, in deference to the notion that $\bar{M}_\Lambda(\mathbf{r})$ is a smoothly varying function of $\mathbf{r}$, the $\sum_{\{S^*\}}$ is written as

$$Z = \int DM_\Lambda\, e^{-\beta L\{M_\Lambda(\mathbf{r})\}}, \tag{5.54}$$

where the notation $\int DM_\Lambda$ denotes a **functional integral**, and we have dropped the overbar. We shall shortly see how this is evaluated in practice, but conceptually, it just means a sum over all smoothly varying functions. Note that $e^{-\beta L\{M_\Lambda(\mathbf{r})\}}$ is proportional to the probability that the configuration $M_\Lambda(\mathbf{r})$ is found.

Once we realise that $L\{M_\Lambda(\mathbf{r})\}$ is a coarse-grained free energy, and not the thermodynamic free energy, there is no theorem requiring that it be convex. Nevertheless, the true *thermodynamic* Gibbs free energy is convex. Integrating over the semi-macroscopic degrees of freedom $\sum_{S^*}$, *i.e.* over the configurations omitted in $L\{M_\Lambda(\mathbf{r})\}$, restores convexity. Our interpretation of $\mathcal{L}$ also explains why we only consider analytic contributions to $\mathcal{L}$: the partial trace involves a finite number of states (for a finite system), and so $\mathcal{L}$ should be analytic.[8]

Let us now make a few general remarks, for a system, with coarse-grained order parameter $\eta_\Lambda(\mathbf{r})$. In practice, $L\{\eta_\Lambda(\mathbf{r})\}$ is never really derived in the way we have discussed, *i.e.* by direct integration of microscopic degrees of freedom; the **real space renormalisation group** method, to be described later, comes close to this, but is hard to implement systematically. There are other ways to try and motivate Landau theory from microscopic considerations, such as the **Hubbard-Stratonovich transformation** used in exercise 3–3. However, this is not a systematic derivation, and the approximations involved become uncontrolled as $T \to T_c$; as we shall see, **perturbation theory** becomes *divergent*, and **renormalisation** and the **renormalisation group** are required to make sense of the entire procedure. In some cases, however, this method can be used without renormalisation as long as T is not too close to T_c. We shall refer to Landau theory motivated in this way as "microscopic Landau theory", to distinguish it from phenomenological Landau theory. An example of the the microscopic Landau theory occurs in the theory of **superconductivity**, where it turns out that the asymptotic critical behaviour occurs so close to T_c as to be inaccessible. The experimentally accessible regime near T_c corresponds to a situation where the fluctuations about mean field theory are small, and it is possible to derive the coarse-grained free energy for superconductors — the so-called **Ginzburg-Landau theory** — from the microscopic theory of superconductivity due to Bardeen, Cooper and Schrieffer.[9]

---

8 For a system which is initially defined on a continuum, a suitable regularisation must first be given. This is a prescription to define the system as the limit of a discrete system. Such a limit may involve technical complications that we will not address here.

9 This was accomplished by L.P. Gorkov, *Sov. Phys. JETP* **9**, 1364 (1959)[*Zh. Eksp. Teor. Fiz.* **36**, 1918 (1959)]. The derivation is outlined by J.R. Schrieffer, *Theory of Superconductivity* (Benjamin, Reading, 1983), p. 248 *et seq.* An instructive functional integral derivation is given by V. Popov, *Functional Integrals in Quantum Field Theory and Statistical Physics* (Reidel, Dordrecht, 1983), p. 191 *et seq.*

In most cases, however, the procedure is to follow the prescription of the phenomenological Landau theory: the terms required by symmetry are written down in $L\{\eta_\Lambda(\mathbf{r})\}$, with more or less *arbitrary* coefficients. We will see later that the coefficients of the starting Landau theory do not matter near T_c, since they acquire corrections dependent upon $T - T_c$: one says that they become *renormalised*. Close to T_c, the coefficients tend towards certain special values, characteristic of the universality class of the starting Landau free energy.

5.7 CORRELATION FUNCTIONS

For definiteness, we consider systems described by the same Landau theory as the Ising model — *i.e.* the Ising universality class. Denote the order parameter by $\eta(\mathbf{r})$; the Landau free energy L is then

$$L = \int d^d\mathbf{r} \left[\frac{\gamma}{2}(\nabla\eta)^2 + at\eta^2 + \frac{1}{2}b\eta^4 - H(\mathbf{r})\eta(\mathbf{r})\right]. \tag{5.55}$$

We begin this section with some mathematical preliminaries.

5.7.1 The Continuum Limit

We will now see how to generate correlation functions from the partition function Z,

$$Z = \mathrm{Tr}_{\eta(\mathbf{r})}\, e^{-\beta L\{\eta(\mathbf{r})\}}, \tag{5.56}$$

in much the same way that we generated correlation functions for the spin degrees of freedom on a lattice in section 3.7.2. The only difference is that here the degrees of freedom η are labelled by a continuous variable $\mathbf{r} \in \Omega$, whereas in section 3.7.2, the degrees of freedom S were labelled by a discrete variable $i = 1, 2, \ldots, N(\Omega)$. In fact, we can simply take over all the results to the present case, by going to the limit that $a \to 0$, $N(\Omega) \to \infty$, but the volume $V = N(\Omega)a^d$ remains constant. This is sometimes referred to as the **continuum limit**. In practice, many of the formulae are unchanged, with the exception that whereas before, in a finite system Ω, there were a finite number of degrees of freedom η_i, now there are an infinite number of degrees of freedom $\eta(\mathbf{r})$.

5.7.2 Functional Integrals in Real and Fourier Space

The functional integral in the partition function looks formal, but in light of the above remarks, it may be thought of as follows. For a finite set of variables η_i, where $-\infty < \eta_i < \infty$, the trace operation is just

$$\mathrm{Tr} \equiv \int \prod_{i=1}^{N(\Omega)} d\eta_i. \tag{5.57}$$

The measure in the functional integral is the continuum limit of eqn. (5.57) *i.e.* is simply an infinite product of integrals. The reader may also be concerned that we have dropped the subscript Λ: how do we implement the notion that $\eta(\mathbf{r})$ is a slowly varying function on the scale of the lattice spacing a? We will see later that in practice, functional integrals are often performed in Fourier space. Then, the measure becomes

$$\mathrm{Tr} \equiv \int \prod_{|\mathbf{k}|<\Lambda} d(\mathrm{Re}\ \eta_{\mathbf{k}})\, d(\mathrm{Im}\ \eta_{\mathbf{k}}). \tag{5.58}$$

We shall sometimes denote the differentials by the shorthand notation $d\eta_{\mathbf{k}}$. Usually, the function $\eta(\mathbf{r})$ is real, and so is equal to its complex conjugate. This means that the Fourier components $\eta_{\mathbf{k}}$ and $\eta_{-\mathbf{k}}$ are not independent, but are related by

$$\mathrm{Re}\ \eta_{\mathbf{k}} = \mathrm{Re}\ \eta_{-\mathbf{k}}; \quad \mathrm{Im}\ \eta_{\mathbf{k}} = -\mathrm{Im}\ \eta_{-\mathbf{k}}. \tag{5.59}$$

This means that an integral with the measure of eqn. (5.58) counts all Fourier components twice. To correctly integrate over each degree of freedom once, the $\mathbf{k}$ vectors should be restricted to span only one half of $\mathbf{k}$ space. Thus, we should sum over Fourier components with (*e.g.*) $k_z > 0$, but $|\mathbf{k}| < \lambda$. We shall denote such an integral by the notation

$$\mathrm{Tr} \equiv \int \prod_{\mathbf{k}}{}' d\eta_{\mathbf{k}}. \tag{5.60}$$

This is also a good point to be precise about Fourier transforms. For a finite system of volume V, we define

$$\eta(\mathbf{r}) = \frac{1}{V} \sum_{\mathbf{k}} \eta_{\mathbf{k}} e^{i\mathbf{k}\cdot\mathbf{r}} \tag{5.61}$$

and

$$\eta_{\mathbf{k}} = \int_V d^d\mathbf{r}\, \eta(\mathbf{r}) e^{-i\mathbf{k}\cdot\mathbf{r}}. \tag{5.62}$$

Substituting eqn. (5.62) into eqn. (5.61) gives the resolution of the delta function

$$\frac{1}{V} \sum_{\mathbf{k}} e^{i\mathbf{k}\cdot(\mathbf{r}-\mathbf{r}')} = \delta(\mathbf{r} - \mathbf{r}'), \tag{5.63}$$

which is correct because converting the sum to an integral, and using the result that the density of states in **k**-space is $V/(2\pi)^d$, we get

$$\sum_{\mathbf{k}} \rightarrow \frac{V}{(2\pi)^d}\int d^d\mathbf{k} \tag{5.64}$$

and

$$\frac{1}{V}\sum_{\mathbf{k}} e^{i\mathbf{k}\cdot(\mathbf{r}-\mathbf{r}')} = \frac{1}{V}\cdot\frac{V}{(2\pi)^d}\int d^d\mathbf{k}\, e^{i\mathbf{k}\cdot(\mathbf{r}-\mathbf{r}')} = \delta(\mathbf{r}-\mathbf{r}'). \tag{5.65}$$

Conversely, substituting eqn. (5.61) into eqn. (5.62) gives the resolution of the Kronecker delta

$$\int d^d\mathbf{r}\, e^{i(\mathbf{k}-\mathbf{k}')\cdot\mathbf{r}} = V\delta_{\mathbf{k}\mathbf{k}'}. \tag{5.66}$$

In the limit that $V \rightarrow \infty$, eqn. (5.63) implies that

$$\int d^d\mathbf{r}\, e^{i(\mathbf{k}-\mathbf{k}')\cdot\mathbf{r}} = (2\pi)^d\delta(\mathbf{k}-\mathbf{k}'), \tag{5.67}$$

and so we have the useful correspondence that as $V \rightarrow \infty$,

$$V\delta_{\mathbf{k}\mathbf{k}'} \longrightarrow (2\pi)^d\delta(\mathbf{k}-\mathbf{k}'). \tag{5.68}$$

5.7.3 Functional Differentiation

In a system with a finite number of degrees of freedom η_i, $i = 1,\ldots,N$, the partition function depends on the N external fields that may couple linearly to the η_i in the Hamiltonian:[10]

$$Z = Z\{H_i\} = \mathrm{Tr}\,\exp\left[-\beta\left(H_\Omega - \sum_i H_i\eta_i\right)\right]. \tag{5.69}$$

Correlation functions may be generated by differentiation: *e.g.*

$$\langle\eta_i\eta_j\rangle = \frac{1}{\beta^2 Z(\{H_k\})}\frac{\partial}{\partial H_i}\frac{\partial}{\partial H_j}Z\{H_k\}, \tag{5.70}$$

[10] Of course the partition function also depends on the coupling constants in the Hamiltonian, but we suppress that dependence for the present argument.

and the left hand side is a function of the $\{H_k\}$, which may be set to zero if desired. In **field theory**, the partition function is sometimes called the generating function(al) for obvious reasons. In a system with an infinite number of degrees of freedom, the partition function becomes a functional of $H(\mathbf{r})$:

$$Z\{H(\mathbf{r})\} = \text{Tr} \exp\left[-\beta\left(H_\Omega - \int d^d\mathbf{r}\, H(\mathbf{r})\eta(\mathbf{r})\right)\right]. \tag{5.71}$$

The partial derivatives of eqn. (5.70) are replaced by **functional derivatives.**

Definition
Let $F\{\eta(\mathbf{r})\}$ be a functional of $\eta(\mathbf{r})$. Then the functional derivative of F with respect to the function $\eta(\mathbf{r}')$ is defined to be

$$\frac{\delta F}{\delta\eta(\mathbf{r}')} \equiv \lim_{\epsilon\to 0} \frac{F\{\eta(\mathbf{r}) + \epsilon\delta(\mathbf{r}-\mathbf{r}')\} - F\{\eta(\mathbf{r})\}}{\epsilon}. \tag{5.72}$$

You should check the following results:

$$\frac{\delta}{\delta\eta(\mathbf{r})}\int d^d\mathbf{r}'\,\eta(\mathbf{r}') = 1, \tag{5.73}$$

$$\frac{\delta}{\delta\eta(\mathbf{r})}\eta(\mathbf{r}') = \delta(\mathbf{r}-\mathbf{r}'), \tag{5.74}$$

$$\frac{\delta}{\delta\eta(\mathbf{r})}\int d^d\mathbf{r}'\,\frac{1}{2}\left(\nabla\eta(\mathbf{r}')\right)^2 = -\nabla^2\eta(\mathbf{r}). \tag{5.75}$$

In the last example, integration by parts and discarding a surface term are necessary.

5.7.4 *Response Functions*

Consider a system described by the Landau theory of eqns. (5.55) and (5.56). From the free energy $F\{H(\mathbf{r})\}$, we generate the expectation value of the order parameter

$$\langle\eta(\mathbf{r})\rangle = -\frac{\delta F}{\delta H(\mathbf{r})} \tag{5.76}$$

and the **generalised isothermal susceptibility**

$$\chi_T(\mathbf{r},\mathbf{r}') = \frac{\delta\delta\langle\eta(\mathbf{r})\rangle}{\delta H(\mathbf{r}')}. \tag{5.77}$$

The definition implies that if a small change $\delta H(\mathbf{r})$ to the external field $H(\mathbf{r})$ is made, the free energy and the expectation value of the order parameter change by

$$\delta F = -\int d^d\mathbf{r}' \, \langle \eta(\mathbf{r}') \rangle \, \delta H(\mathbf{r}') \tag{5.78}$$

and

$$\delta \langle \eta(\mathbf{r}) \rangle \equiv \eta_{H+\delta H} - \eta_H = \int d^d\mathbf{r}' \, \chi_T(\mathbf{r}, \mathbf{r}') \, \delta H(\mathbf{r}'). \tag{5.79}$$

Using the earlier arguments in section 3.7.2, we find

$$\chi_T(\mathbf{r}, \mathbf{r}') = -\frac{\delta^2 F}{\delta H(\mathbf{r}) \delta H(\mathbf{r}')} \tag{5.80}$$

$$= k_B T \left\{ \frac{1}{Z} \frac{\delta^2 Z}{\delta H(\mathbf{r}) \delta H(\mathbf{r}')} - \frac{1}{Z} \frac{\delta Z}{\delta H(\mathbf{r})} \cdot \frac{1}{Z} \frac{\delta Z}{\delta H(\mathbf{r}')} \right\} \tag{5.81}$$

$$= \frac{1}{k_B T} \left\{ \langle \eta(\mathbf{r}) \eta(\mathbf{r}') \rangle - \langle \eta(\mathbf{r}) \rangle \langle \eta(\mathbf{r}') \rangle \right\} = \beta G(\mathbf{r}, \mathbf{r}'). \tag{5.82}$$

For a translationally invariant system,

$$G(\mathbf{r} - \mathbf{r}') = k_B T \chi_T(\mathbf{r} - \mathbf{r}'). \tag{5.83}$$

This is an example of a principal result of **linear response theory**, namely that response functions are proportional to correlation functions. In the translationally invariant case, we can Fourier transform eqn. (5.83) to obtain (using carets to denote Fourier variables)

$$\hat{\chi}_T(\mathbf{k}) = \beta \hat{G}(\mathbf{k}). \tag{5.84}$$

$\hat{\chi}_T(\mathbf{k})$ is the wavevector dependent susceptibility, from which the static susceptibility sum rule follows:

$$\chi_T \equiv \lim_{k \to 0} \hat{\chi}_T(\mathbf{k}) = \beta \hat{G}(\mathbf{k}) \,|_{\mathbf{k}=0} = \beta \int d^d\mathbf{r} \, G(\mathbf{r}). \tag{5.85}$$

5.7.5 *Calculation of Two-Point Correlation Function*

We aim in this section to calculate $G(\mathbf{r})$. We do this in two steps:

(1) Find the equation satisfied by $\eta(\mathbf{r})$ by differentiating the Landau free energy and demanding stationarity.

(2) Differentiate with respect to $H(\mathbf{r})$ to get an equation for $\chi_T(\mathbf{r}-\mathbf{r}')$, *i.e.* an equation for $G(\mathbf{r}-\mathbf{r}')$.

Step 1:

$$\frac{\delta L}{\delta\eta(\mathbf{r})} = 0 \Rightarrow -\gamma\nabla^2\eta(\mathbf{r}) + 2at\eta(\mathbf{r}) + 2b\eta^3(\mathbf{r}) - H(\mathbf{r}) = 0. \tag{5.86}$$

Step 2:

$$\frac{\delta}{\delta H(\mathbf{r}')}\left\{-\gamma\nabla^2\eta(\mathbf{r}) + 2at\eta(\mathbf{r}) + 2b\eta^3(\mathbf{r}) - H(\mathbf{r})\right\} = 0. \tag{5.87}$$

Performing the differentiation, we get

$$(-\gamma\nabla^2 + 2at + 6b\eta(\mathbf{r})^2)\chi_T(\mathbf{r},\mathbf{r}') = \delta(\mathbf{r}-\mathbf{r}'), \tag{5.88}$$

which with eqn. (5.83) gives

$$\beta(-\gamma\nabla^2 + 2at + 6b\eta^2)G(\mathbf{r}-\mathbf{r}') = \delta(\mathbf{r}-\mathbf{r}'). \tag{5.89}$$

Thus the two-point correlation function is actually a **Green function.** For translationally invariant systems, η is just given by the equilibrium value from Landau theory.

For $t > 0$, $\eta = 0$ and

$$\left(-\nabla^2 + \xi_>^{-2}\right)G(\mathbf{r}-\mathbf{r}') = \frac{k_BT}{\gamma}\delta(\mathbf{r}-\mathbf{r}'), \tag{5.90}$$

where

$$\xi_>(t) \equiv \left(\frac{\gamma}{2at}\right)^{1/2}. \tag{5.91}$$

For $t < 0$, $\eta = \pm(-at/b)^{1/2}$ and

$$\left(-\nabla^2 + \xi_<^{-2}\right)G(\mathbf{r}-\mathbf{r}') = \frac{k_BT}{\gamma}\delta(\mathbf{r}-\mathbf{r}'), \tag{5.92}$$

with

$$\xi_<(t) \equiv \left(-\frac{\gamma}{4at}\right)^{1/2}. \tag{5.93}$$

In fact, the correlation length $\xi(t)$ is given by eqns. (5.91) and (5.93) above and below T_c respectively. Since $\xi \sim t^{-1/2}$, we already see that $\nu = \nu' = 1/2$. In summary, the two-point correlation function satisfies

$$\left(-\nabla^2 + \xi^{-2}\right)G(\mathbf{r}-\mathbf{r}') = \frac{k_BT}{\gamma}\delta(\mathbf{r}-\mathbf{r}'). \tag{5.94}$$

There are two obvious ways to solve eqn. (5.94): (1) Fourier transform and (2) Transformation to polar coordinates.

Fourier transform:

$$\hat{G}(k) = \frac{k_B T}{\gamma} \frac{1}{k^2 + \xi^{-2}}. \tag{5.95}$$

At $T = T_c$, $\xi = \infty$, and thus $\hat{G}(k) \sim k^{-2}$, showing that $\eta = 0$. The form of $G(\mathbf{r})$ in real space is obtained by Fourier transforming eqn. (5.95) and using the scaling trick to get

$$G(r, T_c) \propto \frac{1}{r^{d-2}} \quad (d > 2) \tag{5.96}$$

as advertised in (3.162).

For $T \neq T_c$, we can find $\hat{G}(\mathbf{k})$ in terms of measurable quantities, using the static susceptibility sum rule: $\hat{G}(0) = k_B T \chi_T$. Substituting this into eqn. (5.95) we get

$$\chi_T = \frac{\xi^2}{\gamma}, \tag{5.97}$$

which together with the fact that $\nu = 1/2$ shows that the critical exponent $\gamma = 1$. The final form for the two-point correlation function is

$$\hat{G}(\mathbf{k}) = \frac{k_B T \chi_T(t)}{1 + k^2 \xi^2(t)}. \tag{5.98}$$

What is the form of $G(\mathbf{r})$ in real space? We can in principle do the Fourier transform of (5.95) but this is a little tedious because of the d-dimensional angular integrals. The alternative is to go back to the differential equation (5.94).

Polar coordinates:

Without loss of generality, set $\mathbf{r}' = \mathbf{0}$. Since $\delta(\mathbf{r})$ is spherically symmetric, we write ∇^2 in radial polar coordinates in d dimensions:

$$\left[-\frac{1}{r^{d-1}} \frac{\partial}{\partial r} r^{d-1} \frac{\partial}{\partial r} + \xi^{-2} \right] G(r) = \frac{k_B T}{\gamma} \delta(r). \tag{5.99}$$

Again, we use our scaling trick to simplify things: let

$$\rho \equiv r/\xi, \quad \mathcal{G}(\rho) \equiv G(r/\xi). \tag{5.100}$$

Noting that

$$\delta(\rho\xi) = \frac{1}{\xi^d} \delta(\rho), \tag{5.101}$$

eqn. (5.99) becomes:

$$\left[-\frac{1}{\rho^{d-1}}\frac{\partial}{\partial\rho}\rho^{d-1}\frac{\partial}{\partial\rho}+1\right]\mathcal{G}(\rho)=g\delta(\rho), \tag{5.102}$$

where

$$g\equiv\frac{k_BT}{\gamma}\xi(t)^{2-d} \tag{5.103}$$

is a dimensionless measure of the strength of the fluctuations. The equation (5.102) can be solved in terms of the modified spherical Bessel function of the third kind;[11] written as $K_n(\rho)$.

$$\frac{1}{g}\mathcal{G}(\rho)=\begin{cases} e^{-\rho} & d=1; \\ (2\pi)^{-d/2}\rho^{-(d-2)/2}K_{(d-2)/2}(\rho) & d\geq 2. \end{cases} \tag{5.104}$$

The modified spherical Bessel functions of the third kind with $n\geq 0$ have the following asymptotic properties:

$$K_n(\rho)\sim\left(\frac{\pi}{2\rho}\right)^{1/2}e^{-\rho}, \quad \rho\to\infty, \tag{5.105}$$

$$K_n(\rho)\sim\frac{\Gamma(n)}{2}\left(\frac{\rho}{2}\right)^{-n}, \quad \rho\to 0, \tag{5.106}$$

$$K_0(\rho)\sim -\log\rho, \quad \rho\to 0, \tag{5.107}$$

where $\Gamma(n)$ is the Gamma function, equal to $(n-1)!$ for integer n and more generally defined by

$$\Gamma(z)\equiv\int_0^\infty t^{z-1}e^{-t}\,dt. \tag{5.108}$$

Thus we recover the results advertised previously for $T\neq T_c$, $r\gg\xi$, and $d\geq 2$:

$$G(r)=\frac{\pi^{(1-d)/2}}{2^{(1+d)/2}}\cdot\frac{k_BT}{\gamma}\cdot\frac{e^{-r/\xi}}{r^{(d-2)/2}}\cdot\frac{1}{\xi^{(d-3)/2}}. \tag{5.109}$$

For $T=T_c$, the correlation length ξ has diverged to infinity, and the appropriate asymptotic limit to use is $r\ll\xi$: for $d>2$, ξ obligingly cancels out to give

$$G(r)=\frac{k_BT}{\gamma}\cdot\frac{\Gamma(\frac{d-2}{2})}{4\pi^{d/2}}\cdot\frac{1}{r^{d-2}}. \tag{5.110}$$

---

[11] M. Abramowitz and I. Stegun, *Handbook of Mathematical Functions* (Dover, New York, 1970), pp. 375-6, 444.

5.7.6 The Coefficient γ

The strength of the correlations in the system depends on the coefficient γ in $\mathcal{L}$. What physics controls the magnitude of γ? In this section, we will see that the coefficient γ depends on the **range of interactions** in the microscopic physics. We can see this in two ways, one of which is frankly phenomenological, the other of which is based upon a literal interpretation of the 'derivation' of Landau theory from the microscopic theory.

In the first way, we recall the definition of the correlation length

$$\xi = \begin{cases} (\gamma/(-4at))^{1/2} & T < T_c; \\ (\gamma/2at)^{1/2} & T > T_c. \end{cases} \tag{5.111}$$

Let us take, for example, the case $T > T_c$, and write $\xi(t) = \xi(1)|t|^{-1/2}$ where $\xi(1) = (\gamma/2a)^{1/2}$, *i.e.* $\xi(1)$ is the value of the correlation length *extrapolated* from mean field theory to $T = 2T_c$. Thus, $\xi(1)$ gives a measure of the range R of the correlations well away from T_c. Thus, we obtain

$$\gamma \propto R^2. \tag{5.112}$$

Note that this will be qualitatively correct, but the use of the term "extrapolate" in the above paragraph is important. The *actual* temperature dependence of the correlation length away from the critical region may well behave differently from the asymptotic behavior predicted by mean field theory. In other words, when T is not sufficiently close to T_c, the coefficients a_n in Landau theory cannot be simply Taylor expanded to lowest order in $(T - T_c)/T_c$, because this is not a small parameter. Instead, the full functional form of $a_n(T)$ is required.

There is a second way of getting equation (5.112), (*e.g.*) for an Ising-like system, which follows from the microscopic Landau theory. A sketch of the argument follows; the reader will find it most comprehensible if exercise 3–3 has been completed. In the mean field approximation, the Gibbs free energy is given by

$$\Gamma_{\rm MFT} = -\frac{1}{2}\sum_{ij} J_{ij} m_i m_j + \Phi\{m_i\}, \tag{5.113}$$

where Φ is the homogeneous free energy contribution, and the first term represents the coupling between the average magnetisation $m_i \equiv \langle S_i \rangle$ at each site with position $\mathbf{r}_i$. If

$$J_{ij} = \begin{cases} J & 0 < |\mathbf{r}_i - \mathbf{r}_j| < R; \\ 0 & \text{otherwise,} \end{cases} \tag{5.114}$$

then we can approximate the coupling term as

$$\sum_{ij} J_{ij} m_i m_j = -\frac{Ja^2}{4} \sum_i \sum_{|\delta|<R} \left\{ \left(\frac{m_{i+\delta} - m_i}{a} \right)^2 - \left(\frac{m_{i+\delta} + m_i}{a} \right)^2 \right\},$$

$$\approx -\frac{Ja^2}{4} \sum_i \sum_{|\delta|<R} \left(\frac{m(\mathbf{r}_i + \delta) - m(\mathbf{r}_i)}{a} \right)^2 + O(m^2)$$

$$\approx -\frac{Ja^2}{4} \sum_i \sum_{|\delta|<R} \left(\frac{\delta \cdot \nabla m(\mathbf{r}_i)}{a} \right)^2 + O(m^2)$$

$$\approx -\frac{zJR^2}{4} \int \frac{d^d\mathbf{r}}{a^d} \left(\nabla m(\mathbf{r}) \right)^2 + O(m^2), \qquad (5.115)$$

where z is approximately the co-ordination number of the lattice. This gives $\gamma \approx R^2(Jz/4a^d) = R^2 k_B T_c / a^d$ in agreement with eqn. (5.112), where the last equality follows from the result of exercise 5–1(a).

EXERCISES

Exercise 5–1

This question is a continuation of exercise 3–3, in which the mean field theory for the Ising universality class is derived heuristically. In the chapter above, we have mentioned why this should not be taken too seriously. Here, we use the mean field theory calculation of the Gibbs free energy to present the analogue of the Maxwell construction for magnetic systems, and to motivate the Landau free energy. The notation is given in exercise 3–3.

(a) Consider the case of uniform magnetisation $m_i = m$ on a d-dimensional hypercubic lattice, with coordination number $z = 2d$. Expand Γ to quartic order in m and show that there is a second order transition at $T_c = 2dJ/k_B$. From the equation of state, check the values of the critical exponents β and δ.

(b) Sketch the form of $H(m)$ and $m(H)$ above and below the transition, *as given by the mean field theory.* Notice that your answer contains an unphysical portion below T_c. It is tempting to identify the Landau free energy with the function

$$L'(m, \hat{H}) \equiv \Gamma(m) - mN(\Omega)\hat{H}$$

where $\hat{H}$ is a parameter and not the function $H(m)$, and $m(H)$ is given by the mean field theory. Show that the condition that L' be minimised with respect to m implies the equation of state $\hat{H} = H(m)$. Sketch the form of $L'(m)$ above and below the transition for $\hat{H}$ positive, negative and zero. Hence show that the condition that L' be globally minimised removes the unphysical portion of the curve $m(H)$. This is Maxwell's equal area rule for magnetic systems. Notice the correspondence between the variables m and H in the magnetic case and p and V in the fluid case.

Exercise 5–2

Consider a system described by the Landau free energy density

$$\mathcal{L} = \frac{1}{2}a\eta^2 + \frac{1}{4}b\eta^4 + \frac{1}{6}c\eta^6 - h\eta$$

where $c > 0$, and a and b are both linearly proportional to pressure p and temperature T near the point (T_c, p_c): $a = a_1 t + a_2 p$ and similarly for b, with $t = (T - T_c)/T_c$ and $p = (p - p_c)/p_c$. As p and T are varied both a and b can be made to vanish and change sign. Such a system exhibits a **tricritical point**. An example is a mixture of He^3 and He^4. In this question, we will calculate the phase diagram in the $a - b$ plane for $h = 0$, using Landau theory. First, we set $h = 0$.

(a) Consider the case $a < 0$. Find the extrema of the Landau free energy, and study their stability. Hence show that

$$\langle\eta\rangle^2 \equiv \eta_s^2 = \frac{-b + \sqrt{b^2 - 4ac}}{2c}.$$

(b) Now consider the case $a > 0, b > 0$. How does η_s behave in this region?
(c) Lastly, study carefully the case $b < 0$, $a > 0$. What happens here?
(d) Now sketch the phase diagram in the a-b plane, indicating the order of any phase transitions that you have found, and the positions of the phase boundaries. Sketch the form of the Landau free energy in each of the different regions of the phase diagram. The point $a = 0$, $b = 0$ is called the tricritical point. Can you suggest why? (Hint, think of $h \neq 0$.)
(e) Calculate the thermodynamic critical exponents, by approaching the tricritical point along the line $b = 0$. What do you expect ν to be?
(f) Show that for b small but positive, there is a cross-over from the tricritical behaviour which you have found to ordinary critical behaviour when $b^2 \sim -ac$.

Exercise 5–3

In this question, we consider a class of magnetic materials which can exhibit so-called **modulated phases**.[12] For simplicity, we consider the case of a system of length L in one spatial dimension, and use the Landau function:

$$L_{\text{Landau}} = \int_0^L dx \left[atM^2 + \frac{1}{2}bM^4 + \frac{1}{2}\gamma\left(\frac{\partial M}{\partial x}\right)^2 + \frac{1}{2}\sigma\left(\frac{\partial^2 M}{\partial x^2}\right)^2 \right].$$

Here $\sigma > 0$, $b > 0$, $a > 0$, and t and γ may be of either sign. Below, you will work out the phase diagram in the at–γ plane.

(a) Define the Fourier transform

$$M(x) = \frac{1}{L}\sum_{n=-\infty}^{\infty} e^{iq_n x} M_n \qquad q_n = 2\pi n/L.$$

Write out the inverse transform for M_n in terms of $M(x)$, and verify that you obtain the correct expressions for the Kronecker and the Dirac delta functions.

(b) Write out the Landau free energy in terms of the Fourier components M_n.

(c) By minimising with respect to both n and M_n, show that the system exhibits three possible phases: a paramagnetic phase ($M = 0$), a ferromagnetic phase ($M \neq 0$) and a spatially modulated phase ($M_q \neq 0$, $q \neq 0$).

(d) What is the wavelength of the modulation? Find the phase boundaries and draw the phase diagram. What is the order of the different phase boundaries? You should assume that in the modulated phase, you only need to consider one Fourier component.

---

[12] This example is discussed by A. Michelson, *Phys. Rev. B* **16**, 577 (1977).

CHAPTER 6

Fluctuations and the Breakdown of Landau Theory

6.1 BREAKDOWN OF MICROSCOPIC LANDAU THEORY

We have now come to the point where we must ask: is the preceding theory self-consistent? At the beginning of the previous chapter, we motivated Landau theory by the observation that we could replace

$$\sum_{ij} J_{ij} S_i S_j = \sum_i S_i \sum_j J_{ij} [\langle S_j \rangle + (S_j - \langle S_j \rangle)] \tag{6.1}$$

by $\sum_{ij} S_i J_{ij} \langle S_j \rangle$. We can estimate the validity of this step as follows. If we make the replacement $S_i S_j \to S_i \langle S_j \rangle$, then $\langle S_i S_j \rangle \to \langle S_i \rangle \langle S_j \rangle$ on average (translationally invariant case). The fractional error implicit in this replacement can be quantified by the estimate

$$E_{ij} \equiv \frac{|\langle S_i S_j \rangle - \langle S_i \rangle \langle S_j \rangle|}{\langle S_i \rangle \langle S_j \rangle}, \tag{6.2}$$

where all quantities are calculated using Landau theory for $T < T_c$. The numerator is just the two-point correlation function, and in light of eqn. (6.1), we are really interested in the error E_{ij} estimated when $|\mathbf{r}_i - \mathbf{r}_j| \sim R$, the range of the interaction. Let us call this estimate E_R. Then

$$E_R = \frac{|G(R)|}{\eta_s^2}, \tag{6.3}$$

where $\eta_s(T)$ is the equilibrium value of the order parameter in the broken symmetry phase. If $E_R \ll 1$, then mean field theory is a reasonable approximation, and is self-consistent. To calculate E_R, it was assumed that the fluctuations were negligible compared with the mean field, and the result of the calculation does not violate this premise. On the other hand, if it should turn out that $E_R \geq 1$, then mean field theory would fail, because it would not be self-consistent, *i.e.* it would predict that a quantity is large when it was assumed initially to be small. Let us now calculate E_R in two cases: near the transition and well away from the transition.

6.1.1 Fluctuations Away from the Critical Point

The two-point correlation function used in eqn. (6.3) may be written in the form

$$G(R) = g \cdot f(R/\xi), \tag{6.4}$$

where $f(\rho)$ is given by eqn. (5.104). For $T \ll T_c$, the correlation length $\xi \sim R$, using the considerations of section 5.7.6, and hence the order parameter is saturated at the low temperature value: $\eta = O(1)$. Thus

$$\begin{aligned} E_R = g \times O(1) &\sim \frac{k_B T}{\gamma} \xi^{2-d} \\ &\approx \frac{T}{T_c} \cdot \frac{a^d}{R^2} \cdot R^{2-d} \approx \left(\frac{T}{T_c}\right) \left(\frac{a}{R}\right)^d. \end{aligned} \tag{6.5}$$

Now $(R/a)^d$ is essentially the coordination number $z > 1$; hence, $E_R < 1$ and mean field theory is self-consistent.

In summary, mean field theory works well when R/a is large or (equivalently) the co-ordination number is large. In both cases, each degree of freedom is coupled to many neighbors. Mean field theory is exact when each spin is coupled to all other spins, as shown by the exercises at the end of chapters 2 and 3. In some complex situations, a systematic and relatively easy way to generate mean field theory is by extending the Hamiltonian to be infinite range. Another way of generating mean field theory is to allow the order parameter to have n components and then to consider the case $n \to \infty$. In this limit, each degree of freedom experiences a potential due to the other degrees of freedom, which in the large n limit acts like the mean field we have already discussed. Systematic corrections to the $n = \infty$ limit can be worked out in the so-called $1/n$ expansion.

6.1.2 Fluctuations Near the Critical Point

Near the critical point, the correlation length grows towards infinity and, sufficiently close to T_c, $R \ll \xi$. Using eqn. (5.110), and the fact that $\eta \sim |t|^{\beta}$, we obtain

$$E_R \sim \frac{1}{|t|^{2\beta}} \left(\frac{a}{R}\right)^d , \tag{6.6}$$

which tends to infinity as the critical point is approached. For the mean field theory of the ferromagnetic critical point, $\beta = 1/2$; E_R starts to become large when $|t| < 1/z$, and mean field theory fails as $T \to T_c$. Note that we have used the results from eqn. (5.115) which are based on the *microscopic* interpretation of Landau theory, taking seriously the mean field approximation to the Gibbs free energy![1]

6.2 BREAKDOWN OF PHENOMENOLOGICAL LANDAU THEORY

As we have emphasised, the most satisfactory way to regard Landau theory is as phenomenological. Let us now examine the limitations of the phenomenological theory. As previously noted, the order parameter and Landau theory itself are defined with respect to a length scale Λ^{-1}, which is of the order of (but not larger than) ξ. Thus, to make the estimate of validity of Landau theory, we should really evaluate E_{ij} when the quantities in the numerator and denominator have been averaged over a region whose linear dimension is of order ξ. The region must not be larger than ξ, otherwise the fluctuations will become uncorrelated, and we will not obtain a true estimate of their strength. Thus we shall consider the estimate[2]

$$E_{LG} = \frac{|\int_V d^d\mathbf{r}\, G(\mathbf{r})|}{\int_V d^d\mathbf{r}\, \eta(\mathbf{r})^2} , \tag{6.7}$$

where V is taken to be the correlation volume: $V = \xi(T)^d$. The criterion that E_{LG} be small for the applicability of Landau theory is often referred to as the **Ginzburg criterion.**

---

[1] This criterion for the breakdown of the *microscopic* Landau theory is due to R. Brout, and is discussed in R. Brout, *Phase Transitions* (Benjamin, New York, 1965), pp. 33-44 and Chapter 9.

[2] The criterion for the breakdown of phenomenological Landau theory was given by A.P. Levanyuk, *Sov. Phys. JETP* **36**, 571 (1959)[*Zh. Eksp. Teor. Fiz.* **36**, 810 (1959)] and V.L. Ginzburg, *Sov. Phys. Sol. State.* **2**, 1824 (1960)[*Fiz. Tverd. Tela* **2**, 2031 (1960)]. A good discussion is also given by D.J. Amit, *J. Phys. C* **7**, 3369 (1974).

6.2.1 Calculation of the Ginzburg Criterion

The denominator of E_{LG} is simply

$$\frac{a\xi^d|t|}{b} = \frac{a}{b}\xi(1)^d|t|^{1-d/2}. \tag{6.8}$$

For the numerator we use the static susceptibility sum rule

$$\int_V d^d r\, G(\mathbf{r}) \approx k_B T_c \chi_T \approx \frac{k_B T_c}{4a|t|}, \tag{6.9}$$

where, in the sum rule, the integral over all of space is well approximated by the integral over the correlation volume, due to the rapid decay of $G(r)$. Thus,

$$E_{LG} = \frac{k_B T_c}{4a|t|} \cdot \frac{b}{a\xi(1)^d|t|^{1-d/2}} \tag{6.10}$$

$$= \frac{k_B}{4\Delta C\xi(1)^d}\frac{1}{|t|^{2-d/2}}, \tag{6.11}$$

where ΔC is the discontinuity in heat capacity at the transition, as given in eqn. (5.29). The condition that Landau theory be self-consistent, *i.e.* $E_{LG} \ll 1$, requires that

$$|t|^{(4-d)/2} \gg \frac{k_B}{4\Delta C\xi(1)^d} \equiv t_{LG}^{(4-d)/2}, \tag{6.12}$$

where t_{LG} is the value of the reduced temperature which marks the onset of the **critical region**. Within the critical region, $|t| < t_{LG}$, fluctuations dominate the thermodynamics, and the predictions of Landau theory are not valid.

A remarkable consequence of eqn. (6.12) is its dependence on dimensionality. There are three cases to consider:

(a) **d > 4**: As $t \to 0$, the Ginzburg criterion is **always** satisfied. In this case, Landau theory gives the correct exponents and qualitative picture.

(b) **d < 4**: As $t \to 0$, the Ginzburg criterion is **not** satisfied. In this case, Landau theory is not self-consistent as $t \to 0$, and the correct physics is beyond the scope of Landau theory.

(c) In the marginal case $d = 4$, Landau theory is not quite correct, but acquires logarithmic corrections from fluctuations: for example, the isothermal susceptibility in $d = 4$ behaves according to

$$\chi_T \sim \frac{1}{t}|\log t|^{1/3}. \tag{6.13}$$

A useful way of stating the above conclusions is that for $d < 4$, the quartic term proportional to η^4 in $\mathcal{L}$ becomes increasingly important as $t \to 0$, whilst for $d > 4$, the quartic term is negligible as $t \to 0$. We shall see why this statement is true when we discuss dimensional analysis below.

The analysis presented here has been for the Landau theory of the Ising universality class. Although similar qualitative conclusions apply for other Landau theories (systems with different symmetries, critical points with different codimension) the details do differ. In particular, whilst it is often (but not always) true that there is a dimensionality above which Landau theory is self-consistent, that dimension is not always $d = 4$. This special dimension is called the **upper critical dimension**, and in general we can calculate it as follows. For a system whose Landau theory is characterised by exponents γ, β, ν *etc.*, not necessarily with the values 1, 1/2, 1/2, *etc.*, the numerator and denominator of E_{LG} become

$$\int_V d^d\mathbf{r}\, G(\mathbf{r}) \sim k_B T \chi_T \sim t^{-\gamma}, \tag{6.14}$$

$$\int_V d^d\mathbf{r}\, \eta^2 \sim \xi^d |t|^{2\beta} \sim t^{2\beta - \nu d}. \tag{6.15}$$

Suppressing factors of order unity, the criterion that Landau theory applies is that:

$$t^{-\gamma} \ll t^{2\beta-\nu d} \quad \text{as } t \to 0. \tag{6.16}$$

For this to be true, we require that

$$d > \frac{2\beta + \gamma}{\nu} \equiv d_c, \tag{6.17}$$

where d_c is the upper critical dimension.

One final comment. *The Ginzburg criterion serves only to convince a believer in Landau theory that it cannot be correct.* The actual calculation of the critical region, with numerical prefactors evaluated carefully, is *not* reliable. The most compelling reason that this is so is that the calculation uses the transition temperature of Landau theory, which we will now denote by T_c^0. The actual temperature T_c at which the phase transition occurs must be lower than this temperature, as can be seen by the following naïve physical argument. Consider a system at some temperature near zero, and lower than the true T_c. As explained in chapter 2, the transition may be thought of as occuring due to the balance of energy and entropy. In Landau theory, the long wavelength fluctuations are neglected, and the transition occurs at some T_c^0 where the short range fluctuations dominate the energy. Now imagine the same system at the same temperature as before, but including fluctuations of the order parameter *i.e.* the long wavelength fluctuations. Since there are more fluctuations, the entropy will be higher, and the system more disordered. At a given temperature, the real system is more disordered than as described by Landau theory. Consequently, the temperature at which the entropy begins to dominate the energy is *lower* than in Landau theory.

6.2.2 Size of the Critical Region

It is a most enlightening exercise to estimate the size of the critical region for different systems. This was important historically, because it showed why (*e.g.*) the heat capacity in the vicinity of the superconducting transition exhibits the discontinuity expected on the basis of the Ehrenfest classification, whereas (*e.g.*) order-disorder transitions in alloys, also considered to be a "second order transition," usually exhibit divergent heat capacities.[3]

A convenient way to carry out the estimate is to note that the quantity $\Delta C \xi(1)^d$ is the heat capacity of the contents of a region of linear dimension $\xi(1)$, which should be about k_B per particle or degree of freedom. Thus, $(\Delta C/k_B)\xi(1)^d$ is approximately the number of degrees of freedom or particles in a region of linear dimension $\xi(1)$.

Examples:

(a) Magnetic systems: $\xi(1) \sim O(1)$Å. Thus $t_{LG} \sim 1$, and the critical region should be readily observable. In particular, the Landau theory prediction that the heat capacity exhibit a discontinuity will not be correct, and instead the heat capacity diverges near T_c.

(b) The **superfluid transition** in ^{4}He: again, $\xi(1) \sim O(1)$Å and $t_{LG} \sim 1$. The characteristic λ-point behaviour of the heat capacity has already been noted.

(c) Weak-coupling superconductors in three dimensions: $\xi(1)$ is roughly the diameter of a Cooper pair, typically $O(10^3)$Å. Thus, the number of particles in the volume $\xi(1)^3$ is $O(10^8)$. Hence, $t_{LG} \approx 10^{-16}$. The critical region is not generally accessible in superconductors, and the behaviour expected on the basis of Landau theory is observed. Note that eqn. (6.12) implies that in three dimensions $t_{LG} \propto \xi(1)^{-6}$, a very strong dependence! In the copper oxide (high temperature) superconductors, the orientational average of $\xi(1)$ is of order 10 Å, which is up to several hundred times

---

[3] V.L. Ginzburg, *op. cit.*

smaller than the figure in classic superconductors. This increases t_{LG} by a factor of about 10^{12}, making it feasible to observe fluctuation effects.[4]

(d) Weak-coupling superconductors in two dimensions *i.e.* thin films whose thickness ℓ is much less than the correlation length. The Ginzburg criterion then gives $t_{LG} \sim 10^{-5}$, which is still large enough that fluctuation effects in the heat capacity are observable.[5]

The last two examples illustrate the dramatic effect of dimensionality: the size of the critical region *increases* by *lowering* the dimensionality. Finally, some quantities are easier to observe than others, and even in classic superconductors, fluctuation effects, most notably in conductivity, are observable.[6]

In summary, below the upper critical dimension, as $t \to 0$, Landau theory fails. Far away from the critical region, Landau theory gives a qualitatively correct description. As the critical region is approached ($t \to t_{LG}$), fluctuations get more important. In some systems, such as superconductors apparently, there exists a region where fluctuations may be observable, *i.e.* $E_{LG} \leq 1$, but are not yet so large that the interactions between them (namely the η^4 terms) need to be taken into account. The width in t of such a region is, needless to say, dependent upon specific details of the system, and is not universal.

---

[4] S.E. Inderhees, M.B. Salamon, N.D. Goldenfeld, J.P. Rice, B.G. Pazol, D.M. Ginsberg, J.Z. Liu and G.W. Crabtree, *Phys. Rev. Lett.* **60**, 1178 (1988); these data yield $\alpha = 1/2$, a value predicted by the Gaussian approximation (see next section) which neglects the interactions between fluctuations (*i.e.* the quartic terms in $\mathcal{L}$). These authors also attempted to use the heat capacity measurements to estimate the number of components of the order parameter; due to the problems with background subtraction alluded to in section 4.6.2, this has been inconclusive to date.

[5] G.D. Zally and J.M. Mochel, *Phys. Rev. Lett.* **27**, 1710 (1971).

[6] **Paraconductivity** — the precursive drop in the resistivity just above the superconducting transition temperature — was first observed in amorphous bismuth by J.S. Shier and D.M. Ginsberg, *Phys. Rev.* **147**, 384 (1966). Subsequent observations of fluctuations in both thermodynamic and transport properties are reviewed by W.J. Skocpol and M. Tinkham, *Rep. Prog. Phys.* **38**, 1049 (1975).

6.3 THE GAUSSIAN APPROXIMATION

In this section, we will repeat the calculation of the correlation function $G(\mathbf{r}-\mathbf{r}')$, this time using the functional integral of eqn. (5.54) directly. We shall also introduce the **Gaussian approximation** to the functional integral, and show how the critical exponents are calculated, in this, the lowest order systematic correction to mean field theory. The Gaussian approximation allows fluctuations about the spatially uniform mean field, but assumes, in effect, that the fluctuations are distributed normally about the uniform mean field. In this approximation, the fluctuations turn out to be non-interacting *i.e.* independent random variables. This is why the Gaussian approximation is exactly soluble. The Gaussian approximation has counterparts in many areas of physics, being simply the approximation of a functional integral for a partition or generating function by a product of Gaussian integrals. In solid state physics, the analogue is the **random phase approximation**, whereas in field theory, the analogue is free field theory.

6.3.1 One Degree of Freedom

Suppose a random variable q is distributed about some mean q_0, with a probability distribution $P(q)$ sufficiently sharply peaked that q_0 closely corresponds to the value of q which maximises P. Then the **Gaussian approximation** is to simply fit $P(q)$ to the form

$$P(q) \propto e^{-(q-q_0)^2/2\sigma^2}, \tag{6.18}$$

where the standard deviation σ is essentially the half-width of P. As we will see, this approximation is very convenient, because in the generalisation that q is a vector $(q_1, q_2, \ldots, q_N)$, the probability distribution factorises:

$$P(q_1, \ldots, q_N) = P(q_1)P(q_2)\cdots P(q_N). \tag{6.19}$$

For a statistical mechanical system with one degree of freedom q and Hamiltonian $H(q)$, the probability distribution for q is

$$P(q) \propto e^{-\beta H(q)} \approx e^{-\beta H(q_0) - \frac{1}{2}(q-q_0)^2/\lambda^2}, \tag{6.20}$$

where we have performed a Taylor expansion about the maximum of $P(q)$:

$$H(q) = H(q_0) + \frac{1}{2}\frac{\partial^2 H}{\partial q^2}\bigg|_{q_0}(q-q_0)^2, \tag{6.21}$$

and

$$\frac{1}{\lambda^2} = \beta \frac{\partial^2 H}{\partial q^2}(q_0). \tag{6.22}$$

Since $H(q_0)$ is just some constant, which can be absorbed into the normalization of (6.20), we can say that $(q - q_0)$ — the fluctuation (deviation from the mean value) — is distributed normally. Hence,

$$\langle q \rangle = \int_{-\infty}^{\infty} P(q)\, q\, dq = q_0 \tag{6.23}$$

$$\langle (q - q_0)^2 \rangle = \langle q^2 \rangle - q_0^2 = \langle q^2 \rangle - \langle q \rangle^2 = \lambda^2. \tag{6.24}$$

The condition that the probability distribution be sharply peaked about its mean value is just the Ginzburg criterion:

$$E_G \equiv \frac{\langle q^2 \rangle - \langle q \rangle^2}{\langle q \rangle^2} = \frac{\lambda^2}{q_0^2} \ll 1. \tag{6.25}$$

The free energy is easily calculated in the Gaussian approximation. Apart from irrelevant constant factors which simply shift the zero of energy,

$$e^{-F/k_BT} = \int_{-\infty}^{\infty} dq\, e^{-H(q)/k_BT} \tag{6.26}$$

$$= e^{-\beta H(q_0)} \sqrt{2\pi\lambda^2}. \tag{6.27}$$

Therefore

$$F = H(q_0) - \frac{1}{2} k_B T \log(2\pi\lambda^2). \tag{6.28}$$

6.3.2 *N degrees of freedom*

The extension of the above argument is straightforward. For

$$\mathbf{q} = (q_1, q_2, \ldots, q_N)$$

Taylor expansion gives (using Einstein summation convention)

$$H(\mathbf{q}) = H(\mathbf{q}_0) + \frac{1}{2}(q_\alpha - q_\alpha^0) \left.\frac{\partial^2 H}{\partial q_\alpha \partial q_\beta}\right|_{\mathbf{q}=\mathbf{q}_0} (q_\beta - q_\beta^0) + \cdots. \tag{6.29}$$

The fluctuation matrix

$$M_{\alpha\beta} \equiv \left.\frac{\partial^2 H}{\partial q_\alpha \partial q_\beta}\right|_{\mathbf{q}=\mathbf{q}_0} \tag{6.30}$$

can be diagonalized by finding its eigenvalues $\hat{\lambda}_i$ and eigenvectors $\mathbf{v}^{(i)}$.

$$M_{\alpha\beta}v_\beta^{(i)} = \hat{\lambda}_i v_\alpha^{(i)}. \tag{6.31}$$

If we define

$$\frac{1}{\lambda_i^2} \equiv \beta\hat{\lambda}_i, \tag{6.32}$$

then

$$\beta H(\mathbf{q}) = \beta H(\mathbf{q}_0) + \frac{1}{2}\sum_{i=1}^{N} q_i'^2/\lambda_i^2, \tag{6.33}$$

where the q_i' are the components of the normal coordinates, *i.e.* the eigenvectors of eqn. (6.31). The calculation of the free energy is as follows:

$$\begin{aligned}
e^{-\beta F} &= \int_{-\infty}^{\infty}\prod_{i=1}^{N} dq_i\, e^{-\beta H(\mathbf{q})} \\
&\simeq e^{-\beta H(\mathbf{q}_0)}\int_{-\infty}^{\infty}\prod_{i=1}^{N} dq_i\, e^{-\frac{1}{2}\beta(q_\alpha - q_\alpha^0)M_{\alpha\beta}(q_\beta - q_\alpha^0)} \\
&= e^{-\beta H(\mathbf{q}_0)}\int\prod_{i=1}^{N} dq_i'\, e^{-\frac{1}{2}\sum_{i=1}^{N} q_i'^2/\lambda_i^2} \\
&= e^{-\beta H(\mathbf{q}_0)}\prod_{i=1}^{N}\left[\int_{-\infty}^{\infty} dq_i'\, e^{-\frac{1}{2}q_i'^2/\lambda_i^2}\right].
\end{aligned} \tag{6.34}$$

Performing the final integrations yields

$$F = H(\mathbf{q}_0) - \frac{1}{2}k_B T\sum_{i=1}^{N}\log(2\pi\lambda_i^2). \tag{6.35}$$

6.3.3 Infinite Number of Degrees of Freedom

The extension to an infinite number of degrees of freedom is straightforward, and mathematically well-defined.[7] We evaluate the functional integral

$$e^{-\beta F} = \int D\eta\, e^{-\beta L\{\eta(\mathbf{r})\}}, \tag{6.36}$$

[7] J. Glimm and A. Jaffe, *Quantum Physics: A Functional Integral Point of View* (Springer-Verlag, New York, 1981).

with

$$L = \int_V d^d\mathbf{r}\left[\frac{1}{2}\gamma(\nabla\eta)^2 + at\eta^2 + \frac{1}{2}b\eta^4 - H\eta\right] + a_0 V. \quad (6.37)$$

In the spirit of the Gaussian approximation, we shall:

(1) Find the uniform configuration which minimises L. For $T > T_c$, this is $\eta = 0$, whereas for $T < T_c$, this is $\eta = \pm\sqrt{-at/b}$.

(2) Expand L about the appropriate minimum to quadratic order in the fluctuation. Here, we will work above T_c, so we simply neglect the quartic terms in eqn. (6.37).

The basic idea of the calculation is to discretise the functional integral, *i.e.* consider it as the continuum limit of a discrete statistical mechanical system, defined on a hypercubic lattice, for example, as described in section 5.7.1. This can be done in several ways.

We could replace the derivatives in the Landau free energy of eqn. (6.37) by **lattice derivatives** defined on a lattice of points $\{\mathbf{r}_i\}$, *e.g.*

$$\frac{\partial\eta(\mathbf{r})}{\partial x} \to \frac{\eta(\mathbf{r}_i + a\hat{\mathbf{x}}) - \eta(\mathbf{r}_i)}{a}. \quad (6.38)$$

Then we have a statistical mechanical problem of the sort discussed in the preceding section, with the variables $\{q_i\}$ being the set $\{\eta(\mathbf{r}_i)\}$. The zeroth order approximation is the uniform state, and the fluctuations represented by the gradient terms in (6.37) generate a particular form of the fluctuation matrix M, which is block tridiagonal. Although it is not difficult to write out the matrix M explicitly, and even to compute the eigenvalues λ_i, we shall instead compute F by first transforming the functional integral into Fourier space. Due to translational invariance, this leads to a fluctuation matrix M which is *diagonal*.

We begin by writing L in Fourier space, after dropping the quartic terms:

$$L = \int d^d\mathbf{r}\left[\frac{1}{2}\gamma(\nabla\eta)^2 + at\eta^2\right] + a_0 V \quad (6.39)$$

$$= \sum_{\mathbf{k}\mathbf{k}'}\int\frac{d^d\mathbf{r}}{V^2}\left[\left(\frac{1}{2}\gamma(-\mathbf{k}\cdot\mathbf{k}') + at\right)\eta_{\mathbf{k}}\eta_{\mathbf{k}'}\right]e^{i(\mathbf{k}+\mathbf{k}')\cdot\mathbf{r}} + a_0 V \quad (6.40)$$

$$= \sum_{\mathbf{k}\mathbf{k}'}\left[\frac{1}{V}\left(\frac{1}{2}\gamma(-\mathbf{k}\cdot\mathbf{k}') + at\right)\eta_{\mathbf{k}}\eta_{\mathbf{k}'}\,\delta_{\mathbf{k}+\mathbf{k}',0}\right] + a_0 V \quad (6.41)$$

$$= \frac{1}{V}\sum_{\mathbf{k}}\frac{1}{2}|\eta_{\mathbf{k}}|^2\left[2at + \gamma k^2\right] + a_0 V, \quad (6.42)$$

where we have used the results of section 5.7.2. Note that $\mathbf{k}$ plays the rôle of a *label* in these manipulations. In mean field theory, we simply have the uniform state $\mathbf{k} = \mathbf{0}$. States with $\mathbf{k} \neq \mathbf{0}$ represent fluctuations. The sum over $\mathbf{k}$ is really restricted to states with $0 < |\mathbf{k}| < \Lambda$, since $\eta(\mathbf{r})$ is not defined on length scales shorter than the cut-off Λ^{-1}.

The free energy is given by

$$e^{-\beta F} = \int_{-\infty}^{\infty} \prod_{\mathbf{k}}{}' d\eta_{\mathbf{k}}\, e^{-\beta L} \tag{6.43}$$

$$= \prod_{\mathbf{k}}{}' \int_{-\infty}^{\infty} d\eta_{\mathbf{k}}\, e^{-\beta(2at+\gamma k^2)|\eta_{\mathbf{k}}|^2/2V} \cdot e^{-\beta a_0 V}, \tag{6.44}$$

using the notation of section 5.7.2. Writing $x \equiv \mathrm{Re}\ \eta_{\mathbf{k}}$, $y \equiv \mathrm{Im}\ \eta_{\mathbf{k}}$, each integral is of the form

$$\int_{-\infty}^{\infty} dx dy\, e^{-A(x^2+y^2)} = \frac{\pi}{A}. \tag{6.45}$$

Taking account of the fact that $\eta(\mathbf{r})$ is real, as explained in section 5.7.2, we obtain

$$e^{-\beta F} = \left(\prod_{\mathbf{k}}{}' \frac{2\pi V k_B T}{2at + \gamma k^2} \right) e^{-\beta a_0 V}$$

$$= \exp\left\{ \frac{1}{2} \sum_{|\mathbf{k}|<\Lambda} \log\left[\frac{2\pi V k_B T}{2at + \gamma k^2} \right] \right\} e^{-\beta a_0 V}, \tag{6.46}$$

where the $\sum_{\mathbf{k}}$ in eqn. (6.46) is over *all* $\mathbf{k}$-space, and so a factor of $1/2$ has been inserted. Thus

$$F = a_0 V - \frac{1}{2} k_B T \sum_{|\mathbf{k}|<\Lambda} \log \frac{2\pi V k_B T}{2at + \gamma k^2}. \tag{6.47}$$

To evaluate this expression, we will convert the sum into an integral using eqn. (5.64) for the density of states in $\mathbf{k}$-space. Before we do that, and calculate the heat capacity in the Gaussian approximation, it is useful to derive the results for the two-point correlation function directly using functional integration.

6.3.4 Two-point Correlation Function Revisited

We can calculate the two-point correlation function directly, but a quick and dirty way to get the right result is to use the **equipartition of energy**.[8] Thus,

$$\langle |\eta_k|^2 \rangle \frac{2at + \gamma k^2}{V} = 2 \times \frac{k_B T}{2}, \tag{6.48}$$

where the factor of 2 accounts for the fact that $\eta_{\mathbf{k}}$ has both real and imaginary parts. Hence

$$\langle |\eta_k|^2 \rangle = \frac{k_B T V}{2at + \gamma k^2}, \tag{6.49}$$

which we can relate to the two-point correlation function by noting that the definition of $\eta_{\mathbf{k}}$ of eqn. (5.62) implies that

$$\begin{aligned} \langle |\eta_{\mathbf{k}}|^2 \rangle &= \int d^d\mathbf{x}\, d^d\mathbf{y}\, e^{i\mathbf{k}\cdot(\mathbf{y}-\mathbf{x})} \langle \eta(\mathbf{y})\eta(\mathbf{x}) \rangle \\ &= V\hat{G}(\mathbf{k}). \end{aligned} \tag{6.50}$$

Thus we get

$$\hat{G}(\mathbf{k}) = \frac{k_B T}{2at + \gamma k^2} = \frac{k_B T}{\gamma} \cdot \frac{1}{k^2 + \xi_>^{-2}} \tag{6.51}$$

in agreement with eqn. (5.95).

Let us repeat the calculation from first principles, which is really equivalent to proving the equipartition theorem. This time, we will start with

$$\eta(\mathbf{r})\eta(\mathbf{r}') = \frac{1}{V} \sum_{\mathbf{k}\mathbf{k}'} e^{i\mathbf{k}\cdot(\mathbf{r}+\mathbf{r}')} \eta_{\mathbf{k}} \eta_{\mathbf{k}'}. \tag{6.52}$$

Now we must calculate

$$\langle \eta_{\mathbf{k}} \eta_{\mathbf{k}'} \rangle = \frac{\int_{-\infty}^{\infty} d\eta_{\mathbf{k}_1} d\eta_{\mathbf{k}_2} \ldots d\eta_{\mathbf{k}} \ldots d\eta_{\mathbf{k}'} \ldots e^{-\beta L} \eta_{\mathbf{k}} \eta_{\mathbf{k}'}}{\int_{-\infty}^{\infty} d\eta_{\mathbf{k}_1} d\eta_{\mathbf{k}_2} \ldots d\eta_{\mathbf{k}} \ldots d\eta_{\mathbf{k}'} \ldots e^{-\beta L}}, \tag{6.53}$$

where L is given by eqn. (6.42). Except for the terms in $\eta_{\mathbf{k}}$ and $\eta_{\mathbf{k}'}$, all the integrals cancel out between the numerator and denominator. Collecting

---

[8] The theorem of the equipartition of energy states that if a degree of freedom makes only a quadratic contribution to the Hamiltonian, then the average energy of the corresponding term in the Hamiltonian is $k_B T/2$.

both real and imaginary parts (*i.e.* using the fact that $\eta_{\mathbf{k}} = \eta^*_{-\mathbf{k}}$), the right hand side of eqn. (6.53) becomes

$$\frac{\int_{-\infty}^{\infty} d\eta_{\mathbf{k}} d\eta_{\mathbf{k}'} (\eta_{\mathbf{k}}\eta_{\mathbf{k}'}) e^{-\frac{2\beta}{2v}|\eta_{\mathbf{k}}|^2(2at+\gamma k^2)} e^{-\frac{2\beta}{2v}|\eta_{\mathbf{k}'}|^2(2at+\gamma k'^2)}}{\int_{-\infty}^{\infty} d\eta_{\mathbf{k}} d\eta_{\mathbf{k}'} e^{-\frac{2\beta}{2v}|\eta_{\mathbf{k}}|^2(2at+\gamma k^2)} e^{-\frac{2\beta}{2v}|\eta_{\mathbf{k}'}|^2(2at+\gamma k'^2)}}. \tag{6.54}$$

There are now two cases that must be considered:

(i) $\mathbf{k} \neq \pm\mathbf{k}'$. Then $\eta_{\mathbf{k}}$ and $\eta_{\mathbf{k}'}$ are distinct: the integrals in (6.54) factorise, and because they are each of the form

$$\int_{-\infty}^{\infty} d\eta_{\mathbf{k}}\, \eta_{\mathbf{k}}\, e^{-\frac{2\beta}{2V}|\eta_{\mathbf{k}}|^2(2at+\gamma k^2)} = 0 \tag{6.55}$$

the result is that

$$\langle \eta_{\mathbf{k}}\eta_{\mathbf{k}'}\rangle = 0 \quad |\mathbf{k}| \neq |\mathbf{k}'|. \tag{6.56}$$

ii) $\mathbf{k} = \pm\mathbf{k}'$. If $\mathbf{k} = \mathbf{k}'$, then we must calculate $\langle \eta_{\mathbf{k}}^2\rangle$, whereas if $\mathbf{k} = -\mathbf{k}'$, we must calculate $\langle |\eta_{\mathbf{k}}|^2\rangle$. In the former case, it is convenient to write the measure in plane polar coordinates $\eta_{\mathbf{k}} = (|\eta_{\mathbf{k}}|, \theta_{\mathbf{k}})$, so that

$$d\eta_{\mathbf{k}} = |\eta_{\mathbf{k}}|\, d|\eta_{\mathbf{k}}|\, d\theta_{\mathbf{k}}. \tag{6.57}$$

Then

$$\langle \eta_{\mathbf{k}}^2\rangle \propto \int_0^{2\pi} d\theta_{\mathbf{k}}\, e^{2i\theta_{\mathbf{k}}} = 0. \tag{6.58}$$

Thus, only the case $\mathbf{k} = -\mathbf{k}'$ is non-zero.

$$\langle \eta_{\mathbf{k}}\eta_{\mathbf{k}'}\rangle = \delta_{\mathbf{k}+\mathbf{k}',0} \left\langle |\eta_{\mathbf{k}}|^2\right\rangle, \tag{6.59}$$

where, after cancelling out the angular integral over $\theta_{\mathbf{k}}$ in the numerator and denominator, we obtain

$$\left\langle |\eta_{\mathbf{k}}|^2\right\rangle = \frac{\int_0^{\infty} |\eta_{\mathbf{k}}| d(|\eta_{\mathbf{k}}|)\, e^{-\frac{\beta}{V}(2at+\gamma k^2)|\eta_{\mathbf{k}}|^2} |\eta_{\mathbf{k}}|^2}{\int_0^{\infty} |\eta_{\mathbf{k}}| d(|\eta_{\mathbf{k}}|)\, e^{-\frac{\beta}{V}(2at+\gamma k^2)|\eta_{\mathbf{k}}|^2}}. \tag{6.60}$$

Making the change of dummy variable $z \equiv |\eta_{\mathbf{k}}|$, and noticing that $z\,dz = d(z^2/2)$, we find that

$$\left\langle |\eta_{\mathbf{k}}|^2\right\rangle = \frac{k_B T V}{2at + \gamma k^2}, \tag{6.61}$$

as we anticipated from the equipartition theorem eqn. (6.49).

In real space,

$$
\begin{aligned}
\langle \eta(\mathbf{r})\eta(\mathbf{r}')\rangle &= \frac{1}{V^2}\sum_{\mathbf{k}\mathbf{k}'} e^{i(\mathbf{k}\cdot\mathbf{r}+\mathbf{k}'\cdot\mathbf{r}')}\eta_{\mathbf{k}}\eta_{\mathbf{k}'} \\
&= \frac{1}{V^2}\sum_{\mathbf{k}\mathbf{k}'} e^{i(\mathbf{k}\cdot\mathbf{r}+\mathbf{k}'\cdot\mathbf{r}')}\,\delta_{\mathbf{k}+\mathbf{k}',0}\frac{k_B T V}{2at+\gamma k^2} \\
&= \frac{1}{V}\sum_{\mathbf{k}} \frac{k_B T}{2at+\gamma k^2}\, e^{i\mathbf{k}\cdot(\mathbf{r}-\mathbf{r}')}. \qquad (6.62)
\end{aligned}
$$

As expected in a translationally invariant system, $G(\mathbf{r},\mathbf{r}')$ is a function only of the separation. We can also check the static susceptibility sum rule, since

$$
\begin{aligned}
\int_V d^d\mathbf{r}\, G(\mathbf{r}) &= \int_V d^d\mathbf{r}\,\langle \eta(\mathbf{r})\eta(\mathbf{0})\rangle \\
&= \frac{1}{V}\sum_{\mathbf{k}} \frac{k_B T}{2at+\gamma k^2}\cdot \int_V d^d\mathbf{r}\, e^{i\mathbf{k}\cdot\mathbf{r}} \\
&= \frac{1}{V}\sum_{\mathbf{k}} \frac{k_B T}{2at+\gamma k^2}\cdot V\delta_{\mathbf{k},0} \\
&= \frac{k_B T}{2at} = k_B T\chi_T \qquad (6.63)
\end{aligned}
$$

in agreement with eqn. (6.50).

6.4 CRITICAL EXPONENTS

We already saw in computing the correlation function that we recover the results obtained earlier: $\eta = 0$, $\nu = 1/2$, $\gamma = 1$. The only exponent which changes from the mean field prediction is the heat capacity exponent α. Since the free energy is extensive, it is convenient to calculate the specific heat, which is the heat capacity per unit volume:

$$
c_V = -T\frac{\partial^2(F/V)}{\partial T^2}, \qquad (6.64)
$$

where in the Gaussian approximation, F is given by eqn. (6.47). The term $a_0\{K\}$ varies smoothly through the transition, and so will not contribute to singular behavior. It will give a smooth background to c_V.

In mean field theory, we choose $c = 0$ for $T > T_c$, since $\eta = 0$ above the transition. Remember that the sum over $\mathbf{k}$ in eqn. (6.47) does not include $\mathbf{k} = 0$, *i.e.* the mean field uniform state $\eta_{\mathbf{k}=0} = 0$. Below T_c, $\eta_{\mathbf{k}=0} \neq 0$, and there will be an extra term in the free energy below T_c, coming from this uniform mean field contribution.

The term with the sum over $\mathbf{k}$ represents a contribution to the free energy arising from the energy of the fluctuations. To calculate its contribution to c, we just differentiate (6.47):

$$\frac{c}{T} = \frac{\partial^2}{\partial T^2}\left[\frac{1}{2}\frac{k_B T}{V}\sum_{|\mathbf{k}|<\Lambda}\log\left(\frac{2\pi V}{2at+\gamma k^2}\right)\right]. \tag{6.65}$$

Write $c = A + B$, where we define

$$A \equiv \frac{k_B T}{2VT_c^2}\sum_{|\mathbf{k}|<\Lambda}\frac{4a^2}{(2at+\gamma k^2)^2} \tag{6.66}$$

and

$$B \equiv -\frac{k_B}{VT_c}\sum_{|\mathbf{k}|<\Lambda}\frac{2a}{2at+\gamma k^2}. \tag{6.67}$$

Now we shall examine each term A and B as $t \to 0^+$.

We start with A. Replace the summation by an integral, which we write as

$$I = \int_0^\Lambda \frac{d^d\mathbf{k}}{(2\pi)^d}\frac{1}{(2at+\gamma k^2)^2} = \frac{1}{\gamma^2}\int_0^\Lambda \frac{d^d\mathbf{k}}{(2\pi)^d}\frac{1}{(\xi^{-2}+k^2)^2}, \tag{6.68}$$

where the lower and upper limits denote the range of $|\mathbf{k}|$, the angular integrals are over all of the solid angles of d-dimensional space, and for $t > 0$, the correlation length $\xi = \xi_>$, as given in eqn. (5.91). To extract the divergent behavior as $t \to 0^+$, $\xi \to \infty$, we use the scaling trick. Change variables to

$$\mathbf{q} = \xi\mathbf{k}, \tag{6.69}$$

so that the integral (6.68) becomes

$$I = \xi^{4-d}\frac{1}{\gamma^2}\int_0^{\xi\Lambda}\frac{d^d\mathbf{q}}{(2\pi)^d}\frac{1}{(1+q^2)^2}. \tag{6.70}$$

As $t \to 0^+$, $\xi\Lambda \to \infty$. Does the integral converge? Count powers in the numerator and denominator, by writing $d^d\mathbf{q} \propto q^{d-1}dq$. Then in the limits of the integral, the integrand becomes

$$\frac{q^{d-1}}{(1+q^2)^2} \to \begin{cases} 0 & q \to 0 \quad (d > 1); \\ q^{d-5} & q \to \infty. \end{cases} \tag{6.71}$$

In the $q \to 0$ limit for $d > 1$, there is no problem with convergence. At the $q \to \infty$ limit, we use the result that with $a > 0$,

$$\int_a^\infty \frac{dq}{q^\alpha} < \infty \quad \text{for } \alpha > 1, \tag{6.72}$$

which implies that the integral (6.70) converges for $5 - d > 1$, *i.e.* for $d < 4$. The result of the integral is a numerical constant. Hence we see that

$$I \propto \xi^{(4-d)} \quad \text{for } d < 4. \tag{6.73}$$

Since $\xi \to \infty$ as $t \to 0$, we conclude that A makes a contribution to the specific heat c which diverges as the transition is approached. Note the physical origin of the divergence. In the unscaled expression for I, eqn. (6.68), the upper limit of the integral, Λ, is a constant and so the integral, *in its physical units*, cannot diverge because of the large k behaviour. Indeed, if it diverges, it must be because of its behaviour as $k \to 0$. This is the long wavelength behaviour; hence it is sometimes said that the divergence, which we have calculated, is an **infra-red divergence.**

What happens for $d > 4$? From eqn. (6.71), it seems that the *rescaled* integral in eqn. (6.70) diverges, but the prefactor of ξ^{4-d} actually tends to zero as the transition is approached and $\xi \to \infty$. The net result is finite, something that can also be checked from the *unscaled* integral in eqn. (6.68). Here, as $t \to \infty$, the integrand becomes proportional to k^{d-1-4} and since

$$\int_0^\Lambda \frac{dk}{k^\alpha} < \infty \quad \text{for } \alpha < 1, \tag{6.74}$$

the integral I is *finite* for $5 - d < 1$ *i.e.* $d > 4$. Thus, the specific heat does *not* diverge when $d > 4$. Correspondingly, there are no corrections to the critical exponents of mean field theory above four dimensions. In summary we learn from term A that

$$A \propto \begin{cases} \xi^{(4-d)} \sim t^{-(2-d/2)} & \text{for } d < 4; \\ \text{finite} & \text{for } d > 4. \end{cases} \tag{6.75}$$

Now we examine B. Accordingly we consider the integral

$$J \equiv \frac{1}{\gamma} \int_0^\Lambda \frac{d^d\mathbf{k}}{(2\pi)^d} \frac{1}{2at + \gamma k^2} = \int_0^\Lambda \frac{d^d\mathbf{k}}{(2\pi)^d} \frac{1}{\xi^{-2} + k^2}, \tag{6.76}$$

where we are interested in the behavior of J as $\xi \to \infty$. As before, the cause of any potential singular temperature dependence is the limit $k \to 0$

of the integral, since the large k behaviour is cut-off at the scale Λ. To investigate the singular behaviour, rescale $\mathbf{q} \equiv \xi\mathbf{k}$, so that

$$J = \frac{\xi^{2-d}}{(2\pi)^d} \int_0^{\xi\Lambda} \frac{q^{d-1}dq}{1+q^2}. \tag{6.77}$$

The integrand behaves as

$$\frac{q^{d-1}}{1+q^2} \to \begin{cases} q^{d-3} & q \to \infty; \\ 0 & q \to 0 \text{ for } d > 1, \end{cases} \tag{6.78}$$

and so the integral (but not J) converges to some number in the limit $\xi \to \infty$ as long as $3 - d > 1$, *i.e.* $d < 2$. Thus, for $d < 2$, as $t \to 0$, J and hence B diverge:

$$J \sim \xi^{(2-d)} \sim t^{-(1-d/2)}. \tag{6.79}$$

For $d > 2$, it is simplest to consider the unscaled form of J and check finiteness: as $\xi \to \infty$,

$$J \sim \int_0^{\Lambda} \frac{k^{d-1}dk}{k^2}, \tag{6.80}$$

which is finite for $d > 2$ in the $k \to 0$ limit of the integral.

In summary, B does diverge for $d < 2$, but the divergence is less severe than that of A in the same range of d. For $2 \leq d < 4$, only A diverges. Thus we see that for $d < 4$ the fluctuation specific heat behaves as

$$C \sim \begin{cases} t^{-(2-d/2)} & d < 4; \\ \text{finite} & d > 4, \end{cases} \tag{6.81}$$

and in the Gaussian approximation, fluctuations have shifted the exponent α from the mean field value of zero to

$$\alpha = 2 - d/2. \tag{6.82}$$

As noted earlier, the other exponents are unchanged by the inclusion of spatial fluctuations:

$$\gamma = 1; \quad \eta = 0; \quad \nu = \frac{1}{2}; \quad \beta = \frac{1}{2}; \quad \delta = 3. \tag{6.83}$$

This result is quite remarkable. We see explicitly that critical behavior depends on the dimension d, but not on the parameters in the Landau function, a_0, a, b, γ. A similar calculation can be performed for $T < T_c$, which the reader is invited to attempt in the exercises. There, one needs to consider the η^4 term to the extent that $\langle\eta\rangle = \pm(-at/b)^{1/2}$ defines the minimum about which L is expanded, although in the effective Hamiltonian for the fluctuations about $\langle\eta\rangle$, the quartic terms are neglected. Thus, the Gaussian approximation essentially consists in replacing the "potential" for the fluctuations by a harmonic ($y = x^2$) potential. The approximation consists of ignoring the η^4 terms for the fluctuations, *i.e.* the fluctuations are treated as independent harmonic oscillators.

Should we expect to be able to observe the exponents of the Gaussian approximation? In general they are not observed, because as $t \to 0$, the interactions of the fluctuations (the quartic terms) are crucial, as the Ginzburg criterion tells us. However, it is, in principle, possible to observe the Gaussian exponents in a **crossover** region between mean field theory and the true asymptotic critical behaviour, as has been discussed in connection with high temperature superconductivity (section 6.2.2).

EXERCISES

Exercise 6–1

Consider the Landau free energy

$$L = \int d^d x \left\{ \frac{1}{2}\gamma(\nabla\eta)^2 + at\eta^2 + \frac{1}{2}b\eta^4 \right\} + a_0 V$$

discussed above. We studied the specific heat c_+ above T_c in the Gaussian approximation. Here you will examine $T < T_c$.

(a) Writing $\eta = \eta_s + \psi$, where η_s is either one of the degenerate spontaneous mean field values of the order parameter below T_c, calculate the Landau free energy for the fluctuation ψ. Work to quadratic order in ψ, and express your answer in the Fourier components of ψ.

(b) Calculate the most singular contribution to the specific heat below T_c, c_-, and show that whilst the values of c_+ and c_- depend upon the parameters in the Landau free energy, their ratio is universal; find this ratio.

Exercise 6–2

In the text, we presented the Ginzburg criterion for the validity of mean field theory. This determined when the fluctuations were of comparable magnitude to the mean field. This criterion is in principle different from the criterion that the Gaussian approximation is an accurate representation of the fluctuations. Propose a criterion for the validity of the Gaussian approximation, in the same spirit as the Ginzburg criterion, and show explicitly that it leads to the same criterion as the Ginzburg criterion, although with different numerical prefactors. As mentioned in the text, such prefactors should not be taken seriously.

Exercise 6–3

This question requires you to investigate the remarks made in the text about the shift in the transition temperature when Gaussian fluctuations are included. The starting point is again the derivation of mean field theory from the Hubbard-Stratonovich transformation of exercise 3–3. There we showed that the partition function of an Ising system could be written as the functional integral

$$Z = \int_{-\infty}^{+\infty} \prod_{i=1}^{N} d\psi_i \, e^{-\beta S(\{\psi_j\},\{H_j\},\{J_{ij}\})}.$$

In mean field theory, the integral was replaced by the integrand evaluated at the extremum $\overline{\psi}_i$.

(a) Expand the functional S to second order in $\psi_i - \overline{\psi}_i$ and evaluate the resultant Gaussian integral for the partition function. Hence show that the lowest order correction[9] to mean field theory gives for the free energy F

$$F = S(\{\overline{\psi}_i\}) + \frac{1}{2\beta} \log \det \left[\delta_{ij} - \beta\left(1 - \tanh^2(\beta\overline{\psi}_i)J_{ij}\right)\right],$$

where irrelevant constants have been dropped from the partition function.

(b) Calculate the Gibbs free energy in this approximation. You will find that the algebra is eased by writing the second term above as $\epsilon(\delta\Gamma)$, where ϵ is a dummy variable introduced to keep track of orders of approximation, and working to $O(\epsilon)$ only. Also, there is no need to explicitly evaluate $\partial(\delta\Gamma)/\partial H_i$. It actually drops out of the calculation!

(c) In mean field theory, the susceptibility $\chi_T \propto t^{-1}$. From your answer to (b), evaluate χ_T^{-1} (formally) and hence show that it vanishes at a *negative* value of t. This is the shift in the transition temperature due to Gaussian fluctuations.

---

[9] The expansion, which is being performed here, is sometimes known as the loop expansion.

Exercise 6–4

This question is an exercise in multiple Gaussian integrals. The result is known in field theory and statistical mechanics as **Wick's Theorem**; we will use it (and pose it as an exercise again!) in chapter 12.

(a) Prove that

$$\langle x_q x_r \rangle \equiv \frac{\int_{-\infty}^{\infty} d^n x\, x_q x_r e^{-\frac{1}{2}A_{ij}x_i x_j}}{\int_{-\infty}^{\infty} d^n x\, e^{-\frac{1}{2}A_{ij}x_i x_j}} = A^{-1}_{qr}$$

where A is a real symmetric $n \times n$ matrix.

(b) Using the same notation as above, prove that

$$\langle x_a x_b x_c x_d \rangle = \langle x_a x_b \rangle \langle x_c x_d \rangle + \langle x_a x_d \rangle \langle x_b x_c \rangle + \langle x_a x_c \rangle \langle x_b x_d \rangle .$$

CHAPTER 7

Anomalous Dimensions

We have seen that mean field theory in its simplest form can be improved by inclusion of Gaussian fluctuations. The transition temperature is shifted, and the heat capacity does indeed exhibit a divergence rather than the discontinuity predicted for a uniform system. On the other hand, we saw that such a theory predicted its own regime of inapplicability near T_c. The question we now face is: can we continue to improve the theory and extend it into the critical region by going beyond the Gaussian fluctuations? In this chapter, we shall see that such attempts are doomed. Specifically, the expansion, in which the Gaussian and mean field theories are the lowest terms, has an expansion parameter which diverges as the critical point is approached. We shall also find that even if such an expansion were valid, it could not generate critical exponents with values different from the (incorrect) values already derived.

7.1 DIMENSIONAL ANALYSIS OF LANDAU THEORY

We shall now show that for $d > 4$, perturbatively including the interaction of the fluctuations leads to no new singular behaviour as $t \rightarrow 0$. On the other hand, for $d < 4$, this perturbation theory is divergent: the perturbing parameter grows unboundedly as $T \rightarrow T_c$! Our immediate goal, then, is to identify the single parameter in which a systematic expansion can be attempted.

We start with the partition function written in the form of a functional integral:

$$Z = \int D\eta \, e^{-\beta L}, \tag{7.1}$$

where

$$L = \int d^d\mathbf{r} \left\{ \frac{1}{2}\gamma(\nabla\eta)^2 + at\eta^2 + \frac{1}{2}b\eta^4 - H\eta \right\}. \tag{7.2}$$

It is conventional, and convenient, to rescale the order parameter η so that the coefficient of the term $(\nabla\eta)^2$ in βL is just 1/2. This is accomplished by defining

$$\phi \equiv (\beta\gamma)^{1/2}\eta; \quad \frac{r_0}{2} \equiv \frac{at}{\gamma} \equiv \frac{1}{2}\bar{a}t; \quad \frac{u_0}{4} \equiv \frac{1}{2}\frac{b}{\beta\gamma^2}. \tag{7.3}$$

Here we will consider the case $H = 0$. Then

$$H_{\text{eff}}\{\phi\} \equiv L\beta = \int d^d\mathbf{r} \left[\frac{1}{2}(\nabla\phi)^2 + \frac{1}{2}r_0\phi^2 + \frac{1}{4}u_0\phi^4 \right] \tag{7.4}$$

We now proceed in two steps.

(1) Identify the dimensions of the various quantities in eqn. (7.4).

(2) Rewrite (7.4) in terms of dimensionless variables: φ, $\bar{u}_0$, which are ϕ and u_0 scaled by the appropriate powers of r_0.

Step 1

H_{eff} is dimensionless: $[H_{\text{eff}}] = 1$, where we have used the notation $[\ldots]$ to denote the dimensions of the quantity enclosed. What is the dimension of ϕ? Each term separately in eqn. (7.4) must have dimension 1. So

$$\left[\int d^d\mathbf{r} \, (\nabla\phi)^2 \right] = 1 \Longrightarrow L^d \cdot L^{-2}[\phi]^2 = 1 \tag{7.5}$$

Hence,

$$[\phi] = L^{1-d/2}, \tag{7.6}$$

where L is the unit of length. Similarly, we can compute $[r_0]$ and $[u_0]$:

$$\left[\int d^d\mathbf{r} \, r_0\phi^2 \right] = \left[\int d^d\mathbf{r} \, u_0\phi^4 \right] = 1, \tag{7.7}$$

and thus

$$[r_0] = L^{-2}; \quad [u_0] = L^{d-4}. \tag{7.8}$$

Step 2

From eqn. (7.8) we can use r_0 to define a length scale, independent of dimension. Indeed, since $r_0 \propto (T - T_c)$, we know that within the Gaussian approximation $r_0 \propto 1/\xi(T)^2$. Using $r_0^{-1/2}$ as the length scale is equivalent to measuring lengths in units of the correlation length $\xi(T)$.

We have already seen that Gaussian functional integrals are easy to do. So we will write the partition function as a Gaussian functional integral with a modification, which we treat by perturbation theory. We define the following dimensionless variables:

$$\varphi \equiv \frac{\phi}{L^{1-d/2}}; \quad \mathbf{x} \equiv \frac{\mathbf{r}}{L}; \quad \overline{u}_0 \equiv \frac{u_0}{L^{d-4}}; \quad L \equiv r_0^{-1/2}. \tag{7.9}$$

Then eqn. (7.1) becomes, apart from unimportant multiplicative constants,

$$Z(\overline{u}_0) = \int D\varphi \exp\left[-H_0\{\varphi\} - H_{\text{int}}\{\varphi\}\right], \tag{7.10}$$

where

$$H_0 = \int d^d\mathbf{x} \left\{\frac{1}{2}(\nabla\varphi)^2 + \frac{1}{2}\varphi^2\right\} \tag{7.11}$$

$$H_{\text{int}} = \int d^d\mathbf{x} \left\{\frac{1}{4}\overline{u}_0\varphi^4\right\}. \tag{7.12}$$

If $H_{\text{int}} = 0$, the integral (7.10) is just the Gaussian approximation, which is exactly soluble. The partition function has, however, a contribution from the interactions, H_{int}. We might imagine that if $\overline{u}_0 \ll 1$, then we could use perturbation theory:

$$\begin{aligned} Z &= \int D\varphi\, e^{-H_0} e^{-H_{\text{int}}} \\ &= \int D\varphi\, e^{-H_0} \left(1 - H_{int} + \frac{1}{2!}(H_{\text{int}})^2 - \ldots\right). \end{aligned} \tag{7.13}$$

The important point is that the partition function depends on one dimensionless parameter $\overline{u}_0$; this is our perturbation parameter. Written out explicitly,

$$\overline{u}_0 = u_0 r_0{}^{(d-4)/2} = u_0 \bar{a}^{(d-4/2)} t^{(d-4)/2}. \tag{7.14}$$

As $t \to 0$, for $d < 4$, $\overline{u}_0 \to \infty$ and perturbation theory becomes meaningless! On the other hand, for $d > 4$, $\overline{u}_0 \to 0$ as $t \to 0$, and mean field theory becomes increasingly accurate as $T \to T_c^+$. Thus, dimensional analysis already enables us to determine the upper critical dimension, and the importance of fluctuations. Note that all of these considerations are based around the notion that the quadratic (and exactly soluble) terms in the Landau free energy represent a good zeroth order approximation. It is reasonable, but not obviously true, that perturbation theory will break down when the expansion parameter is of order unity. Thus, the critical region is the range of t for which $\overline{u}_0 \geq O(1)$. From eqn. (7.14) we find that

$$t^{(4-d)/2} \geq \frac{b}{a^2}\frac{1}{\xi(1)^d} \tag{7.15}$$

must be satisfied, as $t \to 0$, for perturbation theory to make sense. This is nothing other than the Ginzburg criterion eqn. (6.12).

This argument is, of course, by no means compelling. All we have shown is that the individual terms in the perturbation theory are divergent as $t \to 0$. This does not mean necessarily that the actual perturbation series is divergent, when summed. For example, the function $\exp(-t)$ has the expansion

$$\exp(-t) = \sum_{n=0}^{\infty} \frac{(-t)^n}{n!}, \tag{7.16}$$

in which each term diverges as $t \to \infty$, although the sum is perfectly finite. This example illustrates that care must be taken with manipulating perturbation expansions. Resummations of apparently divergent series form the basis for the most accurate estimates of critical exponents.

Although we have argued plausibly that for $d < 4$ perturbation theory must fail because the expansion parameter diverges as $T \to T_c$, the converse is not necessarily true. For $d > 4$, it is not true that perturbation theory necessarily converges: the smallness of the expansion parameter is not a sufficient condition for convergence. In fact, the perturbation theory is asymptotic, as suggested by a simple physical argument due to Dyson.[1] What happens to Z if we make the transformation $\overline{u}_0 \to -\overline{u}_0$? Then the integral for Z becomes

$$Z(-\overline{u}_0) = \int D\varphi\, e^{-\int d^d\mathbf{x}\,[\frac{1}{2}(\nabla\varphi)^2+\frac{1}{2}\varphi^2]+\int d^d\mathbf{x}\,\overline{u}_0\varphi^4}, \tag{7.17}$$

which is clearly *divergent* due to the sign of the quartic term. On the other hand

$$Z(\overline{u}_0) = \int D\varphi\, e^{-\int d^d\mathbf{x}\,[\frac{1}{2}(\nabla\varphi)^2+\frac{1}{2}\varphi^2]-\int d^d\mathbf{x}\,\overline{u}_0\varphi^4} \tag{7.18}$$

[1] F.J. Dyson, *Phys. Rev.* 85, 631 (1952).

is convergent (at least when regularised as we have discussed). Thus $\overline{u}_0 = 0$ cannot be a regular point, and therefore perturbation theory about it should not converge. Although asymptotic, the perturbation theory is Borel summable, and resummation techniques can be applied to yield accurate estimates of critical exponents.[2]

7.2 DIMENSIONAL ANALYSIS AND CRITICAL EXPONENTS

We can also use dimensional analysis to study the critical exponents themselves. The following naïve argument is intended to bring out the full mystery of critical exponents, and apparently demonstrates that it is impossible for the critical exponents to have values different from those predicted by Landau theory!

As an example, let us consider what dimensional analysis tells us about the two-point correlation function. This is given by

$$G(\mathbf{r}-\mathbf{r}') \equiv \langle \phi(\mathbf{r})\phi(\mathbf{r}')\rangle = \frac{\int D\phi\, e^{-H_{\text{eff}}\{\phi\}}\phi(\mathbf{r})\phi(\mathbf{r}')}{\int D\phi\, e^{-H_{\text{eff}}\{\phi\}}}. \tag{7.19}$$

Thus

$$[G(\mathbf{r}-\mathbf{r}')] = [\phi]^2 = L^{2-d}. \tag{7.20}$$

In k-space

$$\hat{G}(\mathbf{k}) = \frac{1}{V}\int d^d\mathbf{x}\, d^d\mathbf{y}\, e^{-i\mathbf{k}\cdot(\mathbf{x}-\mathbf{y})}G(\mathbf{x}-\mathbf{y}) \tag{7.21}$$

from eqn. (6.50), which means that

$$[\hat{G}(\mathbf{k})] = L^{-d}L^{2d}\cdot L^{2-d} = L^2. \tag{7.22}$$

Thus, if we make a change of the units of length by a factor of ϵ from L to $L' \equiv \epsilon L$, then $\hat{G}$ should transform[3] according to the rule that $\hat{G}'L'^2 = \hat{G}L^2$, which implies that

$$\hat{G}'(\mathbf{k}') = \epsilon^{-2}\hat{G}(\mathbf{k}), \tag{7.23}$$

---

[2] For a detailed discussion of these concepts, see the monograph by J. Zinn-Justin, *Quantum Field Theory and Critical Phenomena* (Clarendon, Oxford, 1989), Chapter 37.

[3] We are using the convention that a physical quantity Q_p represented by the symbol Q is actually given by $Q_p = Q[Q]$. The symbol $[Q]$ designates the units that must be appended to the *number* Q in order to obtain the *dimensionfull* quantity Q_p. Under a change of units, Q changes, whilst Q_p is, of course, invariant.

where $\mathbf{k}' = \epsilon\mathbf{k}$. *This result must always be true*: it follows simply from the definition of the two point function. Let us check it for the Gaussian approximation

$$\hat{G}(\mathbf{k}) = \frac{1}{k^2 + r_0}. \tag{7.24}$$

If we rescale lengths, then from eqn. (7.8), $r_0' = \epsilon^2 r_0$ and the transformed correlation function becomes

$$\hat{G}'(\mathbf{k}') = \frac{1}{\epsilon^2 k^2 + \epsilon^2 r_0} = \epsilon^{-2}\hat{G}(\mathbf{k}) \tag{7.25}$$

in agreement with the general considerations of eqn. (7.23).

But what happens at T_c? In defining the critical exponent η, we have asserted that at long wavelengths ($k \to 0$)

$$\hat{G}(\mathbf{k}, T_c) \sim k^{-2+\eta}, \tag{7.26}$$

which implies that under a change of length scale

$$\hat{G}'(\mathbf{k}') \sim \epsilon^{-2+\eta}\hat{G}(\mathbf{k}). \tag{7.27}$$

This clearly disagrees with our preceding general dimensional considerations, and *must* be incorrect, unless $\eta = 0$, the value given by Landau theory.

We can also use the same argument, perhaps with the flaw somewhat more transparent, for the correlation length exponent ν. The correlation length ξ has dimensions $[\xi] = L$. From eqns. (7.6) and (7.8), the only independent quantity is $r_0 \propto t$, which has dimensions $[r_0] = L^{-2}$. Hence

$$\xi \sim r_0^{-1/2} \sim t^{-1/2}, \tag{7.28}$$

which is indeed the Landau theory result, but in disagreement with the correct result that $\nu \neq 1/2$.

This is, in some sense, the central mystery of critical phenomena: any value for the critical exponents other than that given by Landau theory seems to violate dimensional analysis.

What has happened? How can eqns. (7.26) and (7.23) *both* be correct? The answer is that there *must* be *another* length scale which comes into the dimensional analysis, apart from the correlation length. This seems unreasonable: after all, we have spent considerable effort to show that long wavelength physics is the only dominant physics near the critical point. Nevertheless, the unwelcome conclusion is that the only other length scale in the problem is the microscopic length scale — the lattice spacing a in the original physical problem, or the short distance cut-off given by Λ^{-1} — and this too must be included in the dimensional analysis.

To see how this works, consider first the correlation function at T_c: this must have the form

$$\hat{G}(\mathbf{k}, T_c) \propto a^{\eta} k^{-2+\eta} \tag{7.29}$$

so that under a change of scale

$$\hat{G}'(\mathbf{k}', T_c) = \epsilon^{-2} \hat{G}(\mathbf{k}, T_c) \tag{7.30}$$

as dimensional analysis requires. A calculation of the two-point function in the $k \to 0$ limit at T_c *must* have the form of eqn. (7.29). The exponent η can therefore be obtained by examining the behaviour of $G(\mathbf{k}, T_c)$ at *fixed* k as $a \to 0$: we should obtain $G \sim a^{\eta}$.

Similarly, for the correlation length, if we redo the dimensional analysis taking into account the microscopic length scale a, we have the relations

$$[\xi] = L; \quad [a] = L; \quad [r_0] = L^{-2}. \tag{7.31}$$

Using $r_0 \propto t$, we conclude that

$$\xi = r_0^{-1/2} f(r_0 a^2), \tag{7.32}$$

where $f(x)$ is some function to be determined. Near the critical point $t \to 0$, the argument of f tends to zero. If it so happens that

$$f(x) \sim x^{\theta}, \quad \text{as } x \to 0 \tag{7.33}$$

for some θ to be determined, then as $t \to 0$,

$$\xi \sim t^{-1/2+\theta} a^{2\theta}. \tag{7.34}$$

Thus the critical exponent governing the divergence of the correlation length is

$$\nu = \frac{1}{2} - \theta. \tag{7.35}$$

The difference between this result and that of Landau theory is the so-called **anomalous dimension** θ. In the case of eqn. (7.29), the existence of a non-zero value for η can be considered to come from the fact that ϕ has acquired an anomalous dimension $\eta/2$.

7.3 ANOMALOUS DIMENSIONS AND ASYMPTOTICS

The existence of anomalous dimensions is quite remarkable. Reasoning by "common sense," one would think that since the correlation length is much larger than any microscopic length scale as $T \to T_c$, we could ignore the microscopic details of the theory, such as a. Thus, it would be legitimate to replace $a/\xi \ll 1$ by 0. If we did this, then a would not, of course, appear in any final formula, and the critical exponents would be those of mean field theory. To make it perfectly clear how reasonable this assumption sounds, replace a in the above argument by the radius of the proton! It would be ridiculous to think that the radius of the proton could affect phenomena on the scale of (*e.g.*) microns or larger. Thus, we normally expect that we may disregard physical phenomena which are characterised by widely different length scales. Yet our conclusion above is that this does not apply to critical phenomena! The often-heard statement that "the only important length scale near the critical point is the correlation length" is not only untrue, but misleading — if it were correct, critical phenomena would be very dull indeed.

How, then, does the microscopic length scale affect correlations at macroscopic distances? We have seen that the *existence* of anomalous dimensions follows from the presence of the microscopic length scale. However, the *value* of the anomalous dimension does not in general depend on the microscopic length scale, although it could in principle. Mathematically, the situation can be summarised as follows. Suppose that we are interested in some quantity F that depends in principle on a and ξ. Then, it is only legitimate to replace a/ξ by 0 in the function $F(a/\xi)$ *if* $F(x)$ is not singular in the limit $x \to 0$. There are three possibilities for this limit:

(1) $F(x) \to 0$ as $x \to 0$.

(2) $F(x) \sim x^{-\sigma}\Phi(x)$ as $x \to 0$, with $\sigma > 0$ and $\Phi(x)$ regular in the limit that $x \to 0$.

(3) None of the above.

From the preceding discussion, it follows that critical phenomena are in case (2) above.

Now a very interesting question emerges: are there other phenomena in nature where case (2) occurs? If so, can we conclude that these phenomena too exhibit anomalous dimensions and the analogue of critical exponents, which are apparently at odds with common sense dimensional analysis? The answer to both of these questions is an unqualified yes! Such phenomena are to be found in many areas of physics, although they have continually provoked expressions of surprise from those who have come across them.[4] Relatively recently, Barenblatt has provided an extensive summary of such problems and their solution, accomplished by making the explicit hypothesis that anomalous dimensions exist combined with numerical methods.[5] Historically, however, the problem of anomalous dimensions in critical phenomena was discovered and solved apparently without knowledge or recognition of these other phenomena exhibiting case (2) asymptotics.[6] In later chapters, we shall see how the renormalisation group solves the problem of anomalous dimensions both in critical phenomena, and in the problems discussed by Barenblatt. It would also be of value to apply Barenblatt's methods to critical phenomena, but this has not yet been accomplished.

7.4 RENORMALISATION AND ANOMALOUS DIMENSIONS

To end this short, but important chapter, let us briefly discuss the notion of **renormalisation in field theory**. Field theory is a term used to denote any model system described by a functional integral such as that of eqn. (7.1). We have seen how this arises in statistical mechanics, but such functional integrals may also be used to compute Green's functions (and hence scattering cross-sections) in **quantum field theory**. In quantum field theory, however, the time variable t is first analytically continued from the real axis to the imaginary axis τ ($\tau \equiv -it$), in order that the functional integral be convergent. Such field theories are sometimes known as Euclidean, because after the analytic continuation has been performed, the metric of space-time, with interval s given by $ds^2 = d\mathbf{r}^2 + d\tau^2$, describes a flat (Euclidean) manifold rather than the original curved manifold of Minkowski space, in which $ds^2 = d\mathbf{r}^2 - dt^2$.

In the early days of quantum field theory, it was hoped that the union of quantum mechanics, together with the known classical fields (*i.e.* electromagnetism) would form a self-consistent theory. This hope did not

[4] To the author's knowledge, the first explicit recognition of case (2) was the calculation of how a converging shock wave is focussed; see G. Guderley, *Luftfahrtforschung* **19**, 302 (1942) and L.D. Landau and K.P. Stanyukovich (1944), published in *Unsteady Motion of Continuous Media* (Academic, New York, 1960).

[5] G.I. Barenblatt, *Similarity, Self-Similarity, and Intermediate Asymptotics* (Consultants Bureau, New York, 1979).

[6] It is an interesting historical point that Landau had first-hand knowledge of anomalous dimensions in both critical phenomena and the problem of the converging shock wave, but apparently made no reference to a connection between them.

transpire, because certain quantities turned out to be infinite when calculated. Without going into the technical complications of quantum electrodynamics, we can illustrate the problem with the effective Hamiltonian of eqn. (7.4), for $r_0 > 0$ and $u_0 = 0$. Suppose, for the sake of argument, that eqn. (7.4) represented some "fundamental" physical theory, and we wished to calculate the quantity $\langle \phi(\mathbf{r})^2 \rangle$. Being a fundamental theory, there is no obvious reason for there to be a cut-off in the definition of ϕ or H_{eff}: indeed the most natural assumption is that space-time is a continuum. Thus the cut-off Λ, which we have always stressed is present in critical phenomena, should actually be assumed to be infinite in this model quantum field theory. Hence,

$$\langle \phi(\mathbf{r})\phi(\mathbf{r}') \rangle = \int_0^\infty \frac{d^d\mathbf{k}}{(2\pi)^d} \frac{1}{r_0 + k^2} e^{i\mathbf{k}\cdot(\mathbf{r}-\mathbf{r}')}. \tag{7.36}$$

Setting $\mathbf{r} = \mathbf{r}'$ we find that

$$\langle \phi(\mathbf{r})^2 \rangle = \int_0^\infty \frac{S_d\, k^{d-1} dk}{(2\pi)^d} \frac{1}{r_0 + k^2}, \tag{7.37}$$

where S_d is the surface area of the unit sphere in d dimensions. The integrand behaves as k^{d-3} as $k \to \infty$, and so the integral diverges for $d \geq 2$. This model quantum field theory, and indeed the Landau theory of critical phenomena, is not well-defined without a *finite* length scale Λ. (This is, of course, why we were careful to motivate Landau theory with its associated length scale right at the beginning.)

In quantum field theory, however, where we have assumed that there is no microscopic physics to contribute a cut-off Λ, we must try to make sense of the divergence that we have found. The procedure that is followed — renormalisation — consists of a sequence of steps. First, an artificial cut-off Λ is introduced into the theory; this is sometimes known as **regularisation.** Next, the desired calculation is performed, and is of course, perfectly finite. Finally, the limit $\Lambda \to \infty$ is taken, absorbing the resultant infinities into a redefinition of the parameters in H_{eff}, such as r_0 and u_0. Of course, it is by no means obvious that this can be done, but for a class of quantum field theories, the method can be carried out.

We will describe the whole process in much more detail in a different context in a later chapter, so for now, do not worry about what this renormalisation really involves. The essential point is that the renormalisation procedure introduces a *new* length scale into the problem. Thus, as we have seen in the preceding section, anomalous dimensions may, and indeed do appear. When we come to discuss the physics of problems whose asymptotics is in the category of case (2), we shall explain the precise physical significance of renormalisation.

EXERCISES

Exercise 7–1

Consider the Landau free energy

$$L = \int d^d x \left\{ \frac{1}{2}(\nabla\phi)^2 + \frac{1}{2} r_0 \phi^2 + \frac{u_n}{n!}\phi^n \right\}.$$

(a) Use the Ginzburg criterion or dimensional analysis to find the upper critical dimension.
(b) Comment on the accuracy of the tricritical exponents which were calculated in exercise 5–2, as a function of dimension.
(c) Show that higher powers of $(\nabla\phi)$ and higher derivatives of ϕ are negligible as $T \to T_c$.

CHAPTER 8

Scaling in Static, Dynamic and Non-Equilibrium Phenomena

We have seen that there is a plethora of critical exponents α, β, γ, δ, μ, ν near to the critical point, and we have discussed how in principle they may take on values that differ from those predicted by mean field theory. The next question to address is: are these exponents independent? Remarkably, we will see that in fact, of these six exponents, only two are independent: the rest can all be derived from knowledge of any two through the so-called **scaling laws**. One example of a scaling law is the **Rushbrooke scaling law**[1]

$$\alpha + 2\beta + \gamma = 2, \tag{8.1}$$

whilst another example is

$$\beta\gamma = \beta + \gamma. \tag{8.2}$$

Eqns. (8.1) and (8.2) are examples of thermodynamic scaling laws: they relate exponents describing thermodynamic quantities. There are also scaling laws for the correlation function exponents: (*e.g.*)

$$2 - \alpha = \nu d, \tag{8.3}$$

[1] First conjectured by J.W. Essam and M.E. Fisher, *J. Chem. Phys.* **38**, 802 (1963) and derived as an inequality by G.S. Rushbrooke, *J. Chem. Phys.* **39**, 842 (1963).

which is known as the **Josephson scaling law.**

These and other scaling laws were discovered in a curious fashion, round about 1963. It was noticed that early estimates of the critical exponents obtained from numerical methods approximately satisfied eqn. (8.1), whereupon it was proved that thermodynamics actually implied

$$\alpha + 2\beta + \gamma \geq 2. \tag{8.4}$$

Subsequent work indicated that in all cases examined, with different dimensions and different model systems (in what is now known as the Ising universality class), eqn. (8.1) is satisfied; the inequality of eqn. (8.4) is satisfied as an equality. Many exponent inequalities have now been proved, with varying degrees of rigour. However, the fact remains that they are always satisfied as *equalities.* Finally, **static scaling** was discovered by B. Widom, and this lead to the **static scaling hypothesis**, which is one of the topics of this chapter. From the static scaling hypothesis, the scaling laws in the form of equalities can be derived.

In this chapter, we will explain the static scaling hypothesis, and show how it leads to scaling laws. We will also mention the topic of **dynamical scaling,** in which time-dependent correlations in equilibrium near the critical point exhibit scaling. Finally, we shall introduce the topic of scaling in **non-equilibrium systems.** The latter topic is quite modern, and differs from the preceding ones in that it is not yet understood as a consequence of a detailed theory. Scaling and scaling laws near the critical point of equilibrium systems follow from the renormalisation group theory, which is also capable of calculating the values of the critical exponents and the form of scaling functions when used in combination with approximate methods such as perturbation theory; it remains an open question to what extent these renormalisation group considerations apply to non-equilibrium systems. This will be discussed further in later chapters.

8.1 THE STATIC SCALING HYPOTHESIS

The static scaling hypothesis, written in the form appropriate for a magnetic system, is an attempt to encode two experimental results in one equation. The two results have already been presented: the growth of the magnetisation M slightly below T_c in the absence of an external field H,

$$M(t, h = 0) = \begin{cases} 0 & t > 0; \\ \pm A|t|^{\beta} & t < 0, \end{cases} \tag{8.5}$$

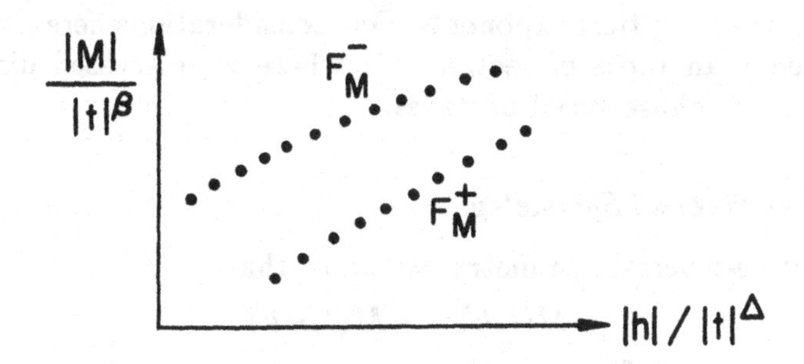

Figure 8.1 Sketch of the magnetisation measured at different temperatures in different external magnetic fields h. When the data is plotted in the manner shown, the data points fall onto only *two* curves, one for the data above T_c, the other for the data below T_c.

and the behaviour along the critical isotherm,

$$M(t=0,h) = \pm B|h|^{1/\delta}, \tag{8.6}$$

where, as usual $t = (T - T_c)/T_c$ and $h = H/k_B T$.

Widom[2] noticed that these results followed from a single formula for $M(t,h)$:

$$M(t,h) = \begin{cases} t^{\beta} F_M^{+}(h/t^{\Delta}) & t > 0; \\ (-t)^{\beta} F_M^{-}(h/(-t)^{\Delta}) & t < 0, \end{cases} \tag{8.7}$$

which is intended to be valid for $|h|, |t| \ll 1$, but for an arbitrary *ratio* of $|h|$ to $|t|$. The exponents β and Δ are assumed to be universal, as are the **scaling functions** $F_M^{+}(x)$ and $F_M^{-}(x)$. Δ is sometimes referred to as the **gap exponent.**

We will investigate the significance of the scaling hypothesis shortly, but first, let us see how it can be tested experimentally. In figure (8.1) is a sketch of experimental data for the magnetisation as a function of reduced temperature t and field h. If the data were plotted against temperature, there would be one curve for each value of the external field h. However, when $M/|t|^{\beta}$ is plotted against $|h|/|t|^{\Delta}$, all of these curves collapse onto two curves, one for the data above T_c, one for the data below T_c. This phenomenon of **data collapse** is the principle significance of scaling. Of course, in order to observe data collapse, it is necessary to choose the correct values of T_c, β and Δ, which are not usually known before hand.

---

[2] B. Widom, *J. Chem. Phys.* **43**, 3898 (1963).

Thus, the observation of scaling requires manipulation of the data, and leads to estimates of T_c and the critical exponents.

Simple conditions on the functions $F_M^{\pm}$ suffice to establish the scaling relations between critical exponents. Our considerations here are no more complicated than those of section 1.1, where we discussed dimensional analysis for the phase speed of waves.

8.1.1 *Time-reversal Symmetry*

From time-reversal symmetry, we know that

$$M(t,h) = -M(t,-h). \tag{8.8}$$

Hence, using eqn. (8.7),

$$F_M^{\pm}(x) = -F_M^{\pm}(-x). \tag{8.9}$$

The scaling functions $F_M^{\pm}$ are odd functions.

8.1.2 *Behaviour as $h \to 0$*

Assuming the smoothness of the limit $h \to 0$, we should recover the zero field result

$$M(t,0) = \begin{cases} (-t)^{\beta} F_M^{-}(0) & t < 0; \\ 0 & t > 0. \end{cases}$$

Thus, we require that

$$F_M^{+}(0) = 0; \quad F_M^{-}(0) = \text{non-zero constant}. \tag{8.10}$$

8.1.3 *The Zero-field Susceptibility*

We can differentiate eqn. (8.7) with respect to H and set $H = 0$ to obtain the isothermal susceptibility.

$$\chi_T(H=0) = \left.\frac{\partial M}{\partial H}\right|_{H=0} = \frac{1}{k_B T}\left.\frac{\partial M}{\partial h}\right|_{h=0}. \tag{8.11}$$

Using eqn. (8.7), we obtain

$$\chi_T = \frac{1}{k_B T}\frac{|t|^{\beta}}{|t|^{\Delta}}\left.\frac{dF_M^{\pm}}{dx}(x)\right|_{x=0} \sim |t|^{\beta-\Delta} F'^{\pm}_M(0). \tag{8.12}$$

Assuming that $F'^{\pm}_M(0) \neq 0$ or ∞, then we can make the identification

$$\beta - \Delta = -\gamma \tag{8.13}$$

from the definition of the exponent γ. So

$$\Delta = \beta + \gamma, \tag{8.14}$$

and is *not* a new critical exponent after all!

8.1.4 The Critical Isotherm and a Scaling Law

Now we consider the limit that $t \to 0$, but h is small but non-zero. The argument of $F_M^{\pm}$, namely $h/|t|^{\Delta}$, tends to infinity. However, we expect that in this limit, $M(0,h)$ is perfectly well-behaved and finite. In fact, from the definition of the critical exponent δ, we know that

$$M(0,h) \sim h^{1/\delta}. \tag{8.15}$$

For this to be the result, we use the same trick that we used between eqns. (1.6) and (1.7). We assume that

$$F_M^{\pm}(x) \sim x^{\lambda} \tag{8.16}$$

as $x \to \infty$, where λ is some power to be determined. Then we have

$$M(0,h) \sim |t|^{\beta} \left(\frac{h}{|t|^{\Delta}} \right)^{\lambda} \sim |t|^{\beta - \Delta\lambda} h^{\lambda}. \tag{8.17}$$

Now, as $t \to 0$, we will either get $M \to 0$ if $\beta - \Delta\lambda > 0$ or $M \to \infty$ if $\beta < \Delta\lambda$. Neither alternative is acceptable physically, and so we conclude that we must have

$$\beta = \Delta\lambda \tag{8.18}$$

in order that the t dependence "cancels out". This implies that

$$M(0,h) \sim h^{\lambda}, \tag{8.19}$$

and hence

$$\lambda = \frac{1}{\delta}, \tag{8.20}$$

$$F_M^{\pm}(x) \sim x^{1/\delta} \quad \text{as } x \to \infty. \tag{8.21}$$

These results imply a scaling law. Combining eqns. (8.18) and (8.20) we find that

$$\Delta = \beta/\lambda = \beta\delta. \tag{8.22}$$

But from eqn. (8.14), $\Delta = \beta + \gamma$. Thus we conclude that

$$\beta\delta = \beta + \delta. \tag{8.23}$$

Does it work? We have not said anything about how we got the exponents. If eqn. (8.23) is true, it should apply to exponents obtained from experiments and exact theoretical calculations; satisfying this criterion is a desideratum of approximate theories. Also note that the scaling law (8.23) is *independent of dimension*. Using the values from table (3.1), one can check that the scaling law is indeed well satisfied, even though the values of the exponents are quite different in the different systems and dimensions.

8.2 OTHER FORMS OF THE SCALING HYPOTHESIS

8.2.1 Scaling Hypothesis for the Free Energy

There is nothing special about writing the scaling hypothesis in terms of the magnetisation. We could equally well start with the free energy per unit volume or at least its singular part $f_s(t,h)$:

$$f_s(t,h) = t^{2-\alpha} F_f\left(\frac{h}{t^\Delta}\right). \tag{8.24}$$

We can derive the Rushbrooke scaling law as follows. The magnetisation is obtained by differentiation:

$$M = -\frac{1}{k_B T}\frac{\partial f_s}{\partial h} \sim t^{2-\alpha-\Delta} F_f'\left(\frac{h}{t^\Delta}\right), \tag{8.25}$$

which must tend towards t^β as $h \to 0$. Hence

$$\beta = 2 - \alpha - \Delta. \tag{8.26}$$

The isothermal susceptibility is obtained by one more differentiation:

$$\chi_T \sim t^{2-\alpha-2\Delta} F_s''\left(\frac{h}{t^\Delta}\right), \tag{8.27}$$

and tends towards $t^{-\gamma}$ as $h \to 0$. Thus

$$2 - \alpha - 2\Delta = -\gamma. \tag{8.28}$$

Eliminating Δ, we obtain the final result

$$\alpha + 2\beta + \gamma = 2. \tag{8.29}$$

8.2.2 Scaling Hypothesis for the Correlation Function

Finally, we remark that it is also possible to write down a scaling form for the two-point correlation function in the form

$$G(\mathbf{r},t,h) = \frac{1}{r^{d-2+\eta}} F_G\left(rt^\nu, \frac{h}{t^\Delta}\right), \tag{8.30}$$

from which more scaling laws follow, involving η, ν:

$$2 - \alpha = d\nu \tag{8.31}$$

$$\gamma = \nu(2-\eta) \tag{8.32}$$

The Josephson relation (8.31) involves d explicitly, and is an example of a **hyperscaling** law — one which involves d. The hyperscaling relations were for a long time on a slightly less secure footing than the thermodynamic scaling laws: numerical work on the $d = 3$ Ising model seemed to indicate minute violations of the hyperscaling laws, and this is discussed at length in some early books and articles. However, the discrepancies have been shown to be artifacts of approximations, and it is generally agreed that the hyperscaling relations are satisfied.[3] For Hamiltonians with long-range power law interactions, however, it is known that the hyperscaling relations are not satisfied.

8.2.3 *Scaling and the Correlation Length*

What is the origin of scaling and the scaling laws? The principle physical idea is that the divergence of $\xi(T)$ as $T \to T_c$ is responsible for **all** the singular behaviour; the *only* length scale which enters dimensional analysis is $\xi(T)$. As we have already discussed, this statement is too naïve, and actually incorrect, in the sense that if it were literally true, the critical exponents would have the mean field theory values. To be precise, let us consider the singular part of the free energy per unit volume f_s, which has dimensions L^{-d}. Then, we may write that

$$\frac{f_s}{k_B T} \sim \xi^{-d}\left(A + B_1\left(\frac{l_1}{\xi}\right)^{\sigma_1} + B_2\left(\frac{l_2}{\xi}\right)^{\sigma_2} + \cdots\right), \tag{8.33}$$

where the l_i are whatever microscopic length scales are present, A, B_i are coefficients with weak (non-singular) temperature dependence, and σ_i are non-negative exponents. In the limit that $t \to 0$, the correlation length ξ diverges, and the corrections to the leading behaviour in eqn. (8.33) may be ignored. Thus

$$\frac{f_s}{k_B T} \sim t^{\nu d}, \tag{8.34}$$

which immediately leads to the Josephson scaling law by differentiation to obtain the specific heat:

$$c_V = -T\frac{\partial^2 f_s}{\partial T^2} \sim t^{\nu d - 2} = t^{-\alpha} \tag{8.35}$$

using the definition of the exponent α, and the Josephson scaling law (8.31) follows immediately.

---

[3] See *Physics Today*, Nov. 1980, p. 18 and the article by B. Nickel in *Phase Transitions*, Proceedings of the Cargèse Summer School 1980 (Plenum, New York, 1982).

Let us close this section by giving another example of this sort of analysis, this time for the two-point correlation function $\hat{G}(\mathbf{k})$ in k-space. For simplicity, we will just write the formulæ as if there were only one microscopic length l. Then, remembering the dimensional considerations of section 7.2, we have

$$\hat{G}(\mathbf{k}) = l^{-2-a}\xi^{a}\left(\hat{g}(k\xi)\left[1 + A\left(\frac{l}{\xi}\right)^{\sigma} + \cdots\right]\right), \tag{8.36}$$

where $\hat{g}$ is a scaling function to be determined, a is an exponent to be determined, σ is non-negative and A is a weakly temperature-dependent quantity. At the critical point, ξ has diverged to infinity; to recover the result that $\hat{G} \sim k^{-2+\eta}$, the factors of ξ must "cancel out" for large ξ. Thus $\hat{g}(x) \sim x^{-a}$ as $x \to \infty$, and $\hat{G} \sim l^{-2-a}k^{-a}$ at the critical point. Thus, we identify $a = 2 - \eta$. On the other hand, we know from the static susceptibility sum rule that $\hat{G}(\mathbf{0}) \propto \chi_T$. Assuming that $\hat{g}(x) \to$ constant as $x \to 0$, we also make the identification that $a = \gamma/\nu$, where we used the facts that $\xi \sim t^{-\nu}$ and $\chi_T \sim t^{-\gamma}$. Hence, we conclude that

$$\frac{\gamma}{\nu} = 2 - \eta. \tag{8.37}$$

In conclusion, the simple rule for constructing scaling hypotheses is to find the dimensions of the dependent and independent variables: as we have seen in chapter 7, everything can be expressed in terms of some given length scale, which we will usually take to be the correlation length. Then, the scaling form is given by writing the desired quantity in terms of combinations of independent variables and the correlation length that are invariant under scale transformations. The principal assumption of the scaling hypothesis is the way in which the correlation length enters the theory. That is, if a quantity Q has the dimensions of L^y, in scaling theory, that quantity appears in the scale invariant combination $Q\xi^{-y}$. However, in principle, this need not be the case if the quantity Q acquires an anomalous dimension. In other words, we do not know *a priori* that the correct dependence is not $Q\xi^{a-y}l^{-a}$, where l is the microscopic length scale. This point was explicitly brought out in the discussion following eqn. (8.36). As we will shortly see, the scaling hypothesis follows from the RG; in particular, the RG gives a criterion for which variables acquire anomalous dimensions, and which do not, and allows the calculation of the anomalous dimensions, when combined with other information, such as perturbation theory.

8.3 DYNAMIC CRITICAL PHENOMENA

Equilibrium statistical mechanics is primarily concerned with static quantities. Nevertheless, time-dependent fluctuations of a system in equilibrium also fall within the scope of equilibrium statistical mechanics, through the use of the **fluctuation dissipation theorem.** We shall not give a comprehensive treatment of this important topic here, but merely present some key points.[4]

8.3.1 Small Time-Dependent Fluctuations

In order to discuss the time-dependent fluctuations about equilibrium, we need to specify the dynamics of the system. The level of description that we have found useful in dealing with static critical phenomena is that of Landau theory and its associated order parameter η. If we wish to preserve this level of generality (and we do!), we must address the question of the dynamics of the order parameter. In general, it is an impossible task to derive this from the dynamics of the microscopic dynamical variables of the system. However, in the spirit of Landau theory, we may guess the form of the result that would be obtained if we were to derive the dynamics from the microscopic physics. This we do as follows.

In equilibrium, the spatial configuration of the order parameter is given by

$$\frac{\delta L}{\delta \eta(\mathbf{r})} = 0, \tag{8.38}$$

where we will assume that L is the by now familiar expression (5.55)

$$L = \int d^d\mathbf{r} \left\{ \frac{1}{2}\gamma(\nabla\eta)^2 + \tilde{a}\eta^2 + \frac{1}{2}b\eta^4 - H\eta \right\}, \tag{8.39}$$

where we have introduced the definition $\tilde{a} \equiv at$ to avoid confusion between the reduced temperature and the time t. If the system is slightly out of equilibrium, it is not unnatural to guess that the rate at which the system relaxes back to equilibrium is proportional to the deviation from equilibrium. This assumption of linear response is purely phenomenological, and leads to the following equation for the rate of change of the order parameter:

$$\frac{\partial \eta(\mathbf{r})}{\partial t} = -\Gamma \frac{\delta L}{\delta \eta(\mathbf{r})}, \tag{8.40}$$

---

[4] For a comprehensive treatment, see the review article by P.C. Hohenberg and B.I. Halperin, *Rev. Mod. Phys.* **49**, 435 (1977).

where Γ is a phenomenological parameter, which we will assume to be independent of η, and weakly temperature dependent (*i.e.* it has no singular temperature dependence).

This equation (known as the **time-dependent Ginzburg-Landau equation** in the theory of superconductivity) cannot possibly be a correct description of the approach to the equilibrium state, because the equilibrium state is actually a *global* minimum of L. Equation (8.40) will cause the order parameter to evolve towards local minima, but not necessarily the global minimum. To ensure that the system approaches the global minimum, we must remember that actually the order parameter dynamics is *not* purely relaxational, but may exhibit fluctuations, arising from the microscopic degrees of freedom. These "thermal fluctuations" will sometimes cause the order parameter to move further away from equilibrium during its time evolution, and thus prevent the system from becoming trapped in any metastable minima of L that may be present. This state of affairs may be modelled by introducing a **noise** term ζ into eqn. (8.40) to give

$$\frac{\partial\eta(\mathbf{r})}{\partial t} = -\Gamma\frac{\delta L}{\delta\eta(\mathbf{r})} + \zeta(\mathbf{r},t), \tag{8.41}$$

where the noise is assumed to be a Gaussian random function. This just means that $\zeta(\mathbf{r},t)$ is chosen at random from an ensemble of space and time-dependent functions, with a probability distribution

$$P_\zeta(\{\zeta(\mathbf{r},t)\}) \propto \exp\left[-\frac{1}{2D}\int dt\, d^d\mathbf{r}\, \zeta(\mathbf{r},t)^2\right] \tag{8.42}$$

with the constant of proportionality formally being equal to the inverse of the functional integral

$$\int D\zeta \exp\left[-\frac{1}{2D}\int dt\, d^d\mathbf{r}\, \zeta(\mathbf{r},t)^2\right], \tag{8.43}$$

and the variance of the distribution being D:

$$\langle\zeta(\mathbf{r},t)\rangle_\zeta = 0; \quad \langle\zeta(\mathbf{r},t)\zeta(\mathbf{r}',t')\rangle_\zeta = D\delta(\mathbf{r}-\mathbf{r}')\delta(t-t'). \tag{8.44}$$

The notation $\langle\cdots\rangle_\zeta$ denotes averaging with respect to the probability distribution P_ζ. In order that the **stochastic differential equation** (8.41), usually known as the **Langevin equation**, leads to the correct equilibrium probability distribution P^e_η for η, the amplitude of the noise must be related to the temperature of the system. This makes sense physically, because the origins of ζ are the microscopic degrees of freedom whose interaction with the order parameter is responsible for equilibration in the first place. The probability of finding the order parameter in the configuration $\eta(\mathbf{r})$ is a function of time, and is given by

$$P_\eta(\{\eta(\mathbf{r})\},t) = \langle \delta[\,\eta(\mathbf{r}) - \overline{\eta}(\mathbf{r},t,\{\zeta\})\,]\rangle_\zeta , \tag{8.45}$$

where $\overline{\eta}(\mathbf{r},t,\{\zeta\})$ is a solution of the Langevin equation (8.41) for a particular realisation of the noise $\zeta(\mathbf{r},t)$. This can be understood as follows: the delta function in the average is zero unless the configuration $\eta(\mathbf{r})$ requested is matched by the solution of the Langevin equation. Within a small region of function space, the probability of finding the desired function is then the integral of the delta function over the small region: this essentially counts the frequency of occurence of the desired event. Finally, we average over all realisations of the noise ζ. The time evolution of P_η may be found by differentiating the definition (8.45), and using the Langevin equation (8.41), as shown in the appendix at the end of the present chapter. The result is the **Fokker-Planck equation**

$$\partial_t P_\eta(\{\eta(\mathbf{r})\},t) = \int d^d\mathbf{r}' \frac{\delta}{\delta\eta(\mathbf{r}')}\left[\Gamma \frac{\delta L}{\delta\eta(\mathbf{r}')} P_\eta + \frac{D}{2}\frac{\delta P_\eta}{\delta\eta(\mathbf{r}')}\right]. \tag{8.46}$$

As $t \to \infty$, the solution to the Fokker-Planck equation approaches the equilibrium solution

$$P^e_\eta\{\eta(\mathbf{r})\} \propto \exp\left(-\frac{2\Gamma L\{\eta(\mathbf{r})\}}{D}\right), \tag{8.47}$$

which should be the Boltzmann distribution. For this to be the case, the strength of the fluctuations D must be related to the temperature T and the strength of the dissipation Γ by

$$D = 2\Gamma k_B T. \tag{8.48}$$

This result is an example of the fluctuation-dissipation theorem. The connection with the dissipation is actually quite natural and rather general. The resistance that a body experiences in moving through a gas is due to the bombardment of gas particles. This same bombardment is also responsible for Brownian motion. We now proceed to discuss the phenomenological consequences of the order parameter dynamics that we have introduced.

8.3.2 The Relaxation Time

Consider the system above T_c, where $\overline{\eta} \equiv \langle\eta\rangle = 0$. Time-dependent fluctuations of the order parameter are generated by the noise term $\zeta(\mathbf{r},t)$

and evolve according to eqn. (8.41). For small deviations from equilibrium $\delta\eta \equiv \eta - \overline{\eta}$, the dynamics can be linearised to yield

$$\frac{\partial \delta\eta}{\partial t} = -\left[\frac{\delta\eta}{\tau_0} - \gamma\Gamma\nabla^2\delta\eta\right] + \zeta \tag{8.49}$$

where the zero wavenumber relaxation time is

$$\tau_0^{-1} \equiv 2\tilde{a}\Gamma. \tag{8.50}$$

Taking the Fourier transform gives

$$\frac{\partial \delta\eta_{\mathbf{k}}}{\partial t} = -\left[\frac{\delta\eta_{\mathbf{k}}}{\tau_{\mathbf{k}}}\right] + \zeta_{\mathbf{k}} \tag{8.51}$$

with the relaxation time of the mode with wavenumber **k** being

$$\tau_{\mathbf{k}}^{-1} = \tau_0^{-1} + \gamma\Gamma k^2. \tag{8.52}$$

The zero wavenumber relaxation time for the uniform component of the order parameter is proportional to $1/\tilde{a}$, and hence diverges as $T \to T_c^+$. As explained earlier, this is the origin of **critical slowing down**. When $\mathbf{k} \neq 0$, $\tau_{\mathbf{k}}$ is finite as $T \to T_c^+$.

The response of the system to an external perturbation is described by the order parameter susceptibility or response function $\widetilde{\chi}_T(\mathbf{k},\omega)$, defined by

$$\widetilde{\chi}_T(\mathbf{k},\omega) \equiv \lim_{\widetilde{h}\to 0} \frac{\delta \langle \widetilde{\eta}(\mathbf{k},\omega)\rangle_\zeta}{\delta \widetilde{h}(\mathbf{k},\omega)}, \tag{8.53}$$

where $h(\mathbf{r},t)$ is a space and time dependent external driving force on the system, added to the right hand side of eqn. (8.41), and the response function is evaluated in the linear response regime where h is small ($h \to 0$). The quantity $\delta \langle \widetilde{\eta}(\mathbf{k},\omega)\rangle_\zeta$ is the difference between the average taken with $h \neq 0$ and the average taken with $h = 0$, and the response function in real space and time is related to $\widetilde{\chi}_T(\mathbf{k},\omega)$ by the Fourier transform

$$\chi_T(\mathbf{r},t) = \frac{1}{V}\sum_{\mathbf{k}} \int \frac{d\omega}{2\pi} e^{i(\mathbf{k}\cdot\mathbf{r}-\omega t)} \widetilde{\chi}_T(\mathbf{k},\omega). \tag{8.54}$$

As we mentioned in section 5.7, response functions are related to correlation functions, and are readily measurable using scattering techniques, using light or neutrons, for example.

When we ignore the non-linearities in the Langevin equation (8.41) and perform the noise average, we simply obtain

$$\widetilde{\chi}_T(\mathbf{k},\omega) = \frac{\gamma}{\tau_{\mathbf{k}}^{-1} - i\omega}. \tag{8.55}$$

However, a calculation of the response function beyond this level of approximation — essentially Landau theory — is more problematic, because it is necessary to average over the noise ζ the non-linear term in eqn. (8.41), which is cubic in η. Attempts to implement this calculation using perturbation theory suffer from the same problems that we found for the statics, in the critical region.

8.3.3 Dynamic Scaling Hypothesis for Relaxation Times

The dynamic scaling hypothesis is a set of assertions about the scaling behaviour of the relaxation time and the response function in the critical region, and implies a relationship between *a priori* independent exponents. First, we consider the relaxation times $\tau_{\mathbf{k}}$. We saw in mean field theory that $\tau_{\mathbf{k}}$ was a function of t and $\mathbf{k}$, with the following limits:

$$\tau_{\mathbf{k}} = \begin{cases} (2a\Gamma t)^{-y} & \mathbf{k}=0; \\ (\gamma\Gamma k^z)^{-1} & T = T_c, \end{cases} \tag{8.56}$$

with the new critical exponents $y = 1$ and $z = 2$. Just as the limiting behaviour of the magnetisation given by eqns. (8.5) and (8.6) followed from the static scaling hypothesis (8.7), so we can generalise eqn. (8.56) for the relaxation time, and derive it from the single scaling hypothesis that

$$\tau_{\mathbf{k}}(t) = t^{-y} F_\tau(k\xi(t)), \tag{8.57}$$

where F_τ is a scaling function, and now we do not expect z and y to have their mean field values necessarily. In order that the scaling hypothesis reproduce the facts summarised in eqn. (8.56), we require that $F_\tau(x)$ is constant when $x \to 0$. In the opposite limit that $T \to T_c$ but $\mathbf{k} \neq 0$, the correlation length diverges as $\xi \sim t^{-\nu}$, and we require that as $x \to \infty$, $F_\tau(x) \sim x^{y/\nu}$ in order to cancel out the singular t dependence and yield a finite result for $\tau_{\mathbf{k}}$. Thus, we find the scaling law

$$z = \frac{y}{\nu}, \tag{8.58}$$

so that

$$\tau_{\mathbf{k}} = t^{-z\nu} F_\tau(\mathbf{k}t^{-\nu}). \tag{8.59}$$

8.3.4 Dynamic Scaling Hypothesis for the Response Function

We can write a scaling form for the response function itself, by assuming that

$$\tilde{\chi}_T(\mathbf{k},\omega) = t^{-v} F_\chi(\mathbf{k}\xi(t),\omega\tau_0), \tag{8.60}$$

where v is a critical exponent and F_χ is a scaling function. In the case that $\mathbf{k} = \mathbf{0}$ and $\omega = 0$, the static susceptibility sum rule implies that $F_\chi(\mathbf{0},0)$ is a constant and that the exponent v is simply the static susceptibility critical exponent γ. Now let us examine the consequences of the dynamic scaling hypothesis when $\mathbf{k} = \mathbf{0}$. As $T \to T_c$, τ_0 diverges like $\sim t^{-z\nu}$. If the scaling function has power law behaviour in the limit that its second argument diverges, we require that $F_\chi(\mathbf{0},x) \sim x^{-\gamma/z\nu}$ in order that the susceptibility remains finite. Hence we find that at the critical point, the zero wavenumber frequency dependent response function has the power law form

$$\tilde{\chi}_T(\mathbf{0},\omega) \sim \omega^{-\gamma/z\nu}. \tag{8.61}$$

Actually, the dependence of the response function on frequency ω must occur through the combination $-i\omega$, because of the form of the Langevin eqn. (8.41). Hence, the correct form for the scaling of the response function at the transition is

$$\tilde{\chi}_T(\mathbf{0},\omega) = A(-i\omega)^{-\gamma/z\nu}, \tag{8.62}$$

with A being a real constant. From this, we can make another striking prediction: the phase lag δ between the driving force h and the response of the order parameter is given by

$$\delta \equiv \tan^{-1}\left(\frac{\mathrm{Im}\,\tilde{\chi}_T(\mathbf{0},\omega)}{\mathrm{Re}\,\tilde{\chi}_T(\mathbf{0},\omega)}\right) \tag{8.63}$$

$$= \frac{\pi\gamma}{2z\nu} \tag{8.64}$$

which is a universal value.[5] This result can also be obtained from eqn. (8.61) using the Kramers-Kronig relations.

8.3.5 Scaling of the Non-linear Response

Scaling ideas can also be used to go beyond linear response theory. The basic idea is to consider the **non-linear response function**

$$\chi^{\mathrm{nl}} = \frac{\delta\eta}{h}, \tag{8.65}$$

[5] See J.P. Clerc *et al.*, *J. Phys. (Paris), Lett.* **45**, L913 (1984) for an application to the elastic response of fractal networks, and A. T. Dorsey, *Phys. Rev. B* **43**, 7575 (1991), for an application to scaling at the superconducting transition.

where all quantities are at zero wavenumber and frequency. Dimensionally, χ^{nl} should have the same dimensions as χ_T, and so we might anticipate the scaling form

$$\chi^{\mathrm{nl}} = \xi^{-\gamma/\nu} F_{\chi}^{\mathrm{nl}}(h\xi^{y}), \tag{8.66}$$

where the exponent y describes the dimensions of h: $[h] = L^{-y}$, and the combination $h\xi^{y}$ does not change under scale transformations. At T_c, the assumed finiteness of χ^{nl} requires that ξ cancel out of eqn. (8.66), leading to the relation

$$\chi^{\mathrm{nl}} \sim h^{\gamma/y\nu}, \tag{8.67}$$

which is analogous to the behaviour of the equation of state on the critical isotherm.

A more interesting example of the scaling of the non-linear response is to the dynamics of currents near a critical point. The transport properties of a system may differ greatly above and below T_c, and this has consequences for the scaling of response functions. Here, we consider the critical dynamics near the superconducting transition[6] and write down the scaling theory for the frequency-dependent conductivity $\sigma(\omega)$. Below T_c, $\sigma(\omega) \sim \rho_s/(-i\omega)$, which follows from the London equation relating the electric field $\mathbf{E}$ to the current $\mathbf{j}$ in a superconductor:

$$\mathbf{E} = \frac{\partial}{\partial t}\frac{m}{\rho_s e^2}\mathbf{j}, \tag{8.68}$$

where ρ_s is the superfluid density *i.e.*, the "number density of superconducting electrons" and e and m are the electron charge and mass respectively.[7] It can be shown[8] that near the superconducting transition, the superfluid density varies as $\rho_s \sim \xi^{2-d}$. Hence, the conductivity should exhibit the scaling form

$$\sigma(\omega) = \xi^{2-d+z} F_{\sigma}(\omega\xi^{z}). \tag{8.69}$$

Our present concern is the non-linear d.c. conductivity, defined by

$$\mathbf{j} = \sigma(E)\mathbf{E}. \tag{8.70}$$

To construct a scaling theory for this quantity, we need to know the dimension of the electric field $\mathbf{E}$. This follows from the fact that $\mathbf{E} \sim \partial\mathbf{A}/\partial t$,

---

[6] S.A. Wolf, D.U. Gubser and Y. Imry, *Phys. Rev. Lett.* **43**, 324 (1979); D.S. Fisher, M.P.A. Fisher and D.A. Huse, *Phys. Rev. B* **43**, 130 (1991); A.T. Dorsey, *op. cit.*

[7] M. Tinkham, *Introduction to Superconductivity* (McGraw-Hill, New York, 1975), p. 4.

[8] B.D. Josephson, *Phys. Lett.* **21**, 608 (1966); see also section 11.1.4.

where **A** is the vector potential. The latter quantity has dimensions of inverse length, as can be seen by the fact that it enters the Ginzburg-Landau theory for the superconducting order parameter in the combination $(-i\nabla - 2e\mathbf{A}/c)$. Thus $[E] = L^{-1}T^{-1}$. Using the dynamic scaling hypothesis that the relaxation time $\tau \sim \xi^z$, the scaling form for $\sigma(E)$ becomes

$$\sigma(E) = \xi^{2-d+z} F_\sigma^{\mathrm{nl}}(\xi^{1+z} E). \tag{8.71}$$

At T_c, using the by now familiar argument, we conclude that

$$\sigma(E) \sim E^{-(2-d+z)/(1+z)}. \tag{8.72}$$

8.4 SCALING IN THE APPROACH TO EQUILIBRIUM

The preceding section dealt with time-dependent phenomena in or very close to equilibrium. In this section, we show briefly how scaling ideas are beginning to be applied to systems far from equilibrium. It is convenient to group these systems into two classes: those systems in a driven or **non-equilibrium steady state**, and those systems which are approaching equilibrium after a non-infinitesimal displacement from the equilibrium state. An example of the former category is the random deposition of particles on a substrate, resulting in a solid phase growing at a steady rate.[9] This simple dynamical process results in a surface that develops fluctuations in its height as it grows, as depicted in figure (8.2a), and the outcome of numerical simulations is well accounted for by simple dynamical scaling ideas. The second class of systems is exemplified by the problem of the **spinodal decomposition** of a binary alloy.[10] At high temperature, the two species A and B are uniformly mixed in the alloy, but below a critical temperature T_c, the uniform state is no longer stable, and the equilibrium state of the system is two co-existing domains of different compositions. It should come as no surprise that this system can be modelled by the Ising model, in a similar way to that in which a fluid may be modelled by a lattice gas. What is of interest here is not the equilibrium statistical mechanics, but the time dependence of the phenomenon. In a solid solution, the time scales are relatively slow, and it is possible to observe the decomposition occuring over several hours. Observations at late times reveal a convoluted domain structure with the

---

[9] Brief reviews are given by F. Family, *Physica A* **168**, 561 (1990) and by B.M. Forrest and L.-H. Tang, *Mod. Phys. Lett. B* **4**, 1185 (1990).

[10] The scaling theory is reviewed by H. Furukawa, *Adv. Phys.* **34**, 703 (1985).

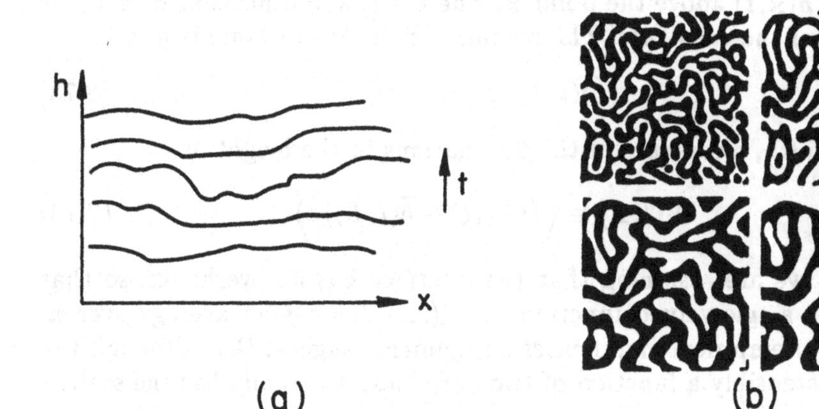

Figure 8.2 (a) Growth of a fluctuating surface. (b) Caricature of the time evolution of domain structure during spinodal decomposition. The dark regions represent the A-rich phase, and the light regions represent the B-rich phase.

two phases closely interlocked, as sketched in figure (8.2b). This structure develops by domain merger and growth, in a way which is statistically self-similar in time, and both experiments and numerical simulation are well described by scaling theory. Although RG ideas have been applied to the first class of problems, there is still no such understanding of the scaling behaviour observed during the approach to equilibrium.

8.4.1 *Growth of a Fluctuating Surface*

Consider a planar substrate in d dimensions of linear dimension L, on which atoms are being deposited at random from above to form a growing aggregate whose surface has dimensionality $d' = d - 1$, as shown in figure (8.2a). There are several ways in which the incoming particles can attach themselves to the existing aggregate, but here we shall just mention **ballistic deposition**, a model situation in which particles fall vertically onto the substrate and stick to the aggregate at the first point of contact. Although it is far from clear that this represents any realistic growth process, it is of interest to examine the behaviour of this model; moreover, it is found empirically that the results exhibit universality in the sense that the scaling and associated exponents observed in computer simulations are common features to a wide variety of growth processes. What, then, is the behaviour of this simple model?

The stochastic element of the growth process — namely the random position of the deposited atom — and the absence of any surface relaxation dynamics suggest that the growing interface will not be completely smooth, but will be rough, in some sense. Let us describe the interface by its height $h(\mathbf{x},t)$ above the point $\mathbf{x}$, where $\mathbf{x}$ is a d'-dimensional vector in the plane of the substrate. The average height at any time is just

$$\overline{h}(t,L) \equiv \langle h(\mathbf{x},t)\rangle \tag{8.73}$$

and the width, a measure of the fluctuations in the height, is

$$W(t,L) \equiv \left\langle \left(h(\mathbf{x},t) - \overline{h}(t,L)\right)^2 \right\rangle . \tag{8.74}$$

Note that we are assuming that the interface has no overhangs, so that $h(\mathbf{x},t)$ is a single-valued function, and $\langle \ldots \rangle$ denotes an average over $\mathbf{x}$. A variety of empirical and heuristic arguments suggest that although the width is ostensibly a function of two variables, it actually has the scaling form

$$W(t,L) = L^{\chi} f(tL^{-z}), \tag{8.75}$$

where χ and z are critical exponents and f is a scaling function. Eqn. (8.75) is presumably valid for times long enough that the influence of the initial conditions has decayed, but still short enough that the width is not time independent. We will refer to such a time regime as one exhibiting **intermediate asymptotics.** On a finite substrate in equilibrium, the width scales as $W \sim L^{\chi}$, showing that the scaling function $f(x)$ tends towards a constant for large values of its argument x. On the other hand, when $L \to \infty$, the length of the intermediate asymptotic regime diverges. Crudely speaking, it takes an infinite amount of time for the system to equilibrate, and the width simply grows as a power law function of time: $W \sim t^{\beta}$, where β is an exponent to be determined. The scaling form (8.75) implies that in order for W to remain finite as $L \to \infty$, $f(x) \sim x^{\chi/z}$ as $x \to 0$. Thus $W(t,L) \sim t^{\chi/z}$ for short times or large systems.

The scaling description above begs the question: how can one substantiate the starting point, eqn. (8.75), and how can one actually compute the exponents that appear? A partial answer to these questions has been obtained by the assumption[11] that the equation of motion for the interface height may be modelled by a Langevin equation of the form[12]

$$\frac{\partial h}{\partial t} = \nu \nabla^2 h + \frac{\lambda}{2} (\nabla h)^2 + \zeta, \tag{8.76}$$

where ν and λ are constants, and $\zeta(\mathbf{x},t)$ is a Gaussian random noise, as in the earlier discussion of dynamic critical phenomena. Note, however,

---

[11] M. Kardar, G. Parisi and Y.C. Zhang, *Phys. Rev. Lett.* **56**, 889 (1986).

[12] This Langevin equation seems to allow for the possibility that $\partial_t h$ can be negative at some time and position where $\zeta(\mathbf{r},t)$ is sufficiently negative.

that the two-point correlation function of the noise is not related to the temperature of the system through the fluctuation-dissipation theorem (8.48), because the system is not in thermal equilibrium. This equation of motion — often referred to as the KPZ equation — is not of the form of the time-dependent Ginzburg-Landau equation (8.41), because the non-linear term cannot be obtained from differentiating a coarse-grained free energy functional. This may be viewed as an additional reflection of the fact that interface growth occurs far from equilibrium, and in a more microscopic description of interface growth, the non-linear term is found to be purely kinetic in origin; in fact, the KPZ equation is nothing more than the assumption that at each point on the interface, the velocity v_n along the outward normal is just

$$v_n = \lambda + \nu\kappa + \zeta, \tag{8.77}$$

where κ is the curvature at that point. Thus λ is the velocity of a flat interface in response to a thermodynamic driving force, ν is a positive constant so that the motion tends to reduce the curvature everywhere, and ζ represents the random deposition of particles. The KPZ description of interface motion is closely analogous to dynamic critical phenomena, and it is this analogy which has been exploited using the methods of critical phenomena, including the renormalisation group, to address the question of scaling.[13]

8.4.2 Spinodal Decomposition in Alloys and Block Copolymers

When a homogeneous binary alloy is quenched rapidly from its spatially uniform state of homogeneous composition at high temperature to a temperature below the critical point, phase separation takes place. Eventually, the system returns to equilibrium, which at this reduced temperature is a state with two coexisting domains. In figure (8.3), a caricature of the phase diagram is shown, with the compositions c_α and c_β of the two domains in equilibrium at temperature T_q indicated schematically. The coexistence curve follows by equating the chemical potentials of the two phases, and is purely thermodynamic in origin. On the other hand, this section is concerned with the *time dependence* of the approach to equilibrium, focussing only on the rôle that scaling arguments have had in describing this phenomenon.

---

[13] For a comprehensive discussion of the attempts to study non-equilibrium growth using the methods of critical phenomena, see the article by B. Grossman, H. Guo and M. Grant, *Phys. Rev. A* **43**, 1727 (1991).

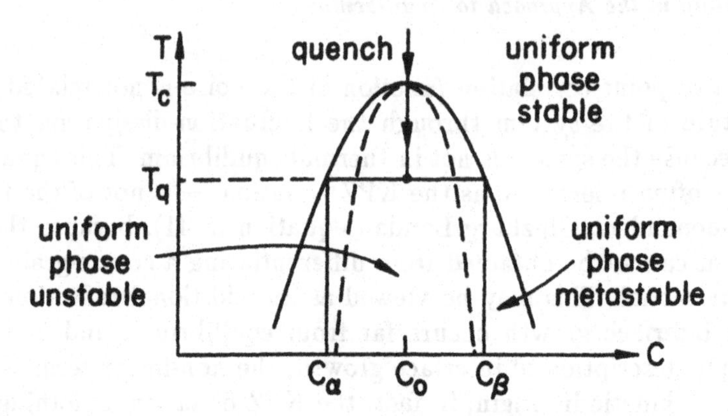

Figure 8.3 Idealised phase diagram of a binary alloy. The full line is the coexistence curve, whilst the dashed line represents the classical spinodal.

There are two principal mechanisms for phase separation, **nucleation** and **spinodal decomposition**, which may be described in a simplified way, as follows. Within the coexistence curve, the homogeneous state is not the *global* minimum of the free energy; however, there may still be a region within the coexistence curve where the homogeneous state is a *local* minimum, and is stable to small local fluctuations in the composition. This is known as **metastability**, and in this region of the phase diagram, the system may proceed from the homogeneous state to the phase separated one only when a sufficiently large thermal fluctuation in the local composition spontaneously occurs. This process is known as **homogeneous nucleation**; the local composition fluctuation may be thought of, for present purposes, as a droplet of one of the two phases coexisting in equilibrium. For small droplets, the surface free energy cost of creating the interface separating the droplet from the rest of the system outweighs the bulk free energy benefit of creating within the volume of the droplet the state which represents the true minimum of the free energy, and thus the homogeneous state is stable to such fluctuations. On the other hand, if the droplet is sufficiently large, the bulk free energy overwhelms the surface energy, and the free energy of the system is reduced by the presence of big enough droplets. Hence the homogeneous state is unstable towards sufficiently large fluctuations. If the surface free energy per unit area of the droplet is σ, and the bulk free energy per unit volume is ϵ, then the difference in free energy between the system with one droplet present of radius R and the system with no droplets is

$$\Delta F = 4\pi\sigma R^2 - \frac{4}{3}\pi\epsilon R^3, \tag{8.78}$$

showing how the bulk and surface free energies compete. Notice that $\Delta F(R)$ is a maximum for $R = R_c = 2\sigma/\epsilon$. The dynamical theory of nucleation describes the likelihood of forming such droplets from the thermal fluctuations in the system, and the subsequent evolution of the droplets; it turns out that droplets with $R < R_c$ shrink after they have been created, whereas droplets with $R > R_c$ grow. This oversimplified picture has many elaborations, but is in essence correct.[14] In summary, nucleation is the decay of the homogeneous state of the system through the development of small-scale but large amplitude fluctuations.

The parameters σ and ϵ depend on the state of the system. If the quench is into the region of the phase diagram where the surface free energy per unit area vanishes, then the homogeneous state is unstable towards arbitrarily small fluctuations in composition. In this case, the decay of the homogeneous state occurs through the development and subsequent growth of infinitesimal long-wavelength fluctuations, and is termed **spinodal decomposition**. The **classical spinodal** is the curve on the phase diagram where the critical droplet radius tends to zero. In practice, it is only a loosely defined concept, because it is hard to give a precise meaning to the droplet concept, when the size of the droplet is of order the width of the interface at the droplet surface.

During spinodal decomposition, initially microscopic composition fluctuations grow in magnitude and develop the interconnected pattern of domains, sketched in figure (8.2b). It is found empirically that there is an intermediate asymptotic regime, where only a single time dependent length scale $L(t)$ characterises the pattern. We can define an order parameter $\psi(\mathbf{r},t) = c(\mathbf{r},t) - c_0$, where $c(\mathbf{r},t)$ is the local concentration of one of the alloy components (let us say A) at time t and c_0 is the average concentration (*i.e.* the concentration in the homogeneous state): then the equal-time two-point correlation function, which is proportional to the neutron or X-ray scattering differential cross section, has the form in real space[15]

$$S(\mathbf{r}-\mathbf{r}',t) \equiv \langle \psi(\mathbf{r},t)\psi(\mathbf{r}',t)\rangle = F_\psi\left(\frac{|\mathbf{r}-\mathbf{r}'|}{L(t)}\right). \tag{8.79}$$

In this expression, the average is taken over the sample in the case of experiment, and in the case of computer simulations, the average is taken

---

[14] For a review of the kinetics of phase transitions, see the article by J.D. Gunton, M. San Miguel and P.S. Sahni in *Phase Transitions and Critical Phenomena*, Volume 8, edited by C. Domb and J.L. Lebowitz (Academic, New York, 1983).

[15] See, for example, the neutron scattering data on CuMn alloys of B.D. Gaulin, S. Spooner and Y. Morii, *Phys. Rev. Lett.* **59**, 668 (887).

over repetitions of the coarsening process, starting with different initial conditions; the function F_ψ is a scaling function. Experiments actually measure the Fourier transform, which then has the form

$$\hat{S}(\mathbf{k},t) = L(t)^d \hat{F}_\psi(kL(t)). \tag{8.80}$$

Although $\hat{S}(\mathbf{k},t)$ is ostensibly a function of two variables, eqn. (8.80) implies that when plotted appropriately, the data at different times and wavenumber should collapse on to a single curve, and this has indeed been convincingly established. The usual interpretation of these findings is that in the intermediate asymptotic regime, the characteristic scale of the spatial pattern is the size of the domains, which for long enough times, becomes much larger than the width of the interface separating A-rich regions from A-depleted regions. The interface width is of order the equilibrium bulk correlation length ξ in the system: the fact that $L \gg \xi$ suggests that thermal fluctuations do not play an important rôle in the coarsening process during the intermediate asymptotic regime, although thermal fluctuations are certainly important in initiating the phase separation process, and in guaranteeing the eventual equilibration of the system.

In addition to the scaling form exhibited in eqns. (8.79) and (8.80), it is found that the domain size grows with a power law:

$$L(t) \sim t^\phi, \tag{8.81}$$

with ϕ being consistent with a value of 1/3. Although there is currently no complete understanding of these findings, it is anticipated that renormalisation group considerations will provide a framework for a successful theory.

Theoretical descriptions of spinodal decomposition have two principle ingredients: first, it is assumed that the long wavelength behaviour of the system can be described by a coarse-grained free energy functional of the form

$$F\{\psi(\mathbf{r})\} = \int d^d\mathbf{r} \left\{ \frac{1}{2}(\nabla\psi)^2 + f(\psi(\mathbf{r})) \right\}, \tag{8.82}$$

which, in the literature, is referred to as the **Cahn-Hilliard free energy functional**. The function $f(\psi)$ is the free energy per unit volume for a spatially homogeneous system, and in the simplest case of a quench through the critical point, has the usual form appropriate to Landau theory. Thus, for temperatures $T > T_c$, there is only a single minimum of the Cahn-Hilliard free energy at $\psi = 0$, whereas for $T < T_c$, there are two degenerate minima, now representing the two phases of the alloy with different compositions, which coexist in equilibrium. The gradient term in (8.82) accounts for the free energy cost due to the presence of a concentration gradient.

The dynamics of the phase separation process is assumed to be governed by diffusion in a chemical potential gradient. Conservation of material implies that

$$\frac{\partial \psi}{\partial t} + \nabla \cdot \mathbf{j} = 0, \tag{8.83}$$

where the current is given phenomenologically by

$$\mathbf{j} = -M\nabla \frac{\delta F}{\delta \psi(\mathbf{r})}, \tag{8.84}$$

and M is a phenomenological constant representing mobility. The resulting dynamical equation is the celebrated **Cahn-Hilliard equation**

$$\frac{\partial \psi}{\partial t} = M\nabla^2 \left[\frac{\partial f}{\partial \psi} - \nabla^2 \psi\right], \tag{8.85}$$

which is to be solved with initial conditions representing the high temperature phase of the system, $\langle \psi \rangle = 0$, but with fluctuations about the mean to initiate growth.

As an example of the application of the use of scaling ideas to non-equilibrium phenomena, we conclude this section by mentioning a determination of the value of the dynamic exponent ϕ using scaling ideas applied to the Cahn-Hilliard equation. The dynamic exponent is difficult to measure by numerical solution of the equation, because the intermediate asymptotic regime is only of appreciable duration for large systems; to ensure that ones data are truly in this regime, computer simulations of phase separation must run to very long times. This dual restriction of large system size and long times, plus the necessity of averaging the data over many repetitions, has meant that considerable effort must be undertaken to obtain reliable results.

An alternative procedure[16] is to consider instead the equation

$$\frac{\partial \psi}{\partial t} = M\nabla^2 \left[\frac{\partial f}{\partial \psi} - \nabla^2 \psi\right] - \epsilon\psi, \tag{8.86}$$

[16] First suggested on heuristic grounds by Y. Oono and M. Bahiana, *Phys. Rev. Lett.* **61**, 1109 (1988). The scaling analysis given here is due to F. Liu and N. Goldenfeld, *Phys. Rev. A* **39**, 4805 (1989).

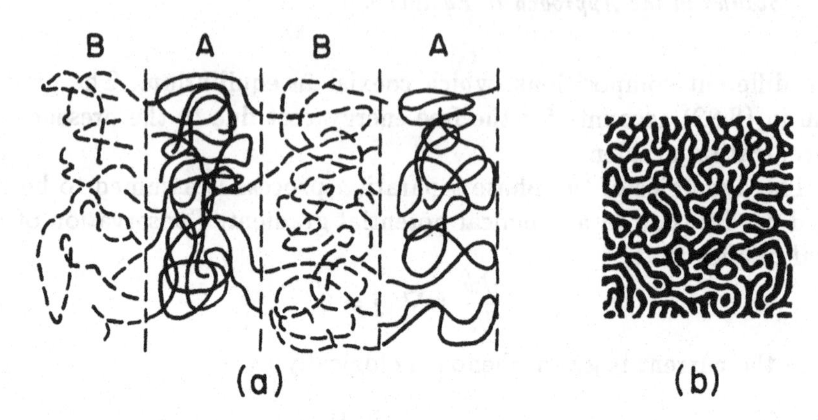

Figure 8.4 (a) Sketch of local phase separation in a block copolymer melt. The monomer A segments of each polymer chain are represented by the full lines, whereas the monomer B segments of each polymer chain are represented by the dashed lines. (b) Morphology as in (a), but shown on a much larger length scale. The dark regions represent the A-rich phase, and the light regions represent the B-rich phase.

which has been proposed to describe phase separation in **block copolymers**.[17] A melt of block copolymers consists of polymer chains, on *each* of which the monomer sequence is

$$\text{A-A-A-}\cdots\text{-A-B-B-B-}\cdots\text{-B}$$

with equal numbers of A and B in the simplest case considered here; the total number of monomers on each chain is N, and it can be shown that in eqn. (8.86), the parameter ϵ is proportional to N^{-2}. Each chain has the character of an alloy! When the temperature is reduced below T_c, the A and B atoms attempt to segregate, but the phase separation is unable to reach completion because the A and B species are constrained to lie on the same polymer chain. The result is that there is *local* phase separation, with domains of A rich and B rich regions of spatial extent governed by the physical dimensions of the polymer chains, as sketched in figure (8.4).

The steady state of alternating domains of A and B rich regions corresponds to periodic steady state solutions of the block coplymer equation (8.86). A detailed analysis shows that there is not a unique periodic steady state solution of the block copolymer equation, but instead, for every value of the parameter ϵ lying in the range $0 \leq \epsilon \leq 1/4$, there is a *band* of linearly stable periodic steady states, parameterised by the wavelength or domain size. Thermal fluctuations will select, in equilibrium, the

---

[17] Y. Oono and Y. Shiwa, *Mod. Phys. Lett. B* **1**, 49 (1987); a coarse-grained free energy for block copolymers was proposed by L. Leibler, *Macromolecules* **13**, 1602 (1980).

steady state which minimises the free energy of the system: numerically, it is found that the domain size λ varies as a power law:

$$\lambda \sim \epsilon^{-\theta/2}, \quad \epsilon \to 0, \tag{8.87}$$

which defines the exponent θ characterising the equilibrium state in the limit of long polymer chains. The value of θ was estimated to be 0.650 ± 0.016.

The interesting question now is: what is the time evolution of the local phase separation? For early times, the A and B monomers diffuse over distances much smaller than the radius of gyration of the polymer chains; hence the constraint imposed by the chain will be inoperative, and we expect the time evolution to follow that of the phase separation of a regular alloy. On the other hand, when the size of the ensuing domains becomes comparable to the chain size, the constraint becomes effective, and the growth of the domains ceases, freezing in eventually to the equilibrium size. In this latter stage, thermal fluctuations are certainly important in facilitating the motion and annealing of defects in the domain structure. The intermediate asymptotic regime, in which the domain growth is similar to that of phase separation in a regular alloy, may be made as large as desired by making the chain length large: thus we are interested in the regime $\epsilon \to 0$. These considerations suggest that the characteristic length scale L in the block copolymer system may be written as

$$L(t, \epsilon) = \epsilon^{-\theta/2} F_L(t\epsilon^{\gamma}), \tag{8.88}$$

where γ is to be determined. We expect that the time to reach equilibrium increases as N increases, and hence as ϵ decreases; thus we anticipate that γ is positive. The scaling function $F_L(x)$ must have the following properties. For ϵ fixed at a non-zero value, but $t \to \infty$, the characteristic length tends towards the equilibrium value; thus $F_L(x) \to$ constant as $x \to \infty$. On the other hand, for large time and $\epsilon \to 0$, the behaviour should be that of regular phase separation, and $L(t) \sim t^{\phi}$. The familiar scaling argument then implies that

$$F(x) \sim x^{\theta/2\gamma}, \quad \text{as } x \to 0 \tag{8.89}$$

to cancel out the ϵ prefactor in $L(t, \epsilon)$. Hence, we conclude that $L \sim t^{\theta/2\gamma}$, and thus

$$\theta = 2\gamma\phi. \tag{8.90}$$

Note that θ is an exponent describing the equilibrium state, whereas ϕ describes the approach to equilibrium. Numerical data are entirely consistent with the scaling law (8.88) for $L(t, \epsilon)$ with $\gamma = 1.0$. Thus the numerical determination of θ implies that the dynamic exponent $\phi = 0.325 \pm 0.008$, which is consistent with experimental and numerical data.

8.5 SUMMARY

The scaling hypothesis is a valuable way to correlate data for systems near the critical point, and for some non-equilibrium systems as they approach equilibrium. These are empirical statements. In the following chapter, we will discuss how the renormalisation group accounts for the success of scaling ideas, for systems in equilibrium. Certain driven non-equilibrium systems, such as models of atomic deposition, exhibit a formal analogy to critical dynamics, and this has proven to be useful in analysing these systems. Finally, there are tantalising suggestions that some systems approaching equilibrium may also be usefully described by renormalisation group ideas. Further discussion of this topic is to be found in chapter 10.

APPENDIX 8 - THE FOKKER-PLANCK EQUATION

In this appendix, we sketch the derivation of the Fokker-Planck equation, starting from the Langevin equation (8.41).

We begin with the definition of the probability distribution for the order parameter $\eta(\mathbf{r})$, and differentiate with respect to time:

$$\begin{aligned}
\partial_t P_\eta(\{\eta(\mathbf{r})\}, t) &= \langle \partial_t \delta\left[\eta(\mathbf{r}) - \overline{\eta}(\mathbf{r}, t, \{\zeta\})\right]\rangle_\zeta \\
&= \int d^d\mathbf{r}' \left\langle \frac{\partial \overline{\eta}}{\partial t} \frac{\delta}{\delta \overline{\eta}(\mathbf{r}', t)} \delta\left[\eta(\mathbf{r}) - \overline{\eta}(\mathbf{r}, t, \{\zeta\})\right] \right\rangle_\zeta \\
&= -\int d^d\mathbf{r}' \frac{\delta}{\delta \eta(\mathbf{r}')} \left\langle \frac{\partial \overline{\eta}}{\partial t} \delta\left[\eta - \overline{\eta}\right] \right\rangle_\zeta \\
&= -\int d^d\mathbf{r}' \frac{\delta}{\delta \eta(\mathbf{r}')} \left\langle \left[-\Gamma \frac{\delta L}{\delta \eta(\mathbf{r}')} + \zeta(\mathbf{r}', t) \right] \delta\left[\eta - \overline{\eta}\right] \right\rangle_\zeta \\
&= \int d^d\mathbf{r}' \frac{\delta}{\delta \eta(\mathbf{r}')} \left[\Gamma P_\eta \frac{\delta L}{\delta \eta(\mathbf{r}')} - \langle \zeta \delta(\eta - \overline{\eta}) \rangle_\zeta \right] \qquad \text{(A8.1)}
\end{aligned}$$

The evaluation of $\langle \zeta \delta(\eta - \overline{\eta}) \rangle_\zeta$ is accomplished by noting the general result

$$\begin{aligned}
\langle F\{\zeta\}\zeta \rangle_\zeta &= \int D\zeta \; (\zeta F) \; P_\zeta \\
&= D \int D\zeta \frac{\delta F}{\delta \zeta} P_\zeta \\
&= D \left\langle \frac{\delta F}{\delta \zeta} \right\rangle_\zeta \qquad \text{(A8.2)}
\end{aligned}$$

where we integrated by parts and used the expression (8.42) for the probability distribution. These operations can be verified, if desired, by writing the functional integral as a multiple integral. In the present case,

$$\begin{aligned}\langle\zeta(\mathbf{r}',t)\delta(\eta-\overline{\eta})\rangle_\zeta &= D\left\langle\frac{\delta}{\delta\zeta(\mathbf{r}',t)}\delta(\eta-\overline{\eta})\right\rangle_\zeta \\ &= D\int d\mathbf{r}''\left\langle\frac{\delta\overline{\eta}(\mathbf{r}'',t)}{\delta\zeta(\mathbf{r}',t)}\frac{\delta}{\delta\overline{\eta}(\mathbf{r}'',t)}\delta(\eta-\overline{\eta})\right\rangle_\zeta \\ &= -D\int d\mathbf{r}''\frac{\delta}{\delta\eta(\mathbf{r}'',t)}\left\langle\frac{\delta\overline{\eta}(\mathbf{r}'',t)}{\delta\zeta(\mathbf{r}',t)}\delta(\eta-\overline{\eta})\right\rangle_\zeta \quad \text{(A8.3)}\end{aligned}$$

The quantity being averaged in the above expression is essentially a response function, and can be evaluated from the formal solution to the Langevin equation (8.41):

$$\overline{\eta}(\mathbf{r},t)=\overline{\eta}(\mathbf{r},0)-\int_0^t dt'\,\Gamma\frac{\delta L}{\delta\eta}\left(\overline{\eta}(\mathbf{r},t')\right)+\int_0^t dt'\,\zeta(\mathbf{r},t'). \quad \text{(A8.4)}$$

Differentiating with respect to $\zeta(\mathbf{r}',t'')$ and noting that causality implies that $\eta(\mathbf{r},t)$ only depends on $\zeta(\mathbf{r},t')$ when $t>t'$, we find that

$$\frac{\delta\overline{\eta}(\mathbf{r},t)}{\delta\zeta(\mathbf{r}',t'')}=\delta(\mathbf{r}-\mathbf{r}')\left[-\int_{t''}^t dt'\left\{\frac{\delta}{\delta\zeta(\mathbf{r}',t'')}\Gamma\frac{\delta L}{\delta\eta}\left(\overline{\eta}(\mathbf{r},t')\right)\right\}+\theta(t-t'')\right]. \quad \text{(A8.5)}$$

In eqn. (A8.3), we require the above quantity evaluated at $t=t''$. The value of the Heaviside function $\theta(t)$ at zero is in this case $\theta(0)=1/2$, as can be seen by repeating the above derivation with the two-point correlation function of ζ being proportional not to $\delta(t-t')$, but to a sharply-peaked even function of $(t-t')$. Thus

$$\frac{\delta\overline{\eta}(\mathbf{r}'',t)}{\delta\zeta(\mathbf{r}',t)}=\frac{1}{2}\delta(\mathbf{r}-\mathbf{r}'). \quad \text{(A8.6)}$$

Substituting into eqn. (A8.3), and collecting results back through eqn. (A8.1), we finally obtain

$$\partial_t P_\eta(\{\eta(\mathbf{r})\},t)=\int d^d\mathbf{r}'\,\frac{\delta}{\delta\eta(\mathbf{r}')}\left[\Gamma\frac{\delta L}{\delta\eta(\mathbf{r}')}P_\eta+\frac{D}{2}\frac{\delta P_\eta}{\delta\eta(\mathbf{r}')}\right], \quad \text{(A8.7)}$$

which is our desired result.

CHAPTER 9

The Renormalisation Group

The preceding chapters have shown that mean field theory does not accurately predict critical exponents, and that attempts to improve mean field theory by perturbation theory fail, due to the presence of divergences as $t \to 0$. We also saw that dimensional considerations imply that the existence of anomalous dimensions can only be accounted for by invoking the presence of a microscopic length scale, such as the lattice spacing, even when the correlation length is very large. By sidestepping these difficulties, the previous chapter showed that experimental data can be correlated by assuming the existence of scaling laws, even though it was recognised that we could justify neither the scaling hypothesis, nor the existence of anomalous dimensions.

In this chapter, we will explain how the scaling hypothesis follows from the presence of a diverging correlation length. The basic argument originates from the insight of L.P. Kadanoff that a diverging correlation length implies that there is a relationship between the coupling constants of an effective Hamiltonian and the length scale over which the order parameter is defined.[1] Kadanoff's ingenious argument is correct in spirit, but not quite right in detail; as we will see, the relationship between coupling constants defined at different length scales is more complicated than assumed.

---

[1] L.P. Kadanoff, *Physics* 2, 263 (1966).

Furthermore, Kadanoff's argument does not enable the critical exponents to be calculated. K.G. Wilson elaborated and completed Kadanoff's argument, showing how the relationship between coupling constants at different length scales could be explicitly computed, at least approximately; Wilson's theory — the **renormalisation group** (RG) — is thus capable of estimating the critical exponents.[2] The RG also provides a natural framework in which to understand **universality**.

9.1 BLOCK SPINS

We begin by describing Kadanoff's basic insight, and showing how it leads to the scaling laws postulated by Widom. We will present the argument in two parts, the first being the thermodynamic scaling laws, and the second being the scaling laws for the two-point correlation function.

9.1.1 Thermodynamics

Consider a system Ω of spins on a d-dimensional hypercubic lattice, with spacing a, and Hamiltonian H_Ω given by

$$\begin{aligned} \beta H_\Omega &= -\beta J \sum_{\langle ij\rangle=1}^{N} S_i S_j - \beta H \sum_i S_i \\ &\equiv -K \sum_{\langle ij\rangle} S_i S_j - h \sum_{i=1}^{N} S_i \end{aligned} \tag{9.1}$$

with

$$\begin{aligned} K &\equiv \beta J; \\ h &= \beta H. \end{aligned} \tag{9.2}$$

Let $f_s(t,h)$ be the singular part of the free energy *per spin* near T_c. Since spins are correlated on lengths of order $\xi(T)$, spins on a length scale ℓa, with $\ell > 1$, act in some sense as a 'single unit' as long as

$$a \ll \ell a \ll \xi(T). \tag{9.3}$$

Thus, we could imagine a coarse-graining procedure, of the form that we invoked in our discussion of Landau theory, in which we replace the spins

---

[2] K.G. Wilson, *Phys. Rev. B* **4**, 3174, 3184 (1971); K.G. Wilson and J. Kogut, *Phys. Rep. C* **12**, 75 (1974).

within a block of side ℓa by a single spin, a **block spin**, which actually contains ℓ^d spins. The total number of blocks, and hence of block spins, is then $N\ell^{-d}$. Now we will examine the consequences of such a coarse-graining procedure, which we will refer to for the moment as a **block spin transformation.**

We define the block spin S_I in block I by

$$S_I \equiv \frac{1}{|\overline{m}_\ell|} \frac{1}{\ell^d} \sum_{i \in I} S_i, \tag{9.4}$$

where the average magnetisation of the block I is

$$\overline{m}_\ell \equiv \frac{1}{\ell^d} \sum_{i \in I} \langle S_i \rangle . \tag{9.5}$$

With this normalisation, the block spins S_I have the same magnitude as the original spins:

$$\langle S_I \rangle = \pm 1. \tag{9.6}$$

Now we make our first significant assumption:

Assumption 1:

Since the original spins interact only with nearest neighbor spins and with the external field, we will make the courageous assumption that the block spins also interact only with nearest neighbor block spins and an effective external field.

Our assumption implies that we should define new coupling constants between the block spins and an effective external field which interacts with the block spins. We will denotes these respectively as K_ℓ and h_ℓ, with the subscript ℓ reminding us that in principle, these coupling constants depend upon the definition of the block spins, and thence depend upon ℓ. The coupling constants of the original Hamiltonian correspond to $\ell = 1$: thus, we have the boundary condition

$$K_1 = K; \quad h_1 = h. \tag{9.7}$$

According to assumption 1, the effective Hamiltonian H_ℓ for the block spins is given by

$$-\beta H_\ell = K_\ell \sum_{\langle IJ \rangle}^{N\ell^{-d}} S_I S_J + h_\ell \sum_{I=1}^{N\ell^{-d}} S_I, \tag{9.8}$$

which, by construction, is of the same form as the original Hamiltonian (9.1). The system described by H_ℓ is identical to that described by H_Ω, except that the lattice spacing between the block spins is ℓa, whereas the spacing between the original spins is a; the former system also has fewer spins. Thus, for the block spins, the correlation length *measured in units of the spacing* ℓa *of the block spins*, ξ_ℓ, is *smaller* than the correlation length ξ_1 of the initial system, *measured in units of the spacing* a *between the original spins.* This follows, because the *actual* physical value of the correlation length ξ, as measured in Å, for example, is of course *unchanged* by our grouping of the spins into block spins. Thus

$$\xi = \xi_\ell(\ell a) = \xi_1 a, \tag{9.9}$$

and hence

$$\xi_\ell = \frac{\xi_1}{\ell}. \tag{9.10}$$

Since $\xi_\ell < \xi_1$, the system with Hamiltonian H_ℓ must be further from criticality than the original system! Thus, we conclude that it is at a new effective reduced temperature, t_ℓ.

Similarly, the magnetic field h has been rescaled to an effective field h_ℓ, when measured in the appropriate units:

$$h \sum_i S_i \cong h\overline{m}_\ell \ell^d \sum_I S_I \equiv h_\ell \sum_I S_I, \tag{9.11}$$

which implies a relationship between the average magnetisation of a block and the effective field:

$$h_\ell = h\overline{m}_\ell l^d. \tag{9.12}$$

The effective Hamiltonian H_ℓ is of the same form as the original Hamiltonian (9.1), and thus the *functional form* of the free energy of the block spin system will be of the same form as that of the original system, albeit with t_ℓ and h_ℓ instead of t and h. In terms of the free energies per spin or block spin,

$$N\ell^{-d} f_s(t_\ell, h_\ell) = N f_s(t, h). \tag{9.13}$$

Hence, we find that

$$f_s(t_\ell, h_\ell) = \ell^d f_s(t, h). \tag{9.14}$$

Although this equation describes how the free energy per spin transforms under a block spin transformation, we still do not have any information on how the reduced temperature and external field have changed during the transformation. Thus, we make

Assumption 2:
Since, we seek to understand the power-law and scaling behaviour in the critical region, we assume that

$$t_\ell = t\ell^{y_t} \quad y_t > 0; \tag{9.15}$$
$$h_\ell = h\ell^{y_h} \quad y_h > 0. \tag{9.16}$$

The exponents y_t and y_h are assumed to be positive, but we cannot say anything more about them at this stage. We will see below that in the full RG theory, the circumstances under which assumption 2 is correct are identified, and then it is possible to perform the block spin transformation approximately, and hence y_t and y_h can be estimated. For now, we substitute into eqn. (9.13) to obtain the central result of this section:

$$f_s(t,h) = \ell^{-d} f_s(t\ell^{y_t}, h\ell^{y_h}). \tag{9.17}$$

Note that we have not specified ℓ, and thus we are at liberty to choose ℓ as we please. Thus, we will choose

$$\ell = |t|^{-1/y_t} \tag{9.18}$$

i.e.

$$\ell^{y_t}|t| = 1. \tag{9.19}$$

With this choice

$$f_s(t,h) = |t|^{d/y_t} f(1, h|t|^{-y_h/y_t}). \tag{9.20}$$

Let us define

$$\Delta \equiv \frac{y_h}{y_t} \tag{9.21}$$

and

$$2 - \alpha \equiv \frac{d}{y_t}, \tag{9.22}$$

for reasons to become obvious. Then eqn. (9.20) becomes

$$f_s(t,h) = |t|^{2-\alpha} F_f(h/|t|^\Delta) \tag{9.23}$$

with

$$F_f(x) \equiv f_s(1,x) \tag{9.24}$$

a function which depends only on x. This is just the basic starting point of the static scaling hypothesis, eqn. (8.24).

9.1.2 Correlation Functions

Now consider the correlation function for the block spin Hamiltonian

$$G(\mathbf{r}_\ell, t_\ell) \equiv \langle S_I S_J \rangle - \langle S_I \rangle \langle S_J \rangle, \tag{9.25}$$

where $\mathbf{r}_\ell$ is the displacement between the centres of blocks I and J *in units of* $a\ell$; if $\mathbf{r}$ denotes the displacement between the centres of blocks I and J in units of a, then $\mathbf{r}_\ell = \mathbf{r}/\ell$. In order that the notion of a block spin correlation function be well-defined, we require that the separation between the blocks be much larger than the block size itself: thus, we are concerned only with the long-wavelength limit $r \gg a$. How is $G(\mathbf{r}_\ell, t_\ell)$ related to $G(\mathbf{r}, t)$? Using the definition of S_I, eqn. (9.4) and eqn. (9.10), the average magnetisation in a block becomes

$$\overline{m}_\ell = h_\ell \ell^{-d}/h = \ell^{y_h - d}. \tag{9.26}$$

Thus the correlation function transforms as

$$G(\mathbf{r}_\ell, t_\ell) = \frac{1}{\ell^{2(y_h-d)} \cdot \ell^{2d}} \sum_{i \epsilon I} \sum_{j \epsilon J} [\langle S_i S_j \rangle - \langle S_i \rangle \langle S_j \rangle] \tag{9.27}$$

$$= \frac{1}{\ell^{2(y_h-d)}} \cdot \frac{1}{\ell^{2d}} \cdot \ell^d \cdot \ell^d \cdot [\langle S_i S_j \rangle - \langle S_i \rangle \langle S_j \rangle] \tag{9.28}$$

$$= \ell^{2(d-y_h)} G(\mathbf{r}, t). \tag{9.29}$$

Including the dependence on h, we have

$$G\left(\frac{\mathbf{r}}{\ell}, t\ell^{y_t}, h\ell^{y_h}\right) = \ell^{2(d-y_h)} G(\mathbf{r}, t, h). \tag{9.30}$$

Again, we can choose ℓ as we please, and so we set $\ell = t^{-1/y_t}$ as before, to obtain

$$G(\mathbf{r}, t, h) = t^{2(d-y_h)/y_t} G(\mathbf{r}t^{1/y_t}, ht^{-y_h/y_t}, 1). \tag{9.31}$$

The prefactor $t^{2(d-y_h)/y_t}$ is unfamiliar, but we can easily re-arrange this expression by writing

$$G(rt^{1/y_t}, ht^{-y_h/y_t}, 1) \equiv \left(rt^{1/y_t}\right)^{-2(d-y_h)} F_G\left(rt^{1/y_t}, ht^{-y_h/y_t}\right), \tag{9.32}$$

which defines the scaling function F_G. Substituting into eqn. (9.31), we obtain our final result

$$G(r, t, h) = \frac{1}{r^{2(d-y_h)}} F_G\left(rt^{1/y_t}, ht^{-y_h/y_t}\right). \tag{9.33}$$

This is just the form assumed by scaling, eqn. (8.30). Comparing with the expectations of the scaling hypothesis we can read off

$$\nu = \frac{1}{y_t} \tag{9.34}$$

$$2(d - y_h) = d - 2 + \eta \tag{9.35}$$

$$\Delta = y_h / y_t. \tag{9.36}$$

We see that, in fact, there are only two independent critical components, corresponding to the two variables t, h.

9.1.3 Discussion

Kadanoff's block spin argument successfully motivates the functional form of the scaling relations; but it gives neither the exponents y_t and y_h, nor the scaling functions themselves. It does not address the issue of universality either, and as we have presented it, applies only to the spin half Ising model, although it is clear that a generalisation to other systems is possible. The most crucial step is the assumption that the block spin Hamiltonian is of the same form as the original Hamiltonian. This cannot possibly be correct, as the following counter-example shows.

Suppose that in the original Hamiltonian the spins did not interact at all with their nearest neighbours, but instead interacted only with the next nearest neighbours. After coarse-graining, it would be ludicrous to assume that the block spins also interacted only with next nearest neighbours. Each block spin contains many original spins, and it is difficult to see why the block spins would not interact with their nearest neighbours. This plausibility argument not only shows that the block spin Hamiltonian is not necessarily the same as the original Hamiltonian, but also provides a precursor to the explanation of universality. The block spin Hamiltonian should be more or less the same, regardless of whether or not the original Hamiltonian involved nearest neighbour interactions only or next nearest neighbour interactions only.

The conceptual importance of the argument is that it suggests how fruitful it might be to get away from the conventional statistical mechanical approach of treating all the degrees of freedom at once. This approach was, of course, rooted in the history of statistical mechanics, where the importance of exact calculations of the partition function was emphasised. In progressively thinning out or coarse-graining the degrees of freedom, Kadanoff's argument focuses on the fact the coupling constants vary with a change of length scale. The question of precisely how the coupling constants vary, under *repeated* elimination of short length scales is the crucial one, and is addressed by the work of Wilson, to which we now turn.

9.2 BASIC IDEAS OF THE RENORMALISATION GROUP

The renormalization group (RG) consists of two principal steps, which address two distinct issues. The first step is a concrete realisation of the coarse-graining transformation used in the preceding section, and we will examine carefully the properties of block spin transformations. The second step is to identify the origin of singular behaviour. The crucial idea is this: after a block spin transformation has been performed, the block spins are separated by a distance ℓa. If we now rescale lengths so that in the new units, the block spins are separated by the original distance a between the microscopic spins, then to all intents and purposes, the system looks like the original system, in terms of the degrees of freedom, but with a different Hamiltonian. Repeating this sequence of steps yields a sequence of Hamiltonians, each describing statistical mechanical systems further and further away from criticality, by the argument in the previous section. We will see that while the construction of a block spin transformation is perfectly analytic, non-analyticities can arise after an infinite number of repetitions, in which all the degrees of freedom in the thermodynamic limit have been integrated.

We consider a system described by the Hamiltonian.

$$\mathcal{H} \equiv -\beta H_\Omega = \sum_n K_n \Theta_n\{S\} \tag{9.37}$$

where K_n are the coupling constants, and $\Theta_n\{S\}$ are the local operators which are functionals of the degrees of freedom $\{S\}$. We will, from this point on, use the term Hamiltonian to denote both H and $\mathcal{H}$; the latter quantity is the more convenient, for present purposes.

9.2.1 Properties of Renormalisation Group Transformations

Let us now consider how $\mathcal{H}$ changes under a transformation which coarse-grains the short wavelength degrees of freedom, leaving an effective Hamiltonian for the long wavelength degrees of freedom. Conceptually, this means that we group together degrees of freedom in a block of linear dimension ℓa, although in practice there are more sophisticated ways of achieving this. The block spin transformation in the spin half Ising model was an example of such a transformation, but now we contemplate a corresponding procedure for a general Hamiltonian $\mathcal{H}$. We will call such a transformation a **renormalisation group transformation** R_ℓ, and we will see explicit examples later. The term 'renormalisation group' has historical origins, which will become clearer in the following chapter: all it means, however, is the change or redefinition of the coupling constants under a change of scale, and a recaling of the degrees of freedom.

Suppose that under R_ℓ, the set of coupling constants $\mathbf{K} \equiv [K]$ become

$$[K'] \equiv R_\ell[K] \quad \ell > 1. \tag{9.38}$$

As before, R_ℓ describes how the coupling constants change as the length scale, over which the local operators are defined, is varied. Equation. (9.38) is sometimes referred to as a **recursion relation**. In general, R_ℓ is a very complicated, non-linear transformation. Since $\ell > 1$, there is no inverse transformation. As we discussed at the end of section 9.1.3, we expect that two different forms of Ising model Hamiltonian can give rise to very similar block spin Hamiltonians, and we anticipate that this is a very general feature.

The transformations R_ℓ for different $\ell > 1$ do indeed form a semi-group: two successive transformations with $\ell = \ell_1$ and $\ell = \ell_2$ should be equivalent to a combined scale change of $\ell_1\ell_2$:

$$[K'] = R_{\ell_1}[K] \tag{9.39}$$

$$[K''] = R_{\ell_2}[K'] \tag{9.40}$$

$$= R_{\ell_2} \cdot R_{\ell_1}[K] \tag{9.41}$$

and thus

$$R_{\ell_1\ell_2}[K] = R_{\ell_2} \cdot R_{\ell_1}[K]. \tag{9.42}$$

How do we calculate R_ℓ? There are many different ways, and indeed, there is no unique RG transformation; many different RG transformations may be constructed for a given problem. All involve a coarse-graining of the degrees of freedom, which we now describe formally here. The discussion here is related to that of section 5.6.2.

We begin by writing down the partition function

$$Z_N[K] = \mathrm{Tr}\, e^{\mathcal{H}}, \tag{9.43}$$

and defining a quantity g, related to the free energy per degree of freedom:

$$g[K] \equiv \frac{1}{N} \log Z_N[K]. \tag{9.44}$$

An RG transformation reduces the number of degrees of freedom by a factor ℓ^d, leaving $N' = N/\ell^d$ degrees of freedom, described by 'block variables' $\{S'_I\}$, $I = 1 \ldots N'$, with an effective Hamiltonian $\mathcal{H}'_{N'}$. This is accomplished by making a partial trace over the degrees of freedom $\{S_i\}$ keeping the block degrees of freedom $\{S'_I\}$ fixed:

$$e^{\mathcal{H}'_N\{[K'],S'_I\}} = \text{Tr}\,'_{\{S_i\}} e^{\mathcal{H}_N\{[K],S_i\}} \tag{9.45}$$

$$= \text{Tr}_{\{S_i\}} P(S_i, S'_I) e^{\mathcal{H}_N\{[K],S_i\}} \tag{9.46}$$

where $P(S_i, S'_I)$ is a **projection operator** which allows the trace in eqn. (9.46) to be unrestricted. The projection operator $P(S_i, S'_I)$ is constructed so that the coarse-grained degrees of freedom S'_I have the same range of values as S_i.

Example:

For Ising spins on a square lattice, we could define the following RG transformation using blocks with linear dimension $(2\ell + 1)a$, so that the number of degrees of freedom within each block is odd. Then, we let

$$S'_I = \text{sign}\left(\sum_{i\epsilon I} S_i\right) = \pm 1. \tag{9.47}$$

The associated projection operator is

$$P(S_i, S'_I) = \prod_I \delta\left(S'_I - \text{sign}\left[\sum_{i\epsilon I} S_i\right]\right). \tag{9.48}$$

Clearly, there are many other ways of defining a RG transformation with the property that $S'_I = \pm 1$.

The projection operator $P(S_i, S'_I)$ must satisfy the following three requirements:

$$(i)\ P(S_i, S'_I) \geq 0; \tag{9.49}$$

$$(ii)\ P(S_i, S'_1) \text{ reflects the symmetries of the system;} \tag{9.50}$$

$$(iii)\ \sum_{\{S'_I\}} P(S_i, S'_I) = 1. \tag{9.51}$$

Condition (i) guarantees that $\exp(\mathcal{H}_{N'}\{[K'], S_I\}) \geq 0$, so that we can safely identify $\mathcal{H}_{N'}$ with the effective Hamiltonian for the degrees of freedom S'_I. Condition (ii) implies that $\mathcal{H}'_{N'}$ possesses the symmetries exhibited by the original Hamiltonian $\mathcal{H}_N$. For example, if $\mathcal{H}_N$ has the form

$$\mathcal{H}_N = NK_0 + h\sum_i S_i + K_1 \sum_{ij} S_i S_j + K_2 \sum_{ijk} S_i S_j S_k + \ldots \tag{9.52}$$

where each multinomial in S_i allowed by symmetry is included in $\mathcal{H}$, with coupling constants that are in general non-zero, then the transformed or renormalised Hamiltonian

$$\mathcal{H}'_{N'} = N'K'_0 + h'\sum_I S'_I + K'_1 \sum_{IJ} S'_I S'_J + K'_2 \sum_{IJK} S'_I S'_J S'_K + \ldots \quad (9.53)$$

has the *same* form, but with transformed, or renormalised coupling constants.[3] If a certain coupling constant K_m happened to be zero in $\mathcal{H}$, then in general, K'_m will be non-zero. A spin model with infinite range coupling between pairs of spins, but no three spin couplings is indeed of the form of eqn. (9.52) with $K_2 = 0$, but after transformation, three spin interactions may be present with $K'_2 \neq 0$. Condition (*iii*) guarantees that

$$\begin{aligned} Z_{N'}[K'] &\equiv \mathrm{Tr}_{\{S'_I\}} e^{\mathcal{H}'_{N'}\{[K'],S'_I\}} & (9.54)\\ &= \mathrm{Tr}_{\{S'_I\}} \mathrm{Tr}_{\{S_I\}} P(S_i, S'_I) e^{\mathcal{H}_N\{[K],S_i\}} & (9.55)\\ &= \mathrm{Tr}_{\{S_i\}} e^{\mathcal{H}_N\{[K],S_i\}} \cdot 1 & (9.56)\\ &= Z_N[K]. & (9.57) \end{aligned}$$

Thus, the partition function is invariant under a RG transformation. What about the 'free energy' g?

$$\begin{aligned} \frac{1}{N} \log Z_N[K] &= \frac{\ell^d}{\ell^d N} \log Z_{N'}[K'] & (9.58)\\ &= \ell^{-d} \frac{1}{N'} \log Z_{N'}[K'], & (9.59) \end{aligned}$$

and thus

$$g[K] = \ell^{-d} g[K']. \quad (9.60)$$

The general formal framework above encapsulates the spirit of the Kadanoff block spin transformation, but differs from it in that at the outset we allow for the possibility that new local operators are generated during the RG transformation. So far, the formalism is exact, but not obviously useful. The only virtue of the presentation that emerges so far is that although the calculation of the $[K']$ as functions of $[K]$ is

---

[3] Note that we have explicitly kept track of the spin independent term K_0 in the Hamiltonian. The recursion relations for K_n ($n \geq 1$) do not involve K_0, but this term must be kept to compute correctly the free energy. Thus, K_0 is often neglected in discussions, because it does not affect critical exponents. See exercise 9–3 for an explicit example, and the discussion by D.R. Nelson, *Phys. Rev. B* 11, 3504 (1975).

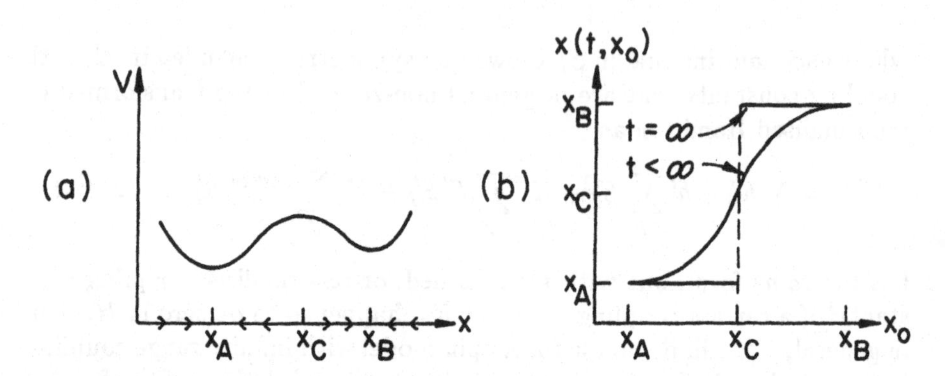

Figure 9.1 (a) The potential $V(x)$. The arrows on the x-axis indicate the direction of motion of the particle as a function of x. (b) Position of the particle after time t as a function of initial position, for finite and infinite times.

in general very difficult, the functions should be *analytic*, because only a finite number of degrees of freedom have been integrated over in the RG transformation. The usefulness of the RG approach derives from the fact that it is thus considerably easier to approximate the $[K']$ than the partition function itself. This remark brings us now to the second part of the RG technique: the origin of singular behaviour.

9.2.2 The Origin of Singular Behaviour

An infinite number of **iterations** of the RG transformations is required in order to eliminate all the degrees of freedom of a thermodynamic system in the thermodynamic limit $N \to \infty$. In this way, singular behavior can occur.

To see this, consider the following simple example of how singular behavior can arise from an analytic transformation. A particle moving in a one-dimensional potential $V(x)$ is subject to damping sufficiently strong that inertia can be neglected. The particle's position $X(t)$ is determined by the equation

$$\frac{dX}{dt} = -V'(X), \tag{9.61}$$

where the units have been chosen so that the coefficient of friction is unity. Suppose that $V(x)$ is as drawn in figure (9.1), and the particle is released from any point $x < x_C$. The particle will roll to x_A and stop. If the particle is released from any point $x > x_C$, it rolls to x_B and stops. Thus, the final position of the particle is a *discontinuous* function of the initial position, x_0. Note that $X(t, x_0)$, the position at time t after release at x_0, is a continuous function of x_0 for finite values of t, but a discontinuous function of x_0 when $t = \infty$. The potential $V(x)$ is perfectly analytic, so the singular behavior is not due to pathologies of $V(x)$. Instead, the origin of the singular behavior is the amplification of the initial condition due to the infinite time limit.

The points x_A, x_B and x_C are **fixed points** of eqn. (9.61). If the particle is at a fixed point at some time t', then it remains there for $t > t'$. We also see that fixed points come in two varieties, repulsive and attractive, with the following properties respectively. If the particle starts of near the fixed point at x_C, it will always end up either at x_A or at x_B, but never at x_C. On the other hand, if the particle starts off near either x_A or x_B, it will end up at that fixed point.

The set of initial conditions $\{x_0\}$ which flows to a given fixed point is called the **basin of attraction** of that fixed point. In the simple example here, the basin of attraction of x_B is $x > x_C$, whilst the basin of attraction of x_A is $x < x_C$. The basin of attraction of x_C is $x = x_C$.

This dynamical system example suggests that singular behavior can arise after an infinite number of RG transformations. The analogue with a dynamical system is quite faithful: after n RG iterations, the coarse-graining length scale is ℓ^n, and the system is described by the coupling constants $K_0^{(n)}$, $K_1^{(n)}$, ... As n varies, the system may be thought of as represented by a point moving in a space whose axes are the coupling constants K_0, K_1, ... On iterating the RG transformation, a given system represented by its initial set of coupling constants, traces out a trajectory in coupling constant space. The set of all such trajectories, generated by different initial sets of coupling constants generates a **renormalisation group flow** in coupling constant space. Although it is possible, in principle, for the trajectory of the representative point to trace out limit cycles, strange attractors, *etc.*, in practice, it is almost always found that the trajectory becomes attracted to fixed points.[4] As we will see, scaling behaviour is invariably associated with the dynamics near a particular sort of fixed point, and the nature of the fixed points and the flows of the representative point in coupling constant space provide important information, allowing the phase diagram of the system to be determined.

---

[4] A discussion of these possibilities is given by D.J. Wallace and R.K.P. Zia, *Ann. Phys. (N.Y.)* **92**, 142 (1975). Exotic trajectories have been reported in some calculations on disordered systems: see A. Weinrib and B.I. Halperin, *Phys. Rev. B* **27**, 413 (1983), and references therein.

9.3 FIXED POINTS

The crucial ingredient of the RG method is the recognition of the importance and physical significance of **fixed points** of the RG transformation. In this section, we will develop these ideas in general terms.

9.3.1 Physical Significance of Fixed Points

Let us suppose that we know the RG transformation $R_\ell[K]$. Then the fixed point of the RG transformation is a point $[K^*]$ in coupling constant space satisfying

$$[K^*] = R_\ell[K^*]. \tag{9.62}$$

Now, under the RG transformation R_ℓ, length scales are reduced by a factor ℓ, as we have discussed. For any particular values of the coupling constants, we can compute the correlation length ξ, which transforms under R_ℓ according to the rule

$$\xi[K'] = \xi[K]/\ell, \tag{9.63}$$

indicating that the system moves further from criticality after a RG transformation has been performed. At a fixed point,

$$\xi[K^*] = \xi[K^*]/\ell, \tag{9.64}$$

which implies that $\xi[K^*]$ can only be zero or infinity.

We will refer to a fixed point with $\xi = \infty$ as a *critical fixed point*, and a fixed point with $\xi = 0$ as a *"trivial" fixed point*. In general, a RG transformation will have several fixed points. Each fixed point has its own basin of attraction or domain: all points in coupling constant space which lie within the basin of attraction of a given fixed point flow towards and ultimately reach the fixed point, after an infinite number of iterations of R_ℓ.

Theorem:

All points in the basin of attraction of a critical fixed point have infinite correlation length.

Proof:

Suppose that we start with a physical system represented by the point in coupling constant space $[K]$. After n iterations of the RG transformation R_ℓ, the system is now at a representative point denoted by $[K^{(n)}]$. Using eqn. (9.63), we have the sequence of identities

$$\xi[K] = \ell\xi[K^{(1)}] = \ell^2\xi[K^{(2)}] = \cdots = \ell^N\xi[K^{(N)}] \tag{9.65}$$

for any N. Taking the limit $N \to \infty$, the right hand side of eqn. (9.65) becomes infinity if $\xi[K^*] = \infty$; *i.e.*, if $[K]$ was in the basin of attraction of a critical fixed point.

Q.E.D.

This set of points — the basin of attraction of a critical fixed point — is often called the **critical manifold**. The fact that *all* points on the critical manifold flow towards the *same* fixed point is the basic mechanism for universality, but is by no means the complete explanation. Universality, after all, involves behaviour exhibited by systems close to, but not at the critical point, and we have so far not said anything about this case. To complete our account, we will need to examine the behaviour off the critical manifold.

We will shortly see that the critical fixed points describe the singular critical behaviour, whereas the trivial fixed points describe the bulk phases of the system. Knowledge of the location and nature of the fixed points of a RG transformation thus enables the phase diagram to be determined, whilst the behaviour of the RG flows near a critical fixed point determines the critical exponents.

Although we have implicitly assumed that the fixed points are isolated points, this is not necessarily the case. It is possible to have lines and surfaces of fixed points, and later, we will classify fixed points according to their **codimension**.

9.3.2 Local Behavior of RG Flows Near a Fixed Point

What can we learn from the behaviour of the flows near a fixed point? Let

$$K_n = K_n^* + \delta K_n, \tag{9.66}$$

so that the starting Hamiltonian is close to the **fixed point Hamiltonian** *i.e.* the Hamiltonian with the coupling constants equal to their fixed point values: $\mathcal{H} = \mathcal{H}[K^*] \equiv \mathcal{H}^*$. Let $\mathcal{H} = \mathcal{H}^* + \delta\mathcal{H}$. Now perform a RG transformation: $[K'] = R_\ell[K]$. Then

$$K_n' = K_n'[K] \equiv K_n^* + \delta K_n', \tag{9.67}$$

with $\delta K_N'$ given by Taylor's theorem:

$$K_n'\{K_1^*+\delta K_1, K_2^*+\delta K_2, \ldots\} = K_n^* + \sum_m \left.\frac{\partial K_n'}{\partial K_m}\right|_{K_m=K_m^*} \cdot \delta K_m + O\left((\delta K)^2\right), \tag{9.68}$$

so that

$$\delta K'_m = \sum_m M_{nm} \delta K_m, \tag{9.69}$$

where

$$M_{nm} \equiv \left. \frac{\partial K'_n}{\partial K_m} \right|_{K=K^*}. \tag{9.70}$$

is the **linearised RG transformation** in the vicinity of the fixed point K^*. The matrix $\mathbf{M}$ is real, but in general it is not symmetric, and we shall have to distinguish between **left eigenvectors** and **right eigenvectors**. Consequently, in general $\mathbf{M}$ is not diagonalisable, and the eigenvalues are not necessarily real. In practice, however, the situation is often more pleasant, and $\mathbf{M}$ is diagonalisable with real eigenvalues.[5] We shall develop the basic ideas of RG assuming that $\mathbf{M}$ is symmetric, for simplicity. At the end, we shall mention the minor modifications that arise for non-symmetric matrices.

Let us now study the RG flows near the fixed point, using the linearised RG transformation $\mathbf{M}^{(\ell)}$, where the superscript ℓ denotes the scale factor involved in the RG transformation R_ℓ. We denote the eigenvalues and eigenvectors by $\Lambda_\ell^{(\sigma)}$ and $e_n^{(\sigma)}$ respectively, where σ labels the eigenvalues and the subscript n labels the component of the vector $\mathbf{e}$. Using Einstein summation convention, we have

$$M_{nm}^{(\ell)} e_m^{(\sigma)} = \Lambda^{(\sigma)} e_n^{(\sigma)}. \tag{9.71}$$

The semi-group property of eqn. (9.42) implies that

$$\mathbf{M}^{(\ell)} \mathbf{M}^{(\ell')} = \mathbf{M}^{(\ell\ell')} \tag{9.72}$$

and thus

$$\Lambda_\ell^{(\sigma)} \Lambda_{\ell'}^{(\sigma)} = \Lambda_{\ell\ell'}^{(\sigma)}. \tag{9.73}$$

One way to solve the functional equation (9.73) is to note that setting $\ell' = 1$ gives $\Lambda_1^{(\sigma)} = 1$. Hence, differentiating eqn. (9.73) with respect to ℓ', setting $\ell' = 1$, and solving the resultant differential equation for $\Lambda_\ell^{(\sigma)}$ gives

$$\Lambda_{(\ell)}^{(\sigma)} = \ell^{y_\sigma}, \tag{9.74}$$

with y_σ being a number to be determined, but independent of ℓ.

---

[5] Certain RG calculations in disordered systems have found complex eigenvalues which may correspond to non-trivial flows. See footnote 4.

How does $[\delta K]$ transform under M? We expand $[\delta K]$ in terms of the eigenvectors of M, and then see how the components of $[\delta K]$ grow or shrink in the eigen-directions.

$$\delta\mathbf{K} = \sum_\sigma a^{(\sigma)}\mathbf{e}^{(\sigma)}, \tag{9.75}$$

writing $[K]$ as a vector $\mathbf{K} = (K_1, K_2, \ldots)$. The coefficients $a^{(\sigma)}$ are obtained from the assumed orthonormality of the eigenvectors:

$$a^{(\sigma)} = \mathbf{e}^{(\sigma)} \cdot \delta\mathbf{K}. \tag{9.76}$$

Note that the orthonormality property does not generally hold when the matrix M is not symmetric.

When we apply the linearised RG transformation M, we find that

$$\delta\mathbf{K}' = \mathbf{M}\delta\mathbf{K} \tag{9.77}$$

$$= \mathbf{M}\sum_\sigma a^{(\sigma)}\mathbf{e}^{(\sigma)} \tag{9.78}$$

$$= \sum_\sigma a^{(\sigma)}\Lambda^{(\sigma)}\mathbf{e}^{(\sigma)} \equiv \sum_\sigma a^{(\sigma)\prime}\mathbf{e}^{(\sigma)}, \tag{9.79}$$

thus defining $a^{(\sigma)\prime}$ as the projection of $\delta\mathbf{K}'$ in the direction $\mathbf{e}^{(\sigma)}$. This equation is very important. It tells us that some components of $\delta\mathbf{K}$ grow under $M^{(\ell)}$ whilst others shrink. If we order the eigenvalues by their absolute value,

$$|\Lambda_1| \geq |\Lambda_2| \geq |\Lambda_3| \ldots \tag{9.80}$$

then we can distinguish three cases:

(1) $|\Lambda^{(\sigma)}| > 1$ *i.e.* $y^\sigma > 0$, which implies that $a^{(\sigma)\prime}$ grows as ℓ increases.
(2) $|\Lambda^{(\sigma)}| < 1$ *i.e.* $y^\sigma < 0$, which implies that $a^{(\sigma)\prime}$ shrinks as ℓ increases.
(3) $|\Lambda^{(\sigma)}| = 1$ *i.e.* $y^\sigma = 0$, which implies that $a^{(\sigma)\prime}$ does not change as ℓ increases.

The significance of these three cases is that after *many* iterations of $\mathbf{M}^{(\ell)}$, only components of $\delta\mathbf{K}$ along directions $\mathbf{e}^{(\sigma)}$ for which case (i) holds, will be important. The projections of $\delta\mathbf{K}$ along the other directions will either shrink or stay fixed at some finite value. The three cases above are given the following terminology:

(1) $\rightarrow$ **relevant** eigenvalues/directions/eigenvectors.
(2) $\rightarrow$ **irrelevant** eigenvalues/directions/eigenvectors.
(3) $\rightarrow$ **marginal** eigenvalues/directions/eigenvectors.

The significance of these distinctions is that if we start at $\mathbf{K}$ near $\mathbf{K}^*$, but *not* on the critical manifold then the flows away from $\mathbf{K}^*$ *i.e.* in directions out of the critical manifold in the vicinity of $\mathbf{K}^*$, are associated with relevant eigenvalues. The irrelevant eigenvalues correspond to directions of flow into the fixed point. The eigenvectors corresponding to the irrelevant eigenvalues span the critical manifold.

The marginal eigenvalues turn out to be associated with logarithmic corrections to scaling, and are important at the upper and lower critical dimensions. The number of relevant eigenvalues must thus be the codimension c of the critical manifold, *i.e.*, the difference between the dimensionalities of the coupling constant space and the critical manifold.

It is very important to remember that the terms relevant, irrelevant and marginal are *always* to be specified with respect to a particular fixed point. A particular term in the Hamiltonian may be relevant at one fixed point, but not at another.

9.3.3 Global Properties of RG Flows

We will soon see that the local behaviour near critical fixed points determines the critical behaviour. The global behaviour of RG flows, however, determines the phase diagram of the system. The basic idea is simple: starting from any point in coupling constant space (*i.e.* in the phase diagram), iterate the RG transformation and identify the fixed point to which the system flows. The state of the system described by this fixed point represents the phase at the original point in the phase diagram. We shall see an explicit example of this later; however, it is useful first to classify the types of fixed points, and to elaborate on the notion of universality. In this subsection, we will describe the global properties of flows in a qualitative fashion. The specific phenomena that we point out will be illustrated later when we explicitly perform a renormalisation group calculation. The purpose of this subsection is to provide the reader with some perspective, before embarking on technical details.

Table (9.1) shows a classification of fixed points by their codimension. Those with codimension 0 have no relevant directions, and therefore trajectories only flow into them, giving rise to the name **sink**. The sinks correspond to stable bulk phases, and the nature of the coupling constants at the sink characterise the phase. For example, a simple three dimensional Ising magnet with nearest neighbour ferromagnetic coupling

Table 9.1 CLASSIFICATION OF FIXED POINTS

Codimension	Value of ξ	Type of Fixed Point	Physical Domain
0	0	Sink	Bulk phase
1	0	Discontinuity FP	Plane of coexistence
1	0	Continuity FP	Bulk phase
2	0	Triple point	Triple Point
2	∞	Critical FP	Critical manifold
Greater than 2	∞	Multicritical point	Multicritical point
Greater than 2	0	Multiple coexistence FP	Multiple coexistence

in an external field H has sinks at $H = \pm\infty$, $T = 0$, corresponding to the fact that in a positive (negative) external field, there is a net positive (negative) magnetisation for all temperatures. Starting at any point in the phase diagram (H,T), successive RG iterations will drive the system to the sink corresponding to the appropriate sign of H. We will see an example of this later.

There are two sorts of fixed point with codimension one: **discontinuity** and **continuity** fixed points. The former correspond to points on a phase boundary and describe a first order phase transition where an order parameter exhibits discontinuous behaviour.[6] An example is the line $H = 0, T < T_c$ in the ferromagnet: all points on that line flow to a discontinuity fixed point at zero temperature and field. A continuity fixed point represents a phase of the system, but nothing interesting happens in its vicinity. An example is the paramagnetic fixed point at $H = 0$, $T = \infty$, which attracts points on the line $H = 0$, $T > T_c$. Both of these codimension one fixed points are unstable towards the sinks: an infinitesimal external field will cause the RG flows to approach the sinks, not the codimension one fixed points. In section 9.6, we calculate these flows explicitly.

Fixed points with codimension greater than or equal to two describe either points of multiple phase coexistence or multicritical points, depending upon the value of the correlation length ξ. The simplest case, codimension two, corresponds to either a triple point ($\xi = 0$) or a critical point ($\xi = \infty$). In each case, *a useful way to interpret the presence of two*

[6] The interpretation of renormalisation group transformations in the vicinity of a first order transition is somewhat delicate, and the picture presented here is somewhat simplified. For a careful discussion see the article by A.C.D. van Enter, R. Fernández and A.D. Sokal, *Phys. Rev. Lett.* **66**, 3253 (1991) and references therein.

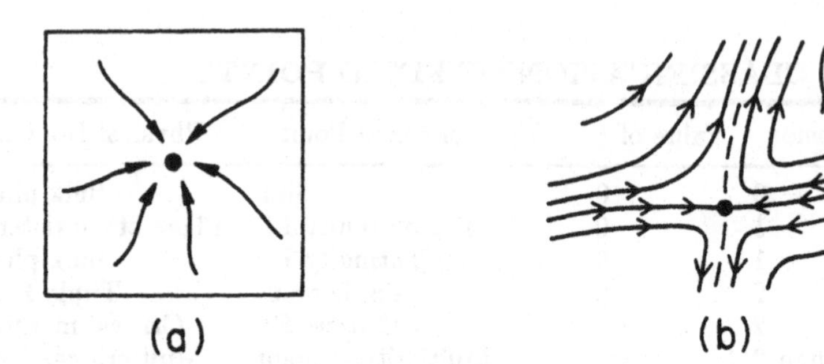

Figure 9.2 Renormalisation group flows near a critical fixed point: (a) View of flows on the critical manifold. (b) View of flows off the critical manifold.

relevant directions is that these represent the two variables that must be tuned in order to place the system at the appropriate point. For example, in order to hold a magnetic system at the critical point, it is necessary to adjust the external field to be zero and the temperature to be the critical temperature.

What happens to a system close to criticality? As depicted in figure (9.2), the trajectories of systems on the critical manifold remain on the manifold and flow to the fixed point. Trajectories which start slightly off the critical manifold initially flow towards the critical fixed point (we expect this to be the case because the only singularities of the flow field are the fixed points themselves), but ultimately are repelled from the critical manifold, because the critical fixed point has two *unstable directions i.e.* the two relevant directions. The fact that it is the *same eigenvalues* which drive *all* slightly off-critical systems away from the fixed point is the origin of universality. Thus, the initial values of the coupling constants do not determine the critical behavior. Only the flow behavior near the fixed point controls the critical behavior.

To illustrate this, consider the flow diagram for an Ising model with nearest neighbour coupling constant $K_1 = J_1/k_BT$, *next nearest neighbour* coupling constant $K_2 = J_2/k_BT$ in an external field $h = H/K_BT$:

$$\mathcal{H} = K_1 \sum_{<ij>} S_iS_j + K_2 \sum_{ij=n.n.n.} S_iS_j + h\sum_i S_i. \tag{9.81}$$

The flow diagram in the $h = 0$ plane is shown in figure (9.3). The arrows indicate the directions of the flows under successive RG transformations. All systems with Hamiltonians of the form of eqn. (9.81) (and many others too!) exhibit critical behaviour governed by the critical fixed point

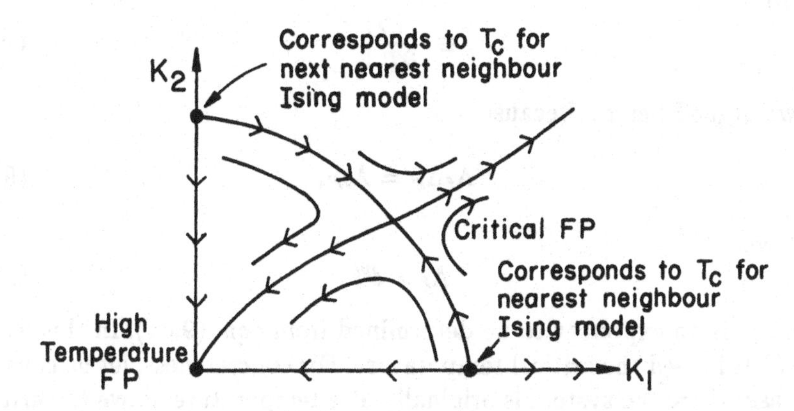

Figure 9.3 Flow diagram for an Ising model with nearest and next nearest neighbour interactions.

shown. Note the following additional features of the diagram. The critical manifold includes the critical fixed point and intersects the K_1 and K_2 axes at the critical value of those couplings K_1^c and K_2^c. The significance of this is that the system described by the Hamiltonian of eqn. (9.81) with $[K] = (K_1^c, 0)$, for example, will undergo a phase transition at a critical temperature $T_c = J_1/k_B K_1^c$, with critical exponents determined by the critical fixed point shown, which has a non-zero value for K_2. Furthermore a system described by a next nearest neighbour Hamiltonian (with $K_1 = 0$) is in the same universality class as the nearest neighbour Ising model. The critical manifold has indeed two unstable directions: one which flows towards the low or high temperature codimension one fixed points, and one out of the $h = 0$ plane towards the sinks.

9.4 ORIGIN OF SCALING

9.4.1 One Relevant Variable

Now that we have some qualitative feeling for RG flows, let us see quantitatively how the RG accounts for scaling behaviour. We start with a simple example, namely a system with only one coupling constant, which can be taken to be temperature (or equivalently $K = J/k_B T$). Then under an RG transformation R_ℓ, T is transformed to $T' = R_\ell(T)$. At a fixed point, $T^* = R_\ell(T^*)$. Linearising in the vicinity of the fixed point, we have

$$T' - T^* = R_\ell(T) - R_\ell(T^*) \tag{9.82}$$

$$\simeq \Lambda_\ell(T - T^*) + O((T - T^*)^2) \tag{9.83}$$

where

$$\Lambda_\ell \equiv \left.\frac{\partial R_\ell}{\partial T}\right|_{T=T^*}. \tag{9.84}$$

As we argued before, because

$$\Lambda_\ell \Lambda_{\ell'} = \Lambda_{\ell\ell'}, \tag{9.85}$$

we have

$$\Lambda_\ell = \ell^{y_t}, \tag{9.86}$$

where y_t is an exponent to be determined from eqn. (9.84). In this example, T^* is indeed the critical temperature. For concreteness, let us consider the case where the system is originally at a temperature above the critical temperature. Defining

$$t = \frac{T - T^*}{T^*}, \tag{9.87}$$

the recursion relation (9.82) becomes

$$t' = t\,\ell^{y_t}, \tag{9.88}$$

which is why we wrote the exponent in eqn. (9.86) as y_t. Now iterate the RG transformation n times, to give

$$t^{(n)} = (\ell^{y_t})^n t. \tag{9.89}$$

This describes how t changes under an n-fold change of scale by a factor ℓ.

In order to make contact with the critical exponents, consider how the correlation length transforms. After one RG transformation, $\xi' = \xi/\ell$, and thus, after n transformations

$$\xi(t) = \ell^n \xi(t^{(n)}). \tag{9.90}$$

But $t^{(n)}$ is given by eqn. (9.89). Hence,

$$\xi(t) = \ell^n \xi(t\,\ell^{n y_t}). \tag{9.91}$$

Although we have proceeded as if ℓ were an integer (to aid our intuition based on block spins), in fact this is not necessary. Let us therefore allow ℓ to be arbitrary and choose it to satisfy

$$\ell^n = (b/t)^{1/y_t}, \tag{9.92}$$

with b being some arbitrary positive number much larger than unity. Thus

$$\xi(t) = \left(b^{-1}t\right)^{-1/y_t} \xi(b) \quad \text{as } t \to 0. \tag{9.93}$$

Note that $\xi(b)$ is the correlation length for temperatures well above T_c, where fluctuations are small, and standard approximation methods, such as perturbation theory work well. Comparing eqn. (9.93) with the definition of the critical exponent ν: $\xi \sim t^{-\nu}$, we read off

$$\nu = \frac{1}{y_t}. \tag{9.94}$$

This is our central result. The exponent y_t is simply given by eqns. (9.86) and (9.84):

$$y_t = \frac{1}{\ell} \log \Lambda_\ell = \frac{1}{\ell} \log \left[\left. \frac{\partial R_\ell}{\partial T} \right|_{T^*} \right]. \tag{9.95}$$

Thus, knowledge of R_ℓ, or of a good approximation to it, enables us to calculate Λ_ℓ, y_t and hence ν. This is the basic idea of the renormalisation group.

We can use a similar calculation to find how the free energy density transforms under the renormalisation group. After one RG transformation, we have

$$f(t) = \ell^{-d} f(t'). \tag{9.96}$$

Iterating n times, and choosing ℓ as before, we obtain

$$f(t) = \left(\frac{t}{b}\right)^{d/y_t} f(b). \tag{9.97}$$

Differentiating twice with respect to t, and comparing with the definition of the specific heat exponent $c \sim t^{-\alpha}$ leads to the result

$$\frac{d}{y_t} = 2 - \alpha. \tag{9.98}$$

Combining with eqn. (9.94), we recover the Josephson scaling law

$$2 - \alpha = \nu d. \tag{9.99}$$

9.4.2 Diagonal RG Transformation for Two Relevant Variables

Let us now consider a slightly more representative example, with two relevant eigenvalues corresponding to t and h. Then we have for the singular part of the free energy density

$$f(t,h) = \ell^{-d} f(t',h'), \tag{9.100}$$

where T and H transform into

$$T' = R_\ell^T(T,H); \tag{9.101}$$
$$H' = R_\ell^H(T,H). \tag{9.102}$$

R_ℓ^T and R_ℓ^H are functions to be determined by the chosen coarse-graining procedure. Let us consider the neighbourhood of the fixed point $T^*_;\ H^*$ given by

$$T^* = R_\ell^T(T^*,H^*); \tag{9.103}$$
$$H^* = R_\ell^T(T^*,H^*). \tag{9.104}$$

We expect that $H^* = 0$ for magnetic systems like the Ising model. Linearising about T^*, H^* in the variables

$$\Delta T = T - T^* \tag{9.105}$$
$$\Delta H = H - H^* \tag{9.106}$$

we obtain the linearised RG transformation

$$\begin{pmatrix} \Delta T' \\ \Delta H' \end{pmatrix} = \mathrm{M} \begin{pmatrix} \Delta T \\ \Delta H \end{pmatrix}, \tag{9.107}$$

with

$$\mathrm{M} = \begin{pmatrix} \partial R_\ell^T/\partial T & \partial R_\ell^T/\partial H \\ \partial R_\ell^H/\partial T & \partial R_\ell^H/\partial H \end{pmatrix}_{\substack{T=T^* \\ H=H^*}}. \tag{9.108}$$

The eigenvectors of M will in general be linear combinations of ΔT and ΔH. In many cases, M is diagonal, and t and h are not mixed. To ease the presentation, we will assume this to be the case, for the time being. Then, writing the eigenvalues of M as

$$\Lambda_\ell^t = \ell^{y_t}; \tag{9.109}$$
$$\Lambda_\ell^h = \ell^{y_h}, \tag{9.110}$$

the RG transformation becomes

$$\begin{pmatrix} t' \\ h' \end{pmatrix} = \begin{pmatrix} \Lambda_\ell^t & 0 \\ 0 & \Lambda_\ell^h \end{pmatrix} \begin{pmatrix} t \\ h \end{pmatrix}. \tag{9.111}$$

After iterating n times, the correlation length transforms as

$$\xi(t,h) = \ell^n \xi(\ell^{ny_t} t, \ell^{ny_h} h). \tag{9.112}$$

It we choose $h = 0$, then we recover the same result as before:

$$\xi \sim t^{-\nu}, \quad \nu = 1/y_t. \tag{9.113}$$

Note that we can also set $t = 0$ and see how ξ diverges as $h \to 0$:

$$\xi(0,h) = \ell^n \xi(0, \ell^{ny_h} h) \tag{9.114}$$

$$\sim h^{-1/y_h} \quad \text{as } h \to 0. \tag{9.115}$$

We will shortly see the physical significance of the exponent y_h. The singular part of the free energy density transforms according to

$$\begin{aligned} f(t,h) &= \ell^{-d} f(t',h') = \ell^{-nd} f(t^{(n)}, h^{(n)}) \\ &= \ell^{-nd} f(\ell^{ny_t} t, \ell^{ny_h} h), \end{aligned} \tag{9.116}$$

which the reader will recognise to be of the form conjectured in the Kadanoff block spin argument, eqn. (9.17). Choosing $\ell^n = bt^{-1/y_t}$, we obtain

$$f(t,h) = t^{d/y_t} b^{-d} f(b, h/t^{y_h/y_t}), \tag{9.117}$$

which is precisely the scaling form (9.20), with

$$2 - \alpha = d\nu = \frac{d}{y_t};$$

$$\Delta = y_h/y_t. \tag{9.118}$$

This is a very important result. Not only have we succeeded in deriving the static scaling hypothesis from the RG, but we also have a way to *calculate* the exponents y_t and y_h, at least approximately, from the RG recursion relations.

9.4.3 Irrelevant Variables

Let us now examine the effect of irrelevant variables, which we neglected in writing down eqn. (9.100). Under an RG transformation, the free energy density transforms according to

$$f(t,h,\tilde{K}_3,\tilde{K}_4,\ldots) = \ell^{-d} f(h\Lambda_\ell^h, \tilde{K}_3\Lambda_\ell^3, \tilde{K}_4\Lambda_\ell^4,\ldots)$$

where

$$\Lambda_\ell^t,\, \Lambda_\ell^h > 1; \tag{9.119}$$

$$\Lambda_\ell^3, \Lambda_\ell^4, \ldots < 1, \tag{9.120}$$

and the irrelevant variables $\tilde{K}_3, \tilde{K}_4, \ldots$ are linear combinations of the original coupling constants, which diagonalise M. Then

$$f(t,h,\tilde{K}_3,\tilde{K}_4,\ldots) = \ell^{-d} f(t\ell^{y_t}, h\ell^{y_h}, \tilde{K}_3\ell^{y_3}, \tilde{K}_4\ell^{y_4}\ldots), \tag{9.121}$$

where the exponents y_3, y_4, *etc.* are negative, in accordance with eqn. (9.120). After n iterations,

$$f(t,h,\tilde{K}_3,\tilde{K}_4,\ldots) = \ell^{-nd} f(t\ell^{ny_t}, h\ell^{ny_h}, \tilde{K}_3\ell^{ny_3}, \tilde{K}_4\ell^{ny_4},\ldots), \tag{9.122}$$

leading to the scaling form

$$f(t,h,\tilde{K}_3,\tilde{K}_4,\ldots) = t^{d/y_t} b^{-d} f(b, ht^{-y_h/y_t}, \tilde{K}_3 t^{-y_3/y_t}, \tilde{K}_4 t^{-y_4/y_t},\ldots). \tag{9.123}$$

Now, as $t \to 0$, the terms in the irrelevant variables become vanishingly small: $\tilde{K}_3 t^{-y_3/y_t} \to 0$ *etc.*, and we obtain

$$f(t,h,\tilde{K}_3,\tilde{K}_4,\ldots) = t^{d/y_t} f(b, ht^{-y_h/y_t}, 0, 0, \ldots). \tag{9.124}$$

In passing from eqn. (9.123) to eqn. (9.124), we have *assumed* that (*e.g.*) the limit $\tilde{K}_3 \to 0$ of $f(t,h,\tilde{K}_3,\ldots)$ is analytic. In fact, this assumption is frequently false! Even, in the simplest situations, described by Landau theory for a scalar order parameter (*i.e.* the Ising universality class), the failure of this assumption has important ramifications, as shown in exercise 9–2 at the end of this chapter. When the free energy density is singular in the limit that a particular renormalised irrelevant variable vanishes, that irrelevant variable is termed a **dangerous irrelevant variable.** The parameter b multiplying the quartic term in the Landau free energy of eqn. (5.55) is a dangerous irrelevant variable.

9.4.4 Non-diagonal RG Transformations

Now that we have explained the basic notions of the RG, let us mention how one proceeds in the general case where the linearised RG transformation $\mathbf{M}$ is not a symmetric matrix. In this case, it is necessary to distinguish between two sets of eigenvectors, **left eigenvectors** and **right eigenvectors.** Right eigenvectors $\mathbf{e}_R$ of a matrix $\mathbf{M}$ satisfy

$$\mathbf{M} \cdot \mathbf{e}_R = \Lambda_R \mathbf{e}_R, \tag{9.125}$$

where Λ_R is the right eigenvalue. Left eigenvectors $\mathbf{e}_L$ and left eigenvalues Λ_L satisfy

$$\mathbf{e}_L \cdot \mathbf{M} = \Lambda_L \mathbf{e}_L. \tag{9.126}$$

In this equation, $\mathbf{e}_L$ is a row vector, multiplying the matrix $\mathbf{M}$ from the left. Taking the transpose of eqn. (9.126), we obtain

$$\mathbf{M}^T \cdot \mathbf{e}_L^T = \Lambda_L \mathbf{e}_L^T. \tag{9.127}$$

Comparing with eqn. (9.125), and using the fact that the determinant of a matrix is the determinant of its transpose, we see that, in fact, the left and right eigenvalues are equal.

The most important result for present purposes is the orthogonality of left and right eigenvectors, which we now demonstrate. Let σ label the eigenvectors. Then, multiplying eqn. (9.125) on the left by $\mathbf{e}_L^{\sigma'}$, we have

$$\mathbf{e}_L^{\sigma'} \cdot \mathbf{M} \cdot \mathbf{e}_R^{\sigma} = \Lambda^{\sigma} \mathbf{e}_L^{\sigma'} \cdot \mathbf{e}_R^{\sigma}. \tag{9.128}$$

Similarly, multiplying eqn. (9.126) on the right by $\mathbf{e}_R^{\sigma}$, we obtain

$$\mathbf{e}_L^{\sigma'} \cdot \mathbf{M} \cdot \mathbf{e}_R^{\sigma} = \Lambda^{\sigma'} \mathbf{e}_L^{\sigma'} \cdot \mathbf{e}_R^{\sigma}. \tag{9.129}$$

Subtracting we find that

$$\mathbf{e}_L^{\sigma'} \cdot \mathbf{e}_R^{\sigma} \left(\Lambda^{\sigma} - \Lambda^{\sigma'} \right) = 0. \tag{9.130}$$

Thus, if the eigenvalues are nondegenerate, $\mathbf{e}_L^{\sigma'}$ is orthogonal to $\mathbf{e}_R^{\sigma}$, for $\sigma \neq \sigma'$, and these eigenvectors can therefore be chosen to be orthonormal. If some eigenvalues are degenerate, then a Gram-Schmidt orthogonalisation must first be performed between the appropriate left or right eigenvectors; however, in this case, it is then possible that the resulting left or right eigenvectors may not span the original space.

For present purposes, these results require that in general, we replace eqn. (9.103) by

$$\delta \mathbf{K} = \sum_{\sigma} a^{(\sigma)} \mathbf{e}_R^{(\sigma)} \tag{9.131}$$

and replace eqn. (9.76) by

$$a^{(\sigma)} = \mathbf{e}_L^{(\sigma)} \cdot \delta \mathbf{K}. \tag{9.132}$$

The subsequent development of the theory is essentially unchanged, and may be carried through in a straightforward manner.

9.5 RG IN DIFFERENTIAL FORM

In this section, we present the RG in a differential form that is very convenient to use in practice, and which avoids various technical complications that can arise in the discrete RG.[7]

Suppose that we start with a system whose lattice spacing is a, and that we renormalise out to blocks of size ℓa. Now construct blocks of size $s\ell a$, with $1 \leq s < \infty$; then

$$[K]_{s\ell} = R_s[K_\ell]. \tag{9.133}$$

The **differential RG transformation** is obtained by choosing $s = 1+\epsilon$:

$$\frac{d[K_\ell]}{dl} = \lim_{\epsilon \to 0} \frac{[K]_{(1+\epsilon)\ell} - [K]_\ell}{\epsilon \ell}. \tag{9.134}$$

The right hand side can also be written as

$$\frac{1}{\ell} \frac{\partial R_s[K_\ell]}{\partial s}\bigg|_{s=\ell} \equiv \frac{1}{\ell} B[K_\ell], \tag{9.135}$$

defining the non-linear transformation $B[K_\ell]$. Thus, the RG recursion relations are sometimes written in a form to emphasise the analogy with dynamical systems:

$$\frac{d[K_\tau]}{d\tau} = B[K_\tau], \tag{9.136}$$

with the time-like variable $\tau \equiv \log \ell$. The fixed points are then the solutions of

$$B[K^*] = 0. \tag{9.137}$$

---

[7] See, for example, M. Nauenberg, *J. Phys. A.* **8**, 925 (1975); D.J. Wallace and R.K.P. Zia, *Rep. Prog. Phys.* **41**, 1 (1978), section 3.2.1.1.

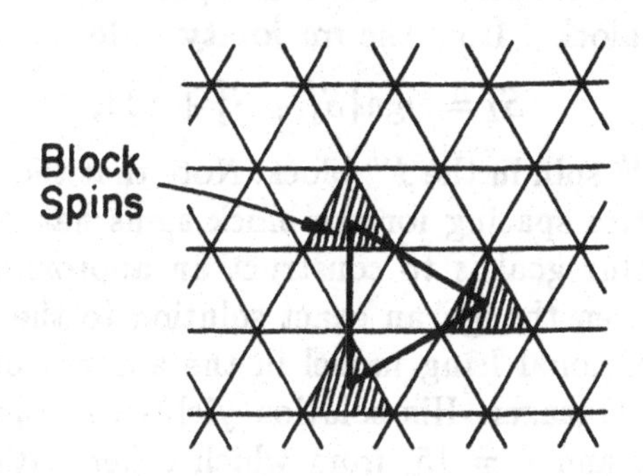

Figure 9.4 Block spin construction for the Ising model on a triangular lattice.

Note that eqn. (9.136) is a coupled set of non-linear ordinary differential equations. We can integrate these equations out to any desired length scale. As one does so, the correlation lengths shrinks, and eventually the system is driven well away from the critical fixed point to a regime where the correlation length is small — *i.e.* $\xi(b)$ in eqn. (9.93) — and the free energy or other quantities may be computed by perturbation theory. Then one can match the solutions of eqn. (9.136) onto the solutions of ordinary perturbation theory outside the critical region.

9.6 RG FOR THE TWO DIMENSIONAL ISING MODEL

Now it is time to put the ideas above into practice. We will study a realisation of the RG which is closest in spirit to the original Kadanoff block spin approach, and which involves only simple computations.[8] We will only present the crudest approximation to the RG transformation itself: more refined and systematic real space procedures are possible, but rapidly become very technical.[9]

We consider the Ising model on a **triangular lattice**, with Hamiltonian

$$\mathcal{H} = K \sum_{\langle ij \rangle} S_i S_j + h \sum_i S_i. \tag{9.138}$$

[8] The approach presented here is due to Th. Niemeijer and J.M.J. van Leeuwen, *Physica* **71**, 17 (1974); see also their chapter in *Phase Transitions and Critical Phenomena*, vol. 6, C. Domb and M.S. Green (eds.) (Academic, New York, 1976).

[9] A useful overview is to be found in *Real Space Renormalization*, T.W. Burkhardt and J.M.J. van Leeuwen (eds.) (Springer-Verlag, New York, 1982).

Block spins are defined by grouping together the spins at the vertices of each triangle of the lattice, as shown in figure (9.4), and computing the block spin S_I in block I from the **majority rule**

$$S_I = \text{sign}\{S_1^I + S_2^I + S_3^I\}, \tag{9.139}$$

where S_j^I is the jth spin in the Ith block. Note that the S_I are normalized, and that the lattice spacing for the block spins has been enlarged by a factor $\ell = \sqrt{3}$. Our goal is to construct an approximate RG with this transformation, even though an exact solution to the partition function for the two dimensional Ising model in the absence of an external field was obtained by Onsager. His solution yields the values of the critical exponents $\alpha = 0$ and $\delta = 15$, from which other critical exponents can be calculated, using scaling laws. We will calculate the critical exponents approximately, but to facilitate the comparison between the RG and the exact result, let us first calculate the *exact* values of the eigenvalues of the RG transformation.

9.6.1 *Exact Calculation of Eigenvalues from Onsager's Solution*

Since $2 - \alpha = \nu d$ and $d = 2$, we expect $\nu = 1$. Thus, the exponent

$$y_t = 1, \tag{9.140}$$

and the eigenvalue Λ_t should have the value

$$\Lambda^t = \ell = \sqrt{3}. \tag{9.141}$$

We can work out what Λ^h should be from eqn. (9.35), which read

$$2(d - y_h) = d - 2 + \eta. \tag{9.142}$$

Using the result proved in the exercises at the end of this chapter that

$$\delta = \frac{d + 2 - \eta}{d - 2 + \eta} \tag{9.143}$$

we find

$$\eta = 1/4 \tag{9.144}$$

and thus

$$y_h = \frac{d + 2 - \eta}{2} = \frac{4 - 1/4}{2} = 15/8. \tag{9.145}$$

Hence,

$$\Lambda^h = \left(\sqrt{3}\right)^{15/8} \cong 2.80. \tag{9.146}$$

Note that Λ^h and Λ^t *do* depend on the block size ℓ, but the critical exponents are *independent* of ℓ.

9.6.2 Formal Representation of the Coarse-grained Hamiltonian

The next step is to write down a formally exact representation of the coarse-grained Hamiltonian $\mathcal{H}$. To ease the notation, we will write σ_I to denote the set of spins which make up the block spin S_I. Thus

$$\sigma_I \equiv \{S_1^I, S_2^I, S_3^I\}. \tag{9.147}$$

Each value of the block spin S_I may arise from 4 configurations of the original spins:

$$\begin{aligned} S_I = +1 \quad \{\sigma_I\} &= \downarrow\uparrow\uparrow \\ & \uparrow\downarrow\uparrow \\ & \uparrow\uparrow\downarrow \\ & \uparrow\uparrow\uparrow; \end{aligned} \tag{9.148}$$

$$\begin{aligned} S_I = -1 \quad \{\sigma_I\} &= \uparrow\downarrow\downarrow \\ & \downarrow\uparrow\downarrow \\ & \downarrow\downarrow\uparrow \\ & \downarrow\downarrow\downarrow\,. \end{aligned} \tag{9.149}$$

The coarse-graining has preserved the total number of degrees of freedom: originally there were $2^3 = 8$ configurations of the spins per plaquette of the lattice, whereas after coarse-graining, for each plaquette there are two values of the block spin, each arising from four configurations of the original spins. Thus in total, there are $2 \times 4 = 8$ configurations per plaquette. The coarse-grained or effective Hamiltonian is given by

$$e^{\mathcal{H}'\{S_I\}} = \sum_{\{\sigma_I\}} e^{\mathcal{H}\{S_I,\sigma_I\}} \tag{9.150}$$

We will estimate $\mathcal{H}'$ using perturbation theory. We begin with the case $h = 0$. The Hamiltonian $\mathcal{H}$ is conveniently split into two parts: interactions between spins within one block spin, and interactions between spins in different block spins. We write

$$\mathcal{H} = \mathcal{H}_0 + V \tag{9.151}$$

where the interactions within a block spin are given by

$$\mathcal{H}_0 = K \sum_I \sum_{i,j\in I} S_i S_j, \tag{9.152}$$

and the interactions between spins in different blocks is

$$V = K \sum_{I \neq J} \sum_{i \in I \, j \in J} S_i S_j. \tag{9.153}$$

$\mathcal{H}_0$ will serve as our zeroth order, unperturbed Hamiltonian, whilst V will be considered to be a perturbation. Define the average of a quantity A with respect to $\mathcal{H}_0$ by

$$\langle A(S_i) \rangle_0 \equiv \frac{\sum_{\{\sigma_I\}} e^{\mathcal{H}_0\{S_I, \sigma_I\}} A(S_I, \sigma_I)}{\sum_{\{\sigma_I\}} e^{\mathcal{H}_0\{S_I, \sigma_I\}}}. \tag{9.154}$$

Then eqn. (9.150) becomes

$$e^{\mathcal{H}'\{S_I\}} = \left\langle e^V \right\rangle_0 \sum_{\{\sigma_I\}} e^{\mathcal{H}_0(S_I, \sigma_I)}. \tag{9.155}$$

If M is the total number of blocks in the system, then

$$\sum_{\{\sigma_I\}} e^{\mathcal{H}_0\{S_I, \sigma_I\}} = Z_0(K)^M, \tag{9.156}$$

where $Z_0(K)$ is the partition function for *one* block, subject to a given value of S_I:

$$Z_0(K) = \sum_{S_1 S_2 S_3} \exp\left\{K\left(S_1^I S_2^I + S_2^I S_3^I + S_3^I S_1^I\right)\right\}. \tag{9.157}$$

Using the spin configurations of eqns. (9.148) and (9.149), we find that $Z_0(K)$ is independent of S_I and has the value

$$Z_0(K) = 3e^{-K} + e^{3K}. \tag{9.158}$$

In summary, we have reduced the problem of calculating $\mathcal{H}'$ to that of evaluating

$$e^{\mathcal{H}'\{S_I\}} = \left\langle e^V \right\rangle_0 Z_0(K)^M. \tag{9.159}$$

9.6.3 Perturbation Theory for the RG Recursion Relation

How do we compute $\left\langle e^V \right\rangle_0$? A useful technique is to use the **cumulant expansion.** We write

$$\left\langle e^V \right\rangle_0 = \left\langle 1 + V + \frac{V^2}{2} + \ldots \right\rangle_0 \tag{9.160}$$

$$= 1 + \langle V \rangle_0 + \frac{\langle V^2 \rangle_0}{2} + \ldots \tag{9.161}$$

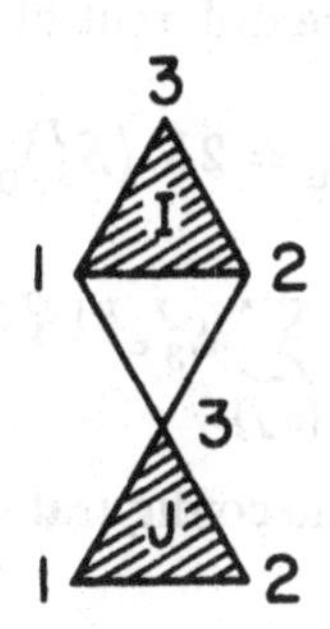

Figure 9.5 Interaction between nearest neighbour block spins to $O(V)$.

and recall that V is a perturbation, and is therefore to be considered small, in some sense. Then, using

$$\log(1+x) = x - x^2/2 + O(x^3), \tag{9.162}$$

we have

$$\log \langle e^V \rangle_0 = \langle V \rangle_0 + \frac{1}{2} \langle V^2 \rangle_0 - \frac{\langle V \rangle_0^2}{2} + O(V^3). \tag{9.163}$$

Re-exponentiating, we finally obtain

$$\langle e^V \rangle_0 = \exp \left\{ \langle V \rangle_0 + \frac{1}{2}[\langle V^2 \rangle_0 - \langle V \rangle_0^2] + O(V^3) \right\}. \tag{9.164}$$

The terms in the exponent on the RHS are called *cumulants*, and the expansion of eqn. (9.164) may be obtained in a more systematic way. Thus

$$\mathcal{H}'\{S_I\} = M \log Z_0(K) + \langle V \rangle_0 + \frac{1}{2}[\langle V^2 \rangle_0 - \langle V \rangle_0^2] + O(V^3) \tag{9.165}$$

The term $M \log Z_0$ is clearly regular, being the partition function for 3 spins and so does not contribute to the singular behavior.

The term $\langle V \rangle_0$ couples nearest neighbor blocks. Writing

$$V = \sum_{I \neq J} V_{IJ}, \tag{9.166}$$

and referring to figure (9.5), we have

$$V_{IJ} = K(S_3^J)(S_1^I + S_2^I). \tag{9.167}$$

Thus

$$\langle V_J \rangle_0 = 2K \left\langle S_3^J S_1^I \right\rangle_0 . \tag{9.168}$$

Since $\mathcal{H}_0$ itself does not couple different blocks, the average in eqn. (9.168) factorises:

$$\langle V_{IJ} \rangle_0 = 2K \left\langle S_3^J \right\rangle_0 \left\langle S_1^J \right\rangle_0 . \tag{9.169}$$

Now

$$\left\langle S_3^J \right\rangle_0 = \frac{1}{Z_0} \sum_{\{\sigma_J\}} S_3^J e^{K[S_1^J S_2^J + S_2^J S_3^J + S_3^J S_1^J]}. \tag{9.170}$$

We evaluate this for each of the configurations of eqns. (9.148) and (9.149). For $S_J = 1$, we find that

$$\left\langle S_3^J \right\rangle_0 = \frac{e^{3K} + e^{-K}}{e^{3K} + 3e^{-K}}, \tag{9.171}$$

whereas for $S_J = -1$, we find that

$$\left\langle S_3^J \right\rangle_0 = - \frac{e^{3K} + e^{-K}}{e^{3K} + 3e^{-K}}. \tag{9.172}$$

Hence

$$\left\langle S_3^J \right\rangle_0 = S_J \left[\frac{e^{-K} + e^{3K}}{e^{3K} + 3e^{-K}} \right]. \tag{9.173}$$

Similarly

$$\left\langle S_1^I \right\rangle_0 = S_I \left[\frac{e^{-K} + e^{3K}}{e^{3K} + 3e^{-K}} \right], \tag{9.174}$$

and thus

$$\langle V \rangle_0 = 2K \Phi(K)^2 \sum_{\langle IJ \rangle} S_I S_J; \tag{9.175}$$

$$\Phi(K) \equiv \frac{e^{3K} + e^{-K}}{e^{3K} + 3e^{-K}}. \tag{9.176}$$

In summary, the effective Hamiltonian is, to first order in V,

$$\mathcal{H}'\{S_I\} = M \log Z_0(K) + K' \sum_{\langle IJ \rangle} S_I S_J + 0(V^2); \tag{9.177}$$

$$K' = 2K \Phi(K)^2. \tag{9.178}$$

This is our goal! We have *calculated* an RG transformation, albeit in a rather uncontrolled way, and derived a crude approximation to the recursion relation for the coupling constant.

9.6.4 *Fixed Points and Critical Exponents*

The next step in the procedure is to find the fixed points of the RG transformation that we have found. The fixed points satisfy

$$K^* = 2K^*\Phi(K^*)^2, \tag{9.179}$$

which gives $K^* = 0, \infty$, or $\Phi(K^*) = 1/\sqrt{2}$. Inverting the latter relation (it helps to use the substitution $x = \exp(4K)$), we find that the non-trivial fixed point, which we will refer to as K_c, is given by

$$K_c = \frac{1}{4}\log(1+2\sqrt{2}) \simeq 0.34. \tag{9.180}$$

The non-trivial fixed point value $K_c \simeq 0.34$ compares reasonably with the exact result of Onsager:

$$K_c = \frac{1}{4}\log 3 = 0.27 \tag{9.181}$$

The eigenvalue

$$\Lambda_t = \left.\frac{\partial K'}{\partial K}\right|_{K_c} = 1.62 \tag{9.182}$$

which is not too far from the exact value $\sqrt{3} \simeq 1.73$ given by eqn. (9.141).

We can improve on this result by going to $O(V^2)$ in the perturbation expansion, requiring us to calculate

$$\langle V^2\rangle_0 - \langle V\rangle_0^2 = K^2\left\langle\sum_{ij}\sum_{mn} S_iS_jS_mS_n\right\rangle - K^2\sum_{ij}\sum_{mn}\langle S_iS_j\rangle_0\langle S_mS_n\rangle_0, \tag{9.183}$$

where spins i and j must be on different blocks from each other and spins m and n must also be on different blocks from each other. Now $\langle S_iS_j\rangle_0 = \langle S_i\rangle_0\langle S_j\rangle_0$ so (9.183) is zero unless either S_i is in the same block as S_m and S_j is in the same block as S_n or S_i is in the same block as S_n and S_j is in the same block as S_m. But since S_i and S_j cannot be in the same block, as cannot be S_m and S_n, we are forced to consider second and third nearest neighbor interactions, as shown in figure (9.6).

The results of this calculation turn out to be

$$\begin{aligned}\Lambda_t &= 1.77\\ \Lambda_2 &= 0.23\\ \Lambda_3 &= -0.12.\end{aligned} \tag{9.184}$$

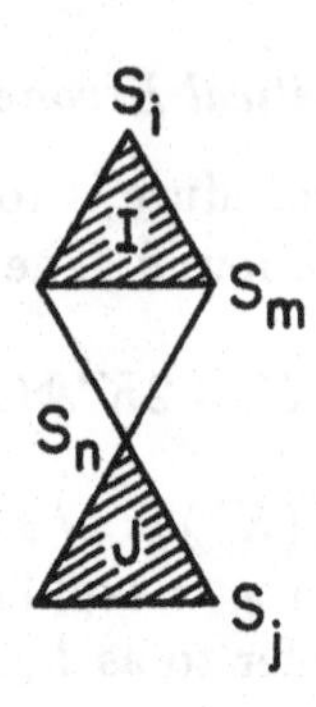

Figure 9.6 Interactions between spins to $O(V^2)$.

The irrelevant eigenvalues with $|\Lambda| < 1$ are associated with second and third nearest neighbor couplings, which are generated under renormalisation, even though absent in the original Hamiltonian. It seems that the results are converging satisfactorily towards $\sqrt{3}$; although systematic extension of the calculations that we have performed do converge to the correct result, the convergence is not uniform.

9.6.5 *Effect of External Field*

Let us now examine the case $h \neq 0$. We can calculate Λ_h by a little trickery, at least in the crudest approximation. We expect $h^* = 0$; how does a small deviation $\delta h = h - h^*$ affect the calculation of $\mathcal{H}'$? Let the change in $\mathcal{H}$ due to a small external field δh be $\delta\mathcal{H}'$. Then, by definition, we have

$$e^{\mathcal{H}'\{S_I\}} = \sum_{\{\sigma_I\}} e^{\mathcal{H}\{S_I,\sigma_I\}} \tag{9.185}$$

$$e^{\mathcal{H}'+\delta\mathcal{H}'} = \sum_{\{\sigma_I\}} e^{\mathcal{H}\{S_I,\sigma_I\}+\delta\mathcal{H}\{S_I,\sigma_I\}}. \tag{9.186}$$

Subtracting and writing $e^x = 1 + x + O(x^2)$ we obtain

$$\delta\mathcal{H}'\{S_I\} = \frac{\sum_{\{\sigma_I\}} e^{\mathcal{H}\{S_I,\sigma_I\}}\delta\mathcal{H}\{S_I,\sigma_I\}}{\sum_{\{\sigma_I\}} e^{\mathcal{H}}}. \tag{9.187}$$

Now, by definition,

$$\delta\mathcal{H}\{S_I,\sigma_I\} = \delta h \sum_i S_i = \delta h \sum_I \sum_{i\epsilon I} S_i^I; \tag{9.188}$$

$$\delta\mathcal{H}' = \delta h' \sum_I S_I. \tag{9.189}$$

But eqn. (9.187) implies that to *zeroth* order in V

$$\delta\mathcal{H}'\{S_I\} = \left\langle \delta h \sum_I \sum_{i\epsilon I} S_i^I \right\rangle_0 \tag{9.190}$$

$$= \delta h \sum_I \langle S_1^I + S_2^I + S_3^I \rangle_0 . \tag{9.191}$$

The right hand side is obtained from eqn. (9.174):

$$\langle S_i^I \rangle_0 = S_I \Phi(K) \quad i = 1, 2, 3. \tag{9.192}$$

Thus, we obtain the recursion relation

$$\delta h' = 3\Phi(K)\delta h, \tag{9.193}$$

which, at the critical fixed point, yields the eigenvalue

$$\Lambda_h = 3\Phi(K_c) = \frac{3}{\sqrt{2}} \simeq 2.12. \tag{9.194}$$

Comparing this zeroth order result with Onsager's exact result $\Lambda_h \simeq 2.8$, we see that the agreement is as good as can be expected. In fact the first order in V calculation for Λ_h gives 3.06, and the second order calculation gives 2.76.

9.6.6 *Phase Diagram*

We can also use RG recursion relations to deduce the phase diagram of the system. Our calculation above did not consistently treat the case of the external magnetic field, but this is straightforward to do, and the reader is invited to perform this calculation in the exercises at the end of this chapter. The resultant flow diagram is sketched in figure (9.7). Some features of the flow diagram are already visible in the calculations that we have done above, however. Along the $h = 0$ axis there are three fixed points, at $K = 0$, $K = K_c$, and $K = \infty$. We have already discussed the local behaviour around the critical fixed point K_c; the fact that $\Lambda_t > 1$ tells us that this fixed point is unstable and that the flows are repelled towards $K = 0$ and $K = \infty$. The line $h = 0$ is sometimes referred to as an **invariant manifold**: if the initial coupling constants are on an invariant manifold, they remain on it after renormalisation. For $h \neq 0$, the flow behaviour is ultimately towards one of the two sinks at $h = \pm\infty$.

What is the nature of the fixed points on the invariant manifold at $K = 0$ and $K = \infty$? The former corresponds to high temperatures: setting

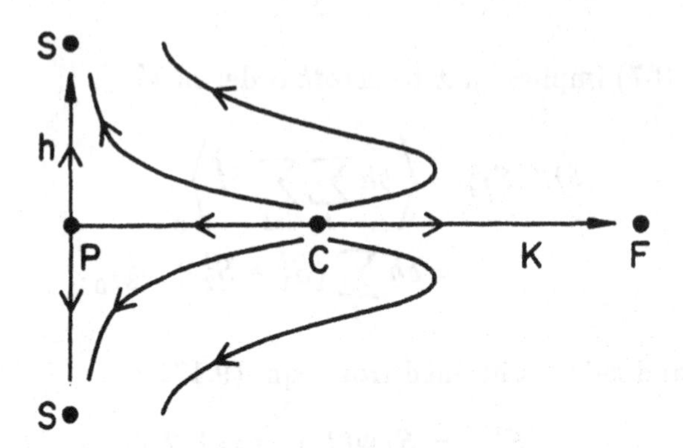

Figure 9.7 Global RG flow diagram for the two dimensional Ising model, according to the calculation in the text. C is a critical fixed point, S are sinks, P is a paramagnetic continuity fixed point, F is a discontinuity fixed point representing the first order transition for $H = 0$, $T < T_c$.

$K = 0$ in the original Hamiltonian shows that at this fixed point, there is no energetic advantage for the spins to align. The probability distribution corresponding to the Hamiltonian $\mathcal{H}^*$ at this fixed point is proportional to $\exp \mathcal{H}^*$, and thus describes a paramagnetic state. At this **paramagnetic fixed point**, the eigenvalues of the linearised RG transformation are, using eqns. (9.178) and (9.193),

$$\Lambda_t = 1/2 \quad \Lambda_h = 3/2. \tag{9.195}$$

On the invariant manifold, the paramagnetic fixed point is stable, but it is unstable in the magnetic field direction. The fixed point at $K = \infty$ is also an attractor on the invariant manifold, and has eigenvalues

$$\Lambda_t = 2 \quad \Lambda_h = 3. \tag{9.196}$$

We refer to this as the **ferromagnetic fixed point**, because the nearest neighbour interaction is so strong there that in the probability distribution for the spin configurations, states with spins all aligned are energetically favourable. The eigenvalue $\Lambda_t = 2$ is apparently a signature of an unstable fixed point, but this is an artifact of the position of the fixed point being at $K = \infty$: near the fixed point at ∞, we can write the recursion relation on the invariant manifold as

$$(K^{-1})' = \frac{1}{2}\,(K^{-1}) \tag{9.197}$$

showing that the fixed point at $K^{-1} = 0$ is actually attractive. Another way to see this is that eqn. (9.196) implies that as $K \to \infty$, K' is always bigger than K, showing that the flow is towards $K = \infty$. The most interesting aspect of this fixed point is the value of the magnetic eigenvalue Λ_h. Recalling that the scale factor of the RG transformation was $\ell = \sqrt{3}$, we see that we could write $\Lambda_h = \ell^d$, where the dimensionality $d = 2$ of course. In the next section, we will show that this value for the exponent signifies a **discontinuity fixed point**, and is the manifestation in the RG of a *first order phase transition* in the basin of attraction of the ferromagnetic fixed point. We will also see that one can calculate the order parameter, *i.e.* the magnetisation, using the RG.

9.6.7 *Remarks*

This example of the RG that we have just worked through shows explicitly the construction of the projection operator and the analyticity of the RG transformation. In general, one must be very careful about preserving the analyticity of the RG transformation.[10] It is also very easy to be drawn to erroneous conclusions by not properly respecting the symmetries of the problem, as the following examples illustrate.

Consider a Heisenberg model with spins that are three component vectors — the so-called $O(3)$ Heisenberg model: $\mathbf{S} = (S_x, S_y, S_z)$, $|S|^2 = 1$, and Hamiltonian

$$\mathcal{H}_{\text{Heis}} = K \sum_{\langle ij \rangle} \mathbf{S}_i \cdot \mathbf{S}_j. \tag{9.198}$$

Suppose that we tried to use the RG transformation

$$S'_I = \text{sign}\left(\sum_{i \in I} S_i^z \right) = \pm 1. \tag{9.199}$$

This would map

$$\mathcal{H}_{\text{Heis}}\{K, \mathbf{S}_i\} \longrightarrow \mathcal{H}_{\text{Ising}}\{K', S'_I\}, \tag{9.200}$$

suggesting that the exponents y_t and y_h of the Heisenberg model are the same as those of the Ising model. This is wrong! The problem is that the RG transformation (9.199) is not analytic: the rotational invariance (*i.e.* $O(3)$ symmetry) of the Heisenberg Hamiltonian is destroyed by this projection.

---

[10] See R.B. Griffiths and P.A. Pearce, *Phys. Rev. Lett.* **41**, 917 (1978); R.B. Griffiths, *Physica* **106A**, 59 (1981); see also the article by van Enter *et al.* referenced under footnote 3.

Another instructive example to consider is the antiferromagnetic Ising model: $K < 0$. A naïve application of (*e.g.*) the real space calculation performed above yields a renormalised nearest neighbour coupling constant K' which is *positive*. Thus, one is led to conclude that the Ising antiferromagnet and ferromagnet are in the same universality class, since they map onto each other under renormalisation. This conclusion is correct in zero field, as we showed in section 2.7.2, but is wrong when $h \neq 0$. The antiferromagnetic state orders at a wavevector $q = \pi/a$, where a is the lattice spacing; however, the renormalisation group transformation constructed above *eliminates* modes on this scale, in favour of the long wavelength $q \approx 0$ modes which are important for the ferromagnetic transition. Thus, the transformation that we used is physically wrong for the antiferromagnetic transition.[11] In summary, it is dangerous to proceed without thinking about the physics!

9.7 FIRST ORDER TRANSITIONS AND NON-CRITICAL PROPERTIES

We have seen how the global behaviour of the RG flows is related to the phase diagram of the system. The RG can also be used to calculate quantitatively physical properties over the entire phase diagram, and in this section we will specifically focus on the order parameter, showing how the line of first order phase transitions for $h = 0$, $T < T_c$ is accounted for by the properties of the low temperature fixed point.[12] The basic idea is that the RG is a set of transformations on $\mathcal{H}$ which drive the system to one of the stable fixed points at low or high temperature. Sufficiently close to these fixed points, it is expected that the usual variety of approximation schemes will work, in contrast to the situation near a critical fixed point.

Consider, for concreteness, how the magnetisation $M[K]$ of the Ising model changes under a RG transformation, starting at a point in coupling constant space $[K^{(0)}]$:

$$M[K^{(0)}] = \left.\frac{\partial g}{\partial h}\right|_{h=0+} \tag{9.201}$$

$$= \ell^{-d}\left.\frac{\partial g[K']}{\partial h}\right|_{h=0+} \tag{9.202}$$

$$= \ell^{-d}\left.\frac{\partial h'}{\partial h}\frac{\partial g[K']}{\partial h'}\right|_{h'=0+} . \tag{9.203}$$

[11] J.M.J. van Leeuwen, *Phys. Rev. Lett.* **34**, 1056 (1975).

[12] B. Nienhuis and M. Nauenberg, *Phys. Rev. Lett.* **35**, 477 (1975).

In the above, $g[K]$ is as defined in eqn. (9.44). Let

$$a[K] \equiv \left.\frac{\partial h'}{\partial h}\right|_{h'=0^+} \tag{9.204}$$

so that

$$M[K^{(0)}] = \frac{a[K^{(0)}]}{\ell^d} M[K']. \tag{9.205}$$

Iterating n times, we obtain

$$M[K^{(0)}] = \prod_{i=0}^{n} \frac{a[K^{(i)}]}{\ell^d} M[K^{(i+1)}], \tag{9.206}$$

where $[K^{(i)}]$ are the coupling constants after i iterations. Suppose we reach a fixed point at $M[K^{(\infty)}]$ after an infinite number of iterations. Then $M[K^{(\infty)}] = 0, -1$ or $+1$, depending upon in which basin of attraction was the initial point $[K^{(0)}]$ in coupling constant space.

The first-order transition for $h = 0, T < T_c$ is manifested as a *discontinuity* $\Delta M[K]$ in the magnetisation across $h = 0$ for $T < T_c$. *i.e.*

$$\Delta M[K] \equiv M(T, 0^+) - M(T, 0^-) \neq 0. \tag{9.207}$$

The necessary condition for $\Delta M[K] \neq 0$ is that $M[K^{(0)}] \neq 0$, and thus

$$\prod_{i=0}^{\infty} \frac{a[K^i]}{\ell^d} \neq 0. \tag{9.208}$$

If $M[K^{(0)}]$ is to be bounded and non-zero, we also require that

$$\lim_{i\to\infty} \frac{a[K^i]}{\ell^d} = 1. \tag{9.209}$$

But $a[K^{(\infty)}]$ is nothing other than $\partial h'/\partial h$ evaluated at the fixed point $[K^{(\infty)}]$, which in this case is the low temperature ferromagnetic fixed point. Thus, at this fixed point we must have

$$\Lambda_h \equiv \frac{\partial h'}{\partial h} = \ell^d. \tag{9.210}$$

This criterion — that one of the eigenvalues of the RG has $\Lambda = \ell^d$ — is sometimes known as the Nienhuis-Nauenberg criterion, and is the manifestation of a first-order or discontinuity fixed point. All other fixed points have $y_\sigma < d$. In the example of the preceding section, we directly found that $\Lambda_h = \ell^d$, showing that the low temperature ferromagnetic fixed point was a discontinuity fixed point associated with the first order transition in the two dimensional Ising model. The formula (9.206) can also be used to calculate the magnetisation at any point in coupling constant space, and in particular shows that for $h = 0$, the magnetisation is zero for $T > T_c$ but non-zero for $T < T_c$. The spontaneous magnetisation as a function of temperature below T_c is in qualitative agreement with the exact result, when computed using the lowest order recursion relations derived earlier in this section, and the agreement can be systematically improved using more elaborate real space RG approximation schemes.

9.8 RG FOR THE CORRELATION FUNCTION

For completeness, we briefly sketch the RG derivation of the scaling form for the two-point correlation function. As we saw in section 5.7.4,

$$G(\mathbf{r}-\mathbf{r}') = \frac{\delta^2 \log Z[h(\mathbf{r})]}{\delta h(\mathbf{r})\delta h(\mathbf{r}')}, \tag{9.211}$$

where $h(\mathbf{r})$ is now a spatially dependent external magnetic field (divided by k_BT) which couples to the magnetisation or order parameter. We assume that $h(\mathbf{r})$ varies slowly over a block size.

Now consider the effect of a RG transformation:

$$Z\{K',h',\ldots\} = Z\{K,h,\ldots\}. \tag{9.212}$$

Differentiating with respect to $h(\mathbf{r})$, we obtain

$$\frac{\delta^2 \log Z\{h'(\mathbf{r})\}}{\delta h(\mathbf{r})\delta h(\mathbf{r}')} = \frac{\delta^2 \log Z\{h(\mathbf{r})\}}{\delta h(\mathbf{r})\delta h(\mathbf{r}')}. \tag{9.213}$$

The fact that the external field is assumed to vary slowly over the scale of a block implies that if $\mathbf{r}$ and $\mathbf{r}'$ in eqn. (9.213) are taken to be the centres of two different blocks of spins, then the derivative simply generates an average over all the spins in these blocks:

$$\frac{\delta^2 \log Z\{h\}}{\delta h(\mathbf{r})\delta h(\mathbf{r}')} = \left\langle \left(\sum_{i\in\mathbf{r}}^{\ell^d} S_i\right)\left(\sum_{i\in\mathbf{r}'}^{\ell^d} S_i\right)\right\rangle - \left\langle \sum_{i\in\mathbf{r}}^{\ell^d} S_i\right\rangle^2 \tag{9.214}$$

$$= \ell^{2d} G(|\mathbf{r}-\mathbf{r}'|, \{\mathcal{H}\}). \tag{9.215}$$

The $\mathcal{H}$ in the equation above simply indicates that the correlation function is evaluated with the coupling constants appropriate to $\mathcal{H}$. Near a fixed point, we can linearise the RG to obtain

$$G\left(\frac{|\mathbf{r}-\mathbf{r}'|}{\ell}, \mathcal{H}'\right) = \frac{\delta^2 \log Z\{h', K'\}}{\delta h'(\mathbf{r})\delta h'(\mathbf{r}')} \tag{9.216}$$

$$= \frac{1}{\Lambda_h^2}\frac{\delta^2 \log Z\{h', K'\}}{\delta h(\mathbf{r})\delta h(\mathbf{r}')} \tag{9.217}$$

$$= \frac{1}{\Lambda_h^2}\ell^{2d} G(|\mathbf{r}-\mathbf{r}'|, \{\mathcal{H}\}). \tag{9.218}$$

Explicitly writing in the dependence of the relevant variables yields

$$G\left(\frac{|\mathbf{r}-\mathbf{r}'|}{\ell}, t\ell^{y_t}, h\ell^{y_h}\right) = \left(\frac{\ell^d}{\ell^{y_h}}\right)^2 G(|\mathbf{r}-\mathbf{r}'|, t, h), \tag{9.219}$$

which is eqn. (9.30), derived from Kadanoff's argument.

9.9 CROSSOVER PHENOMENA

In the remaining sections of this chapter, we discuss some of the experimental consequences of the existence of scaling. In this section, we focus on **crossover phenomena**, the term given to phenomena associated with the failure of a system to attain its asymptotic scaling regime. This can occur for several reasons, the most important of which are:

(*i*) Small residual external fields may be present.

(*ii*) Weak interactions, neglected in writing down $\mathcal{H}$, may break the symmetry of $\mathcal{H}$, and thus generate relevant directions at fixed points which are not associated with the external variables such as temperature, magnetic field *etc.* These relevant variables can drive the system from the neighbourhood of one fixed point to that of another. How do we describe this, and what are the experimental consequences?

(*iii*) The effects of disorder. We will examine under what circumstances disorder can be expected to influence critical behaviour.

9.9.1 *Small Fields*

In their simplest form, crossover phenomena can be thought to be nothing more than the fact that as the external parameters on the system are varied, such as h or t, different asymptotic regimes of the scaling form of the free energy density are encountered. To see this, consider the scaling form for the free energy density

$$f_s = |t|^{2-\alpha} F_\pm(h/|t|^\Delta), \tag{9.220}$$

where, according to eqn. (8.10),

$$F_\pm(0) = \text{constant}. \tag{9.221}$$

How does $F_\pm(x)$ behave as $x \to \infty$? We know that

$$M = -\frac{1}{k_B T}\frac{\partial f_s}{\partial h} \sim t^{2-\alpha-\Delta} F'_\pm(h/|t|^\Delta). \tag{9.222}$$

On the critical isotherm $t = 0$, $M \sim h^{1/\delta}$, by definition; hence, assuming that $F'_\pm(x) \to x^\lambda$ as $x \to \infty$, we can find λ by eliminating the t dependence from eqn. (9.222):

$$M \sim \frac{h^\lambda}{t^{\Delta\lambda}} t^{2-\alpha-\Delta} \sim t^{\beta-\Delta\lambda} h^\lambda \tag{9.223}$$

using eqn. (8.26). Thus, we require that

$$\lambda = \beta/\Delta = 1/\delta, \tag{9.224}$$

and conclude that as $x \to \infty$

$$F_\pm(x) \sim x^{\lambda+1}. \tag{9.225}$$

In an actual experiment on (*e.g.*) magnetic critical phenomena, there is always a small residual magnetic field h_0 present, perhaps due to the Earth or to impurities in the sample. The experimentally relevant question is this: for given values of h and t is the system in the limit $x \to 0$ or in the limit $x \to \infty$? The answer depends on whether or not $h/|t|^\Delta$ is much larger than unity or much less than unity. The crucial observation is that the actual asymptotic regime depends upon how close the system is to criticality.

For example, in figure (9.8), a system is cooled in the field h_0. In the critical regime, but yet not too close to $t = 0$, $h/|t|^\Delta$ is much smaller than unity, and one observes the critical behaviour implied by the $x \to 0$ limit of the scaling function; this is the behaviour one would expect if the external field were *exactly* zero. However, when t is reduced sufficiently that the system is well to the left of the solid line, then $h/|t|^\Delta$ is much larger than unity, and the $x \to \infty$ behaviour of the scaling function is probed. This is,

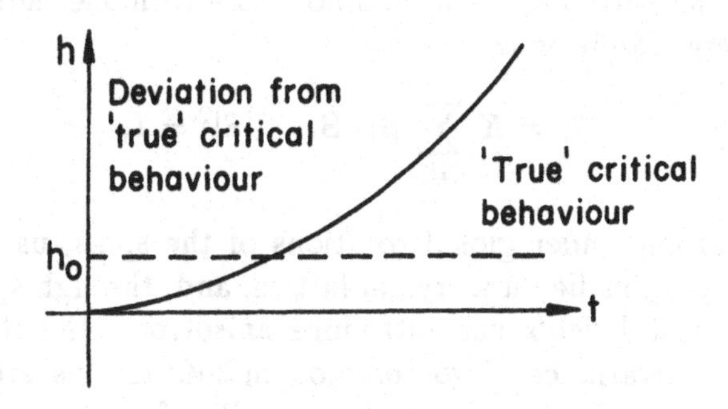

Figure 9.8 Crossover in a magnetic system. The curve represents the points satisfying $h/|t|^\Delta = 1$. Above the curve, the $x \to \infty$ regime of the scaling function is probed, whereas below the curve, the $x \to 0$ regime is probed. An experiment in a small, but non-zero, field h_0 will pass from one regime to the other as t is reduced to zero along the dashed trajectory.

of course, the completely wrong limit if one is interested in determining the critical properties in zero field. Thus, close to the transition, the scaling behaviour *deviates* from the correct critical behaviour if there is a small non-zero field present. It should be clear from the above discussion that the transition between the two regimes is by no means sharp, because in general, one does not know how rapidly the scaling function approaches its limiting behaviours. Thus the line $h/|t|^\Delta = 1$ is just a rough guide to where behaviour intermediate between the two different scaling regimes will be found.

These remarks have important experimental consequences, which augment the difficulties in determining critical exponents, which were pointed out in section 4.6. Within the critical regime we might observe the specific heat c behaving as follows: there exists a reduced temperature t_x, not sharply defined, but where for $t_x < t \ll 1$, $c \sim t^{-\alpha}$, but for $t < t_x$, it is found that there is a cross-over to a different power law $c \sim t^{-\tilde{\alpha}}$, with $\tilde{\alpha}$ *not* being the true α.

9.9.2 Crossover Arising From Anisotropy

A prevalent and therefore important example of crossover is that arising from terms neglected in the original Hamiltonian $\mathcal{H}$. As we have seen, the critical behaviour is governed by fixed points in the space of all possible coupling constants *consistent with the symmetry of the problem.* In reality, our identification of the symmetry of the problem represents an idealisation, and there are often small, and therefore neglected, interactions which can break these idealized symmetries.

For example, although it is commonplace to model ferromagnets by the Heisenberg Hamiltonian

$$\mathcal{H} = K \sum_{<ij>} \mathbf{S}_i \cdot \mathbf{S}_j, \quad |\mathbf{S}|^2 = 1, \tag{9.226}$$

which is invariant under global rotations of the spins, usually it is the case that these spins lie on a crystal lattice, and, through spin-orbit coupling, local crystal fields can introduce anisotropy, thereby destroying the rotational invariance. Two common manifestations are **single ion anisotropy** and **anisotropic exchange**. The former corresponds to a perturbation of the Heisenberg Hamiltonian, such as

$$\mathcal{H} = K \sum_{<ij>} \mathbf{S}_i \cdot \mathbf{S}_j + g \sum_i (S_i^z)^2, \tag{9.227}$$

whereas the latter corresponds to the case when the exchange interactions are different in different lattice directions; a common situation (yet still an idealisation) is **uniaxial anisotropy**, represented by the Hamiltonian

$$\mathcal{H} = K_z \sum_{<ij>} S_i^z S_j^z + K_{xy} \sum_{<ij>} (S_i^x S_j^x + S_i^y S_j^y) \tag{9.228}$$

These modifications turn out to have qualitatively the same effect.[13]

How do we expect the phase diagram to be modified? At high temperature, we still expect a paramagnetic phase. But at low temperatures, the spins order and now the energy of the system depends upon the direction in which the spins align. Let us just consider the case of single ion anisotropy, eqn. (9.227). If $g > 0$, then the spins can lower the physical energy $H = -k_B T \mathcal{H}$ by aligning along the z axis, and at low enough temperatures

$$\langle S_i^z \rangle = \pm 1, \tag{9.229}$$

a state characteristic of the low temperature behaviour of the Ising model. On the other hand, if $g < 0$, then the spins order in the $x-y$ plane; writing $\mathbf{S}_\perp \equiv (S^x, S^y)$, we have at low enough temperatures

$$\left\langle |\mathbf{S}_\perp|^2 \right\rangle = 1, \tag{9.230}$$

---

[13] A detailed discussion is given by A. Aharony in *Phase Transitions and Critical Phenomena*, vol. 6, C. Domb and M.S. Green (eds.) (Academic, New York, 1976).

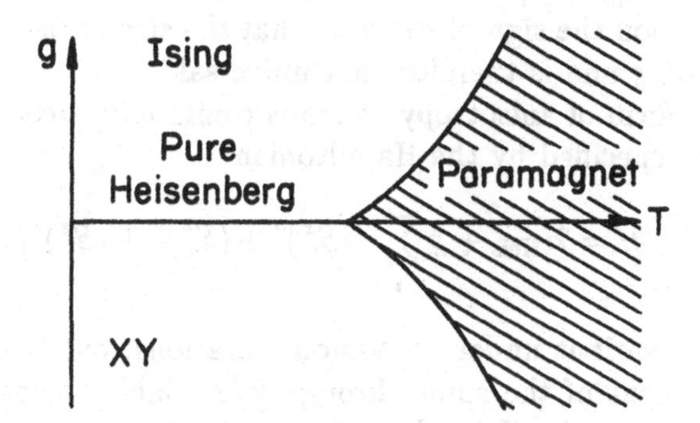

Figure 9.9 Crossover behaviour for single ion anisotropy, showing different regimes of behaviour as the temperature and anisotropy strength g are varied.

which is behaviour characteristic of the XY model.

A consequence of these considerations is that, for *large* g, the phase transition from the paramagnetic phase as T is reduced will be in either the Ising or XY universality class, *depending upon the sign of* g. For $g = 0$ the transition is in the Heisenberg universality class, but for small values of g, the RG flows will end up at the stable low temperature fixed points, namely an Ising fixed point or an XY fixed point. Thus we conclude (heuristically) that the symmetry breaking term must be *relevant* at the Heisenberg fixed point. Small but non-zero values of g will grow under renormalisation, and will carry the system towards a new fixed point, either Ising or XY, depending upon the sign of g. Around each of these fixed points, there will be scaling phenomena different from those of the Heisenberg fixed point. In general, scaling fields which break a symmetry are relevant operators. A high symmetry fixed point is usually (but not always) unstable with respect to a fixed point of lower symmetry.

We can describe this crossover behaviour, in terms of the RG, with two relevant scaling fields near the Heisenberg fixed point. One is t, the other is g, and the singular part of the free energy density will transform as

$$f_s(t,g) = \ell^{-d} f_s(t\ell^{y_t}, g\ell^{y_g}) \tag{9.231}$$

$$= |t|^{2-\alpha} F_{\pm}(g|t|^{-\phi}), \tag{9.232}$$

where the **crossover exponent** ϕ is given by

$$\phi = y_g / y_t > 0. \tag{9.233}$$

The exponent ϕ is positive, because $y_g > 0$. The reader will notice at once the similarity between this description and that of the crossover phenomena in a magnetic field. Thus, we expect to see effective Heisenberg behaviour when $|g|t|^{-\phi}| \ll 1$ and Ising or XY behaviour when $|g|t|^{-\phi}| \gg 1$ (depending upon the sign of g). Note that the size of the crossover region depends upon g and is therefore not universal.

Another form of anisotropy which is commonly encountered is **cubic anisotropy**, specified by the Hamiltonian

$$\mathcal{H} = \mathcal{H}_{\text{Heis}} + g \sum_i \left[(S_i^x)^4 + (S_i^y)^4 + (S_i^z)^4 \right]. \tag{9.234}$$

This is not invariant under continuous rotations, but is invariant under discrete rotations of the spins through $\pi/2$. Cubic anisotropy turns out to be *irrelevant* at the Heisenberg fixed point in $d = 3$, but is *relevant* in $d = 4$. In general, one expects that there exists a whole series of crossovers, determined by arranging the terms of the Hamiltonian in order of *decreasing* symmetry.

In practice, fixed points such as the Ising or Heisenberg fixed point may be unstable not only to symmetry reducing interactions, but also to long-range interactions, such as dipolar interactions.[14]

9.9.3 Crossover and Disorder: the Harris Criterion

Real systems are almost always impure, and it is therefore important to determine whether or not quenched disorder affects the critical behaviour. In the language of the RG, is disorder a relevant variable at the critical fixed point of the pure system? To answer this, we must first quantify the disorder, and then estimate the corresponding eigenvalue of the linearised RG, taking disorder into account. In general, the critical behaviour of disordered systems is very complicated, but there is a simple heuristic criterion, due to Harris,[15] for when the critical behaviour of the disordered system does not differ from that of the pure system.

We consider a system with quenched disorder, such as the presence of impurities at random sites in a crystal lattice, which when pure undergoes a continuous phase transition at a temperature T_c. If the strength of the disorder (*i.e.* the mean impurity concentration in our example) is denoted by p, then in general the critical temperature is $T_c(p)$, and the correlation length diverges as

$$\xi \sim |T - T_c(p)|^{-\nu(p)}. \tag{9.235}$$

---

[14] A map of the different possible crossover phenomena is given by M.E. Fisher, *Rev. Mod. Phys.* 46, 597 (1974).

[15] A.B. Harris, *J. Phys. C* 7, 1671 (1974).

We have assumed that the critical exponent ν in the presence of disorder depends continuously on the strength of the disorder, and that as $p \to 0$, $\nu(p)$ tends smoothly towards its value for the pure system $\nu(0)$. The effect of the disorder may be viewed as changing the local co-ordination number or exchange interaction, and thus, from eqn. (3.139), may be expected to cause the transition temperature to vary from point to point within the impure sample. We model the distribution of

$$\delta T_c(\mathbf{r}) \equiv T_c(\mathbf{r}) - T_c(p) \tag{9.236}$$

throughout the sample as a Gaussian random function, with two-point correlation function

$$W(\mathbf{r} - \mathbf{r}') \equiv \langle \delta T_c(\mathbf{r}) \delta T_c(\mathbf{r}') \rangle \tag{9.237}$$

and mean zero.

Having modelled the disorder, we now attempt to assess its effects on the critical behaviour of the *pure system*. In essence, we are estimating the relevance or irrelevance of disorder *at the critical fixed point of the pure system*, although we may still allow a non-universal quantity such as T_c to have the modified value $T_c(p)$. To this end, consider the fluctuation in $T_c(p)$, averaged over a region of linear dimension L much greater than the lattice spacing a, namely

$$\Delta T_c(p) = \left[\int \frac{d^d\mathbf{r}}{L^d} \int \frac{d^d\mathbf{r}'}{L^d} W(\mathbf{r} - \mathbf{r}') \right]^{1/2} . \tag{9.238}$$

Assuming that $W(\mathbf{r})$ decays faster than $|\mathbf{r}|^{-d}$ for large $|\mathbf{r}|$, then

$$\Delta T_c(p) \sim L^{-d/2}. \tag{9.239}$$

To examine the stability of the pure fixed point to disorder, we require that over a correlation volume, the fluctuations in T_c should be small compared with $|T - T_c(p)|$ as $T \to T_c(p)$ in order that the transition be well-defined. At the pure fixed point, $\xi \sim |T - T_c(p)|^{-\nu}$, so that the criterion for the well-defined transition becomes

$$|T - T_c(p)|^{\nu d/2} \ll |T - T_c(p)| \tag{9.240}$$

as $|T - T_c(p)| \to 0$. Thus we require for self-consistency that $\nu d/2 > 1$, which is equivalent to

$$\alpha < 0, \tag{9.241}$$

using the Josephson scaling relation (8.3). This result is known as the **Harris criterion.**

The physical argument given above is another example of crossover; as we have seen, the disorder introduces a new variable into the thermodynamic description of the system, which we may take to be $w \equiv \Delta T_c(p)$. This is a "temperature-like" variable, and therefore has associated with it the eigenvalue $y_w = y_t$. Hence, the crossover exponent associated with the disorder is $\phi_w = y_w/y_t = 1$. The crossover from the pure fixed point to the disorder fixed point (if it exists) occurs approximately when $w|t|^{-\phi_w} \geq 1$. By virtue of eqn. (9.239), $w \sim \xi^{-d/2}$, so that the crossover point is when $|t|^{\nu d/2-1} \geq 1$. If $\nu d/2 - 1 > 0$, then there is never a crossover from the pure fixed point *i.e.* the pure fixed point is stable if $\nu > 2/d$, as we obtained before.

What happens when the pure fixed point is indeed unstable? Detailed calculations[16] show that the disorder fixed point is a critical fixed point with new critical exponents; in particular, the value of α at the disorder fixed point is never positive.[17]

9.10 CORRECTIONS TO SCALING

In practice, it is very difficult to access the asymptotic critical regime $t \to 0$. Irrelevant variables which can be ignored in the $t \to 0$ limit may not therefore be negligible in practice. As an example, consider the susceptibility

$$\chi_T(t,h) = |t|^{-\gamma} F_\chi^{\pm}\left(\frac{h}{t^\Delta}, \tilde{K}_3 t^{-y_3/y_t}, \ldots\right) \tag{9.242}$$

where $\tilde{K}_3$ is an irrelevant scaling field and $y_3 < 0$. In zero field ($h = 0$), we expect $F_x^{\pm}$ to be an analytic function of the irrelevant variables $\tilde{K}_3, \ldots$ and so we expand $F_x^{\pm}$ for small values of its arguments:[18]

$$\chi_T(t,0) = |t|^{-\gamma}\left(A_\pm + B_\pm \tilde{K}_3 |t|^{-y_3/y_t} + \ldots\right), \tag{9.243}$$

---

[16] See the review by T.C. Lubensky in *Ill-Condensed Matter*, R. Balian, R. Maynard and G. Toulouse (eds.) (North-Holland, Amsterdam, 1979); for an account of the effects of disorder with long-range correlations, see the article by A. Weinrib and B.I. Halperin, *Phys. Rev. B* **27**, 413 (1983).

[17] A rigorous argument, applicable to many situations of practical interest, is given by J.T. Chayes, L. Chayes, D.S. Fisher and T. Spencer, *Phys. Rev. Lett.* **57**, 2999 (1986).

[18] This is valid as long as the variables in question are not dangerous irrelevant variables.

where A and B are non-universal constants. As expected, the leading behavior as $t \to 0$ is $|t|^{-\gamma}$, but there is a first correction of order $|t|^{-y_3/y_t}$. If $|y_3|/y_t > 1$, then this correction becomes smaller as $t \to 0$, whereas if $|y_3|/y_t < 1$, the correction may not be negligible for small but non-zero t. Most importantly, when $|y_3|/y_t < 1$, the first order correction is actually *singular*, giving rise to a cusp at $t = 0$: to appreciate this, it is instructive to sketch $|t|^n$ for two different values of n, one larger than unity, the other smaller than unity. Due to this behaviour, such a correction is often referred to as a **confluent singularity**.

For example, in the Ising universality class, the leading correction term due to irrelevant variables is

$$y_3/y_t \simeq -0.5 \tag{9.244}$$

and this correction to scaling is indeed observed in numerical calculations and experiments involving the superfluid transition, and must be taken into account when attempting to extract critical exponents from data.[19]

9.11 FINITE SIZE SCALING

We have seen in chapter 2 that strictly speaking, there are no phase transitions in a finite system at non-zero temperature. Experiments on real systems, as well as numerical calculations, using either transfer matrix techniques or **Monte Carlo simulation**[20] all use finite systems. How does the failure of a finite size system to exhibit a phase transition manifest itself in the RG, and what can one learn about phase transitions from studies on finite systems? The answers to these questions form the topic of **finite size scaling.** Although first hypothesised before the advent of the RG, finite size scaling is conceptually clearer within the framework of the RG, and that is how we will present this subject here.

Consider a system with linear dimension L and volume $V = L^d$. The singular part of the free energy density scales like

$$f_s([K], L^{-1}) = \ell^{-d} f_s([K'], \ell L^{-1}). \tag{9.245}$$

Here, the free energy density is written as a function not only of the coupling constants, but also of the inverse size of the system. The last

[19] See, for example, the article by G. Ahlers in *Phase Transitions*, Proceedings of the Cargèse Summer School 1980 (Plenum, New York, 1982), p. 1.

[20] A pedagogical introduction to this important technique is given by K. Binder and D.W. Heermann, *Monte Carlo Simulation in Statistical Physics* (Springer-Verlag, Berlin, 1988).

argument on the RHS of eqn. (9.245) comes from the fact that lengths are reduced by a factor ℓ during a renormalisation group transformation R_ℓ. The RG transformation is a local transformation, and therefore it does not matter if it is performed on an infinite system or a finite system.

Close to a fixed point of the RG, we can write eqn. (9.245) in terms of the right eigenvectors of the linearised RG, as before:

$$f_s(t,h,\tilde{K}_3,\ldots,L^{-1}) = \ell^{-d} f_s(t\ell^{y_t}, h\ell^{y_h}, \tilde{K}_3\ell^{y_3},\ldots,\ell L^{-1}). \tag{9.246}$$

We see that L^{-1} behaves like a relevant eigenvector with eigenvalue

$$\Lambda_L = \ell, \tag{9.247}$$

and thus

$$y_L = 1. \tag{9.248}$$

This is in accord with our interpretation of relevant variables as those parameters that must be adjusted by the experimenter in order to place the system at the critical point. A phase transition occurs only when the parameter L^{-1} is zero, along with the other parameters that must be set to zero, such as t and h in Ising systems. Thus, in the formulæ of this section, the variable t is defined with respect to the critical temperature in the thermodynamic limit.

A corollary of the above is that crossover effects become important for finite L. To see this, let us consider the scaling form for the singular part of the free energy density at $h = 0$, for simplicity:

$$f_s(t,L^{-1}) = |t|^{2-\alpha} F_f^{\pm}(L^{-1}|t|^{-y_L/y_t}) \tag{9.249}$$

$$= |t|^{2-\alpha} F_f^{\pm}(L^{-1}|t|^{-1/y_t}). \tag{9.250}$$

A useful way to write this is in terms of the bulk correlation length of the infinite system (*i.e.* with $L = \infty$), which we denote now by $\xi_\infty(t)$:

$$f_s(t,L^{-1}) = |t|^{2-\alpha} F_f^{\pm}(\xi_\infty L^{-1}). \tag{9.251}$$

We can analyze this result in terms of crossover phenomena, as follows. For the true critical behavior, we want $L^{-1} = 0$, *i.e.* the limit $x \to 0$ of $F_f^{\pm}(x)$. Thus the $x \to \infty$ limit of $F_f^{\pm}(x)$ does not correspond to the correct critical behaviour. When $L^{-1}t^{-\nu} \ll 1$, or equivalently $L \gg \xi_\infty(t)$, then the correlation length is not affected by the boundaries of the system, and the thermodynamic properties are those of the infinite system. In the opposite limit, encountered sufficiently close to $t = 0$, $L \ll \xi_\infty(t)$ or equivalently $L^{-1}t^{-\nu} \gg 1$, and the system is no longer governed by the

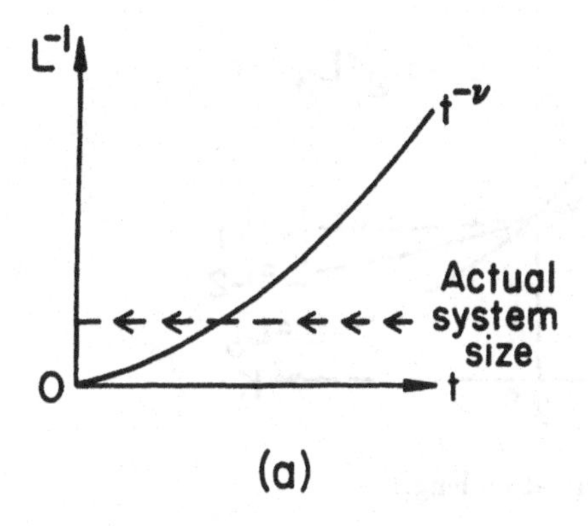

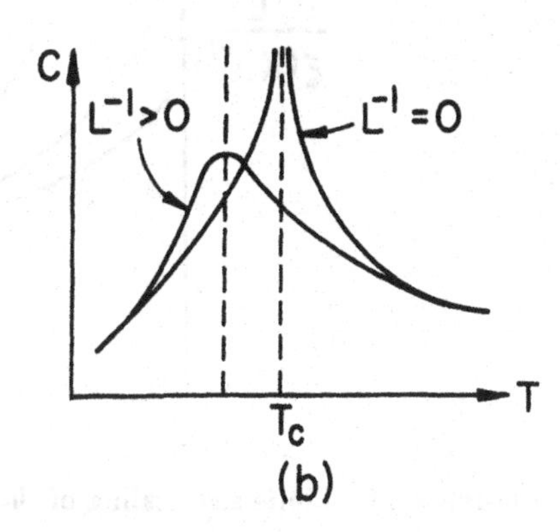

Figure 9.10 (a) Crossover regimes in finite size scaling. (b) Resultant heat capacity curve.

critical fixed point. In this case, the actual correlation length cannot grow beyond L as $t \to 0$, and the transition appears rounded.

The specific heat has the form, given by eqn.(9.250)

$$c(t, L^{-1}) = |t|^{-\alpha} F_f^{\pm}(L^{-1}t^{-\nu}) \tag{9.252}$$

$$= |t|^{-\alpha}(L^{-1}t^{-\nu})^{-\alpha/\nu} D^{\pm}(tL^{1/\nu}) \tag{9.253}$$

$$= L^{\alpha/\nu} D^{\pm}(tL^{1/\nu}), \tag{9.254}$$

where $D(x)$ is a new scaling function, defined by eqn. (9.253), with a maximum at $x = x_0$. Thus the specific heat peak occurs at a reduced temperature shifted from that in the infinite system by an amount

$$t_L = x_0/L^{1/\nu} \propto L^{-1/\nu}. \tag{9.255}$$

Similarly, the maximum height of the specific heat is

$$c(t_L, L^{-1}) = L^{\alpha/\nu} D(x_0) \propto L^{\alpha/\nu}. \tag{9.256}$$

We can exploit these phenomena in practice to obtain estimates of the true critical behavior. As an example, consider the finite size scaling of the correlation length itself.

$$\xi(t, L^{-1}) = \ell\xi(t\ell^{y_t}, \ell L^{-1}) \tag{9.257}$$

$$= t^{-\nu} F_\xi(L^{-1}t^{-\nu}) \tag{9.258}$$

$$= t^{-\nu}(Lt^{\nu})\bar{F}(Lt^{\nu}) \tag{9.259}$$

$$= L\bar{F}(Lt^{\nu}). \tag{9.260}$$

In eqn. (9.259), we have defined a new scaling function $\bar{F}(x)$, which must have the following limiting behaviour. For $L \to \infty$ at fixed $t \ll 1$, we

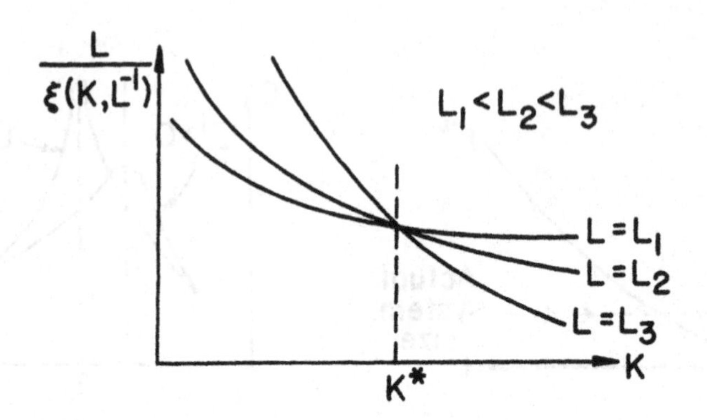

Figure 9.11 Finite size scaling of the correlation length.

expect $\xi(t,0) \sim t^{-\nu}$. Thus $\bar{F}(x) \to x^{-1}$ as $x \to \infty$. For L finite and $t \to 0$, $\bar{F}(x)$ tends towards a constant. It is perfectly analytic in this limit and $\xi \sim L$. Thus, at finite L we can expand about $t = 0$:

$$\frac{L}{\xi(t,L^{-1})} = A + BtL^{1/\nu} + 0(t^2), \tag{9.261}$$

where A and B are constants. The beauty of this form is that if we plot L/ξ *versus* the coupling constant K, for different values of L, all the curves will pass through the same point when $t = 0$ (or equivalently $K = K^*$). Thus we can determine K^*.

We can also compute the critical exponents, using the fact that

$$\frac{\partial}{\partial K}\left(\frac{L}{\xi(t,L^{-1})}\right) = BL^{1/\nu} \tag{9.262}$$

as $T \to 0$. In practice, one computes ν by taking the logarithm:

$$\log\frac{\partial}{\partial K}\left(\frac{L}{\xi}\right) = \log B + \frac{1}{\nu}\log L. \tag{9.263}$$

EXERCISES

Exercise 9–1

This question concerns the static scaling hypothesis.

(a) Derive the relation

$$\delta = \frac{d+2-\eta}{d-2+\eta}.$$

A good starting point is the set of expressions derived from the correlation function scaling in the Kadanoff argument, relating y_t and y_h to ν, η and Δ.

(b) The critical exponents could conceivably be different above and below the transition. Here we will study ν ($T > T_c$) and ν' ($T < T_c$), and show that they are equal. The scaling hypothesis or the Kadanoff argument leads to the relation for the singular part of the free energy density

$$f_s(t,h) = |t|^{d\bar{\nu}} F_f^{\pm}\left(\frac{h}{|t|^{\bar{\nu}y_h}}\right),$$

with $\bar{\nu} = \nu$ ($t > 0$) and $\bar{\nu} = \nu'$ ($t < 0$). For fixed $h \neq 0$, $f_s(t,h)$ should be a smooth function of t, because the only singularity which we expect is at $t = h = 0$. Show that $f_s(t,h)$ can be written in the form

$$f_s(t,h) = h^{d/y_h}\phi_{\pm}\left(\frac{h}{|t|^{\bar{\nu}y_h}}\right)$$

and explain how the smoothness assumption mentioned above constrains the analytic form of the functions $\phi_{\pm}$. Hence show that $\nu = \nu'$.

Exercise 9–2

This question concerns **dangerous irrelevant variables**, and shows how to reconcile the apparent contradiction between the statements "mean field theory works above d=4" and "mean field theory violates hyperscaling because $\alpha = 0$ and $\nu = 1/2$".

(a) Briefly: what is the evidence for the first statement, and what is the violation referred to in the second statement?

(b) Consider the singular part of the free energy density

$$f_s(t,h,\tilde{K}_3) = t^{d/y_t} f_s(1, ht^{-y_h/y_t}, \tilde{K}_3 t^{-y_3/y_t}).$$

$\tilde{K}_3$ is an irrelevant variable, so $y_3 < 0$. In principle there are other irrelevant scaling fields in this expression, but we have suppressed them for simplicity. As $t \to 0$, we expect that

$$f_s(t,h,\tilde{K}_3) \to t^{d/y_t} f_s(1, ht^{-y_h/y_t}, 0)$$

and the usual scaling results follow. However the limit may not be well-defined, in the way discussed in section 7.3. Namely, it might be the case that

$$\lim_{z\to 0} f_s(x,y,z) = z^{-\mu}\bar{f}(x,y) \qquad \mu > 0.$$

Show that this leads to a violation of the Josephson hyperscaling law.

(c) In the Landau theory discussed in lectures, the coefficient of the quartic coupling, u_0, is relevant for $d < 4$ and irrelevant for $d > 4$, as you might expect. It can be shown that in this case $\mu = 1$ and the crossover exponent $y_3/y_t = -(d-4)/2$. Hence, show that for $d > 4$, Landau theory does satisfy hyperscaling after all.

Exercise 9–3

In this question, you are asked to perform a variant of the real space renormalisation group method – decimation – on the $d = 1$ Ising model, and to compare the results with exact results which you obtained from the transfer matrix method. We work with a system of N sites, N may be taken to be even, and we assume periodic boundary conditions $S_{N+1} = S_1$. The Hamiltonian is

$$\mathcal{H} = K\sum_{i=1}^{N} S_i S_{i+1} + h\sum_{i=1}^{N} S_i + NK_0.$$

The partition function is

$$Z_N\{\mathcal{H}\} = \text{Tr } e^{\mathcal{H}}.$$

(a) The partition function can be expressed as a trace over all spins of a product of transfer matrices. Write down the transfer matrix for this case. Now perform the sum over the even numbered spins only, to obtain a new effective transfer matrix for a system with twice the lattice spacing as the original system. Hence calculate the renormalised coupling constants K', h' and K_0' in terms of the original coupling constants. Verify that your expression for the renormalisation group transformation preserves the symmetries of the original problem. Why did we need K_0?

(b) Set the field $h = 0$. Show that the recursion relation for K is

$$e^{4K'} = \cosh^2(2K)$$

and find the fixed points and sketch the flow. Work with the variable $w \equiv e^{-2K}$.

(c) Linearise the recursion relation and find y_t.
(d) Now set $h \neq 0$ and find the fixed points, the flow diagram and the exponent y_h.
(e) What do your results mean for the behaviour of the $d = 1$ Ising model? Use the exact results from the transfer matrix calculations done in class to calculate y_h and y_t.

Exercise 9–4

Perform the real space RG calculation for the Ising model in $d = 2$ in a non-zero external magnetic field, using the scheme described in section 9.6. Construct the RG flow diagram.

Exercise 9–5

This question concerns the use of finite size scaling to estimate critical exponents and transition temperatures from transfer matrix calculations in a strip. Consider the $d = 2$ Ising model on a square lattice. There are N rows parallel to the x axis and M rows parallel to the y axis. We will require that $N \to \infty$ whilst we will calculate the transfer matrix for $M = 1$ and $M = 2$. Periodic boundary conditions apply in both directions, so that our system has the topology of a torus. The Hamiltonian is

$$\mathcal{H} = K \sum_{n=1}^{N} \sum_{m=1}^{M} S_{mn} S_{m+1n} + S_{mn} S_{mn+1}.$$

In exercise 3–2, you constructed the transfer matrix for this problem, and calculated the eigenvalues λ_1 and λ_2. The correlation length is given by

$$\xi^{-1} = \log \lambda_1 / \lambda_2.$$

Use finite size scaling and the results from exercise 3–2 to estimate the critical value of K and the exponent ν.

CHAPTER 10

Anomalous Dimensions Far from Equilibrium

10.1 INTRODUCTION

The solution to the problem of critical exponents given in the previous chapter may not seem very satisfying. Although we appear to have gone beyond mean field theory by the process of successively integrating out short wavelength degrees of freedom, we seem to have finessed the question: where do anomalous dimensions come from?

The present chapter addresses this question directly, in the terms of reference of chapter 7. There, we saw that anomalous dimensions reflect the presence of a microscopic length scale, which affects the behaviour of thermodynamic and correlation functions asymptotically close to the critical point. However, we also pointed out that the mathematical mechanism by which this occurs is simply that, in general, a function $F(x)$ may not be replaced by $F(0)$ as $x \to 0$. Thus, anomalous dimensions may occur not just in the equilibrium statistical mechanics of critical points, but in other areas of physics too.

We will explain the calculation of anomalous dimensions by renormalisation group techniques taking as our examples certain non-linear diffusion problems that arise in fluid dynamics. Although far from equilibrium, and thus formulated as partial differential equations (PDEs) rather than in terms of a partition function, these non-linear diffusion problems also lead to anomalous diffusion laws.[1]

In the conventional case of one-dimensional diffusion, the root mean square displacement of a particle from the starting point as a function of time t is

$$\langle x^2 \rangle \sim t. \tag{10.1}$$

The problems described below lead to laws of the form

$$\langle x^2 \rangle \sim t^{1-\alpha}, \tag{10.2}$$

where α is an anomalous dimension, which may be positive or negative. The key point about these problems is that the long time behaviour is governed by a **similarity solution**, but the similarity variables cannot be deduced from dimensional analysis.

Despite the apparent dissimilarity between critical phenomena and fluid flow far from equilibrium, the techniques involved in calculating anomalous dimensions are actually identical. Thus, we will encounter the problem of divergent perturbation series, just as we did in chapters 6 and 7; the problem is resolved by the procedure of **renormalisation**, first developed by S. Tomonaga, R.P. Feynman and J. Schwinger. The physical interpretation is quite straightforward in the present context. Renormalisation cures the problem of divergences, but still does not lead to anomalous dimensions. We will see that the renormalisation procedure introduces an arbitrary length scale μ, which was not present in the original problem, and thus cannot appear in the solution. In other words the solution possesses an invariance with respect to variations of μ, and it is this invariance which the renormalisation group expresses, and which may be exploited in conjunction with any approximation scheme to calculate anomalous dimensions.[2]

Although this sounds very technical, it is conceptually very simple, as we will see.

---

[1] Note that the PDEs in question have no noise source or other explicit stochastic element in them, unlike the Langevin equation and KPZ equation, discussed in section 8.3. Such stochastic differential equations may be formulated in terms of functional integrals and were shown to be equivalent to field theories by P.C. Martin, E.D. Siggia and H.A. Rose, *Phys. Rev. A* **8**, 423 (1978). See also the article by H.K. Janssen in *Dynamical Critical Phenomena and Related Topics*, Springer Lecture Notes in Physics vol. 104, Ch.P. Enz (ed.) (Springer-Verlag, Berlin, 1979).

[2] This form of the RG, which we will present in this chapter, is due to M. Gell-Mann and F.E. Low, *Phys. Rev.* **95**, 1300 (1964).

10.2 SIMILARITY SOLUTIONS

10.2.1 *Long-time Behaviour of the Diffusion Equation*

We begin with a very simple and familiar example: the problem of solving the one-dimensional **diffusion equation** starting from a localised initial condition.

$$\partial_t u(x,t) = \frac{1}{2}\kappa\partial_x^2 u(x,t), \quad -\infty < x < \infty, \tag{10.3}$$

with diffusion coefficient κ and initial condition

$$u(x,0) = \frac{A_0}{\sqrt{2\pi\ell^2}}e^{-x^2/2\ell^2}. \tag{10.4}$$

Here ℓ is the width of the initial distribution, and A_0 is the area under the initial distribution, which we will refer to as the mass m of the distribution:

$$m \equiv \int_{-\infty}^{\infty} u(x,t)\,dx \tag{10.5}$$

$$= A_0 \quad \text{for } t \geq 0. \tag{10.6}$$

On physical grounds, we are interested only in solutions that vanish sufficiently fast at spatial infinity that m is well-defined, and which have continuous spatial derivatives, at least up to second order. Thus

$$u(x,t) \to 0 \quad \text{as } |x| \to \infty. \tag{10.7}$$

The diffusion equation was derived from a conservation law, and therefore m is conserved. After time t, the solution is

$$u(x,t) = \frac{A_0}{\sqrt{2\pi(\kappa t + \ell^2)}}e^{-x^2/2(\kappa t+\ell^2)}, \tag{10.8}$$

which for long times $t \gg \ell^2/\kappa$ reduces to

$$u(x,t) \sim \frac{A_0}{\sqrt{2\pi\kappa t}}e^{-x^2/2\kappa t} \quad \text{as } t \to \infty. \tag{10.9}$$

This result can also be obtained by keeping the value of t fixed, but taking ℓ small, *i.e.* $\ell^2 \ll \kappa t$:

$$u(x,t) \sim \frac{A_0}{\sqrt{2\pi\kappa t}}e^{-x^2/2\kappa t} \quad \text{as } \ell \to 0. \tag{10.10}$$

In the limit $\ell \to 0$ the initial condition reduces to

$$u(x,0) = A_0\delta(x), \tag{10.11}$$

and we conclude that the long time behaviour of the initial value problem with $\ell \neq 0$ is given by the degenerate limit of the initial value problem with $\ell \to 0$. The solution in this limit is of course the Green function for the diffusion equation.

10.2.2 Dimensional Analysis of the Diffusion Equation

For present purposes, the most important aspect of the similarity solution is that it is dimensionally deficient: whereas the original problem was specified by the variables x, t, ℓ and A_0, the degenerate problem is specified only by x, t and A_0. The fact that ℓ is no longer a parameter is very powerful, and enables us to construct a similarity solution, using dimensional analysis.

The dimensions of the various quantities above, in the system of units ULT, where $U \equiv [u]$ is independent of LT, are:

$$[u] = U; \quad [x] = [\ell] = L; \quad [t] = T; \quad [A_0] = UL; \quad [\kappa] = L^2T^{-1}. \tag{10.12}$$

The dimensionless groups, conventionally denoted by Π, Π_1, *etc.*, may be chosen to be

$$\Pi = \frac{u}{A_0}\sqrt{\kappa t}; \quad \Pi_1 = \frac{x}{\sqrt{\kappa t}}; \quad \Pi_2 = \frac{\ell}{\sqrt{\kappa t}}. \tag{10.13}$$

All other dimensionless groups of parameters may be constructed from these, if desired. Since the Π's are dimensionless, the following relation must be correct:

$$\Pi = f(\Pi_1, \Pi_2), \tag{10.14}$$

where f is a function to be determined, and represents the solution of our problem. We will sometimes refer to f as a scaling function, to emphasise the connection to the static scaling hypothesis. For the similarity solution, ℓ does not appear in the list of parameters of the problem, and so

$$\Pi = f_s(\Pi_1), \tag{10.15}$$

where f_s is the scaling function for the similarity solution. It is straightforward to determine f_s: let $\xi \equiv x/\sqrt{\kappa t}$. Then our dimensional analysis shows that the solution is of the form

$$u(x,t) = \frac{A_0}{\sqrt{\kappa t}} f_s(\xi), \quad \xi \equiv x/\sqrt{\kappa t}. \tag{10.16}$$

Substituting this assumed form into the diffusion equation (10.3), we obtain an ordinary differential equation (ODE) for the unknown function f_s:

$$f_s'' + \xi f_s' + f_s = 0, \tag{10.17}$$

which must satisfy

$$f_s(x) \to 0, \quad \text{as } |x| \to \infty, \tag{10.18}$$

by virtue of eqn. (10.7). A second condition on the scaling function f_s comes from the conservation law (10.6): substituting in the form (10.16), we obtain

$$\int_{-\infty}^{\infty} f_s(\xi)\, d\xi = 1. \tag{10.19}$$

In addition, f_s must be an even function of ξ, by virtue of the relation (10.17) and an even initial condition, and thus

$$f_s'(0) = 0. \tag{10.20}$$

We solve the ODE for $f_s(x)$ by noticing that eqn. (10.17) is an exact differential equation and therefore can be integrated to give

$$f_s' + (\xi f_s) = C, \tag{10.21}$$

where C is a constant of integration. The constant C must be zero because when $\xi = 0$, $f_s'(\xi) = 0$. Integrating eqn. (10.21) gives

$$f_s(\xi) = Be^{-\xi^2/2}, \tag{10.22}$$

where B is determined by the condition (10.19) to be

$$B = 1/\sqrt{2\pi}. \tag{10.23}$$

Hence, the similarity solution is

$$f_s(x,t) = \frac{A_0}{\sqrt{2\pi\kappa t}} e^{-x^2/2\kappa t}. \tag{10.24}$$

10.2.3 Intermediate Asymptotics of the First Kind

We saw that, for the similarity solution,

$$\Pi = f_s(\Pi_1), \tag{10.25}$$

where f_s is the scaling function for the similarity solution. On the other hand, for *any* solution whose initial condition was $\ell \neq 0$, the solution must be of the form

$$\Pi = f(\Pi_1, \Pi_2). \tag{10.26}$$

For long times, $\Pi_2 \to 0$, and 'common sense' dictates that for sufficiently long times, we can regard Π_2 to be small enough that it may be set equal to zero in eqn. (10.14). Hence

$$f_s(\Pi_1) = f(\Pi_1, 0). \tag{10.27}$$

The 'common sense' intuition used above is the statement that a diffusion process loses all memory of the initial distribution once $\langle x^2 \rangle \gg \ell^2$. Although this seems an innocuous assumption, it remains an assumption, and remarkably, may not be correct for some problems, as we will see. In section 10.2.1, we explicitly verified this assumption, for the case where the initial condition was a Gaussian profile, and the reader is invited to perform the corresponding calculation for an arbitrary initial condition in the exercises at the end of this chapter. The assumption that we may safely take the limit to obtain eqn. (10.27) allows us to determine the long time behaviour of the diffusion equation, starting from initial conditions with $\ell \neq 0$, from the similarity solution. We will refer to the asymptotics of the long time limit as **intermediate asymptotics**, the term 'intermediate' denoting the fact that the ultimate behaviour is $u = 0$; the case, where the 'commonsense' assumption is correct, is referred to as *intermediate asymptotics of the first kind*.[3] An immediate consequence of the assumption of intermediate asymptotics of the first kind is that the asymptotic behaviour of the diffusion equation for long times is given by

$$u(x,t) \sim t^{-1/2} g\left(\frac{x}{\sqrt{\kappa t}}\right), \tag{10.28}$$

with $g(x)$ being a function satisfying $g(x) \to$ constant as $x \to 0$, which is indeed what we found in the previous sections.

Intermediate asymptotics arise in many situations in science, and the most common, even unconscious, assumption is that the asymptotics is of the first kind. We may formulate this, in general, as follows. Given a physical problem expressed in dimensionless variables $\Pi, \Pi_1, \Pi_2, \ldots$, the solution of the problem takes the form

$$\Pi = f(\Pi_1, \Pi_2, \ldots), \tag{10.29}$$

with f a function that can only be determined from a detailed analysis of the problem. If the limiting behaviour as (*e.g.*) $\Pi_2 \to 0$ is of the form

$$\Pi \sim f(\Pi_1, 0, \ldots), \tag{10.30}$$

then the asymptotics is said to be of the first kind.

The example given here demonstrates the importance of similarity solutions: they are easy to construct, because they satisfy an equation with one less independent variable than the original PDE, and, most importantly, the intermediate asymptotics of solutions that are themselves *not* similarity solutions are nevertheless often given by similarity solutions. Thus, these solutions have physical significance.

---

[3] G.I. Barenblatt, *Similarity, Self-Similarity and Intermediate Asymptotics* (Consultants Bureau, New York, 1979).

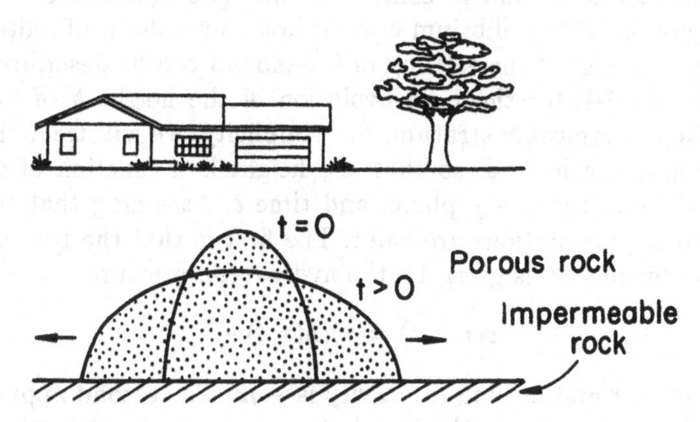

Figure 10.1 Spreading of a groundwater mound in a porous rock.

10.3 ANOMALOUS DIMENSIONS IN SIMILARITY SOLUTIONS

It is not always possible to obtain similarity solutions using dimensional analysis, because anomalous dimensions may occur. In this section, we will present an example of how this happens, by examining the long time behaviour of a modification to the diffusion equation. We begin by motivating the equation to be studied. Next, we find that a naïve application of dimensional analysis fails to obtain the long time behaviour in this case, although a modified similarity solution, with an anomalous dimension, does represent the long time behaviour. In this section, we will closely follow the presentation of these topics due to Barenblatt, postulating the form of the similarity solution, and subsequently determining the anomalous dimension by requiring that this postulated form satisfy all the boundary conditions of the problem. The following section approaches the same problem using the RG.

10.3.1 The Modified Porous Medium Equation

Consider a **porous medium** overlying an impermeable horizontal stratum, as depicted in figure (10.1), in which is present a groundwater mound. Within the mound, the pores of the rock are partially filled with water, whereas outside the mound, the pores are empty of water, but filled with gas. The groundwater mound is unstable in the Earth's gravitational field, and settles by spreading parallel to the plane of the impermeable stratum. We wish to describe the approach to the equilibrium state, where the groundwater mound is completely flat (we ignore the existence of atoms here, so the equilibrium state is not a monolayer of finite extent).[4]

The dynamics of the settling of the mound can be described by writing down a PDE for the time evolution of the height h of the mound above the impermeable stratum. For simplicity, we will treat the case of an axisymmetric mound, so that the height is a function of the radial coordinate r in the $x-y$ plane, and time t. Assuming that the flow is slow, two approximations are valid. The first is that the pressure in the groundwater mound is given by the hydrostatic pressure

$$p(r,z,t) = \rho g(h(r,t) - z), \tag{10.31}$$

where the acceleration due to gravity is $-g\hat{\mathbf{z}}$. The second approximation is that of **D'arcy's Law**: the flux j of groundwater is proportional to the gradient of the so-called pressure head $p + \rho g z$. Thus

$$j = -\frac{\kappa}{\mu}\partial_r(\rho g h), \tag{10.32}$$

where κ is the permeability (typically 10^{-8} cm^2) and μ is the kinematic viscosity of the ground water. It remains to specify the physics associated with the transport of groundwater into and out of the pores in the rock. A given physical volume of rock cannot be completely occupied by fluid: only the pores can be filled by fluid. The fraction of space occupied by the pores is known as the porosity m, and is typically about 10^{-1}. When the groundwater occupies a pore, only a fraction σ is occupied, the remaining volume being filled with gas. When the groundwater exits a pore, a thin wetting layer is retained on the walls of the pore, and so a fraction σ_0 of the pore remains filled with groundwater. We will assume that pores are in one of these two states of occupation only, and for brevity, we shall denote these two states respectively as occupied or unoccupied. This physics introduces a fundamental asymmetry into the dynamics of the groundwater mound: the rate at which empty pores are filled with groundwater is not the same as that at which initially occupied pores become depleted of groundwater. For the groundwater profile indicated in figure (10.1), there is a certain time-dependent radius $r_0(t)$ beyond which $\partial_t h > 0$ and behind which $\partial_t h < 0$. Thus, for $r > r_0(t)$, previously unoccupied pores are becoming occupied with groundwater, whereas for $r < r_0(t)$, previously occupied pores are becoming depleted.

---

[4] This problem is described by G.I. Barenblatt, *Dimensional Analysis* (Gordon and Breach, New York, 1987), and was studied by I.N. Kochina, N.N. Mikhailov and M.V. Filinov, *Int. J. Eng. Sci.* **21**, 413 (1983).

To derive the equation of motion for h, we consider the conservation of water in a shell of radius r, height h and thickness dr. For $r > r_0$, we have

$$m\sigma 2\pi r\, \partial_t h\, dr = \frac{\kappa}{\mu}\partial_r \left[2\pi rh\, \partial_r(\rho gh)\right] dr \tag{10.33}$$

and for $r < r_0$, we have

$$m(\sigma - \sigma_0)2\pi r\, \partial_t h\, dr = \frac{\kappa}{\mu}\partial_r \left[2\pi rh\, \partial_r(\rho gh)\right] dr, \tag{10.34}$$

where the term on the left hand side is the rate at which groundwater is entering/leaving the shell. Thus

$$\partial_t h = \frac{D}{r}\partial_r \left(r\partial_r h^2\right), \tag{10.35}$$

where

$$D = \begin{cases} \kappa_1 & \partial_t h < 0; \\ \kappa_2 & \partial_t h > 0. \end{cases} \tag{10.36}$$

with

$$\kappa_1 \equiv \frac{\kappa\rho g}{2m(\sigma - \sigma_0)\mu}; \quad \kappa_2 \equiv \frac{\kappa\rho g}{2m\sigma\mu}. \tag{10.37}$$

The PDE (10.35) is to be solved subject to the initial condition that the water was confined within a finite region, as depicted in figure (10.1). The requisite boundary conditions are that h and the groundwater current at radius r, namely $2\pi rhj$, are continuous. Thus, both h and $\partial_r h^2$ are continuous: only at $h = 0$ (*i.e.* at the front of the groundwater mound), is it therefore permissible for $\partial_r h$ to exhibit a discontinuity.

Now we have completed our description of the approach to equilibrium of a groundwater mound. Equation (10.35) is the expression in cylindrical polar coordinates of a more general equation, which we shall refer to as the **modified porous medium equation:**

$$\partial_t u(\mathbf{r}, t) = D\Delta_d\, u(\mathbf{r}, t)^{1+n}. \tag{10.38}$$

Here D is a discontinuous function of $\partial_t u$, as in eqn. (10.36), and Δ_d denotes the Laplacian operator in d-dimensions. This equation, when D is a constant, is known as the **porous medium equation**, and models a wide variety of non-equilibrium phenomena in, *inter alia*, fluid dynamics, plasma physics and gas dynamics[5] depending upon the value of n.

---

[5] For a brief summary, see (*e.g.*) W.L. Kath, *Physica D* **12**, 375 (1984); D.G. Aronson, in *Non-linear Diffusion Problems*, A. Fasano and M. Primicerio (eds.), Lecture Notes in Mathematics vol. 1224 (Springer-Verlag, Berlin, 1986).

In the following sections, we will focus on the case $n = 0$, describing the pressure u in an *elasto-plastic porous medium* — one which may expand and contract irreversibly in response to the flow of a fluid through it.[6] We will refer to this equation as **Barenblatt's equation.** Note that the form of D implies that this equation is non-linear. When D is a constant, rather than being a discontinuous function of $\partial_t u$, Barenblatt's equation reduces to the familiar diffusion equation. Needless to say, our discussion can be readily extended to arbitrary n and d; see the exercises at the end of this chapter.

10.3.2 Dimensional Analysis for Barenblatt's Equation

Let us now proceed blindly, and attempt to determine the long time behaviour of Barenblatt's equation in one dimension, which we will write in the form

$$\partial_t u(x,t) = D\kappa \partial_x^2 u(x,t), \tag{10.39}$$

with

$$D = \begin{cases} 1/2 & \partial_t u > 0; \\ (1+\epsilon)/2 & \partial_t u < 0. \end{cases} \tag{10.40}$$

The constants κ and ϵ can be related to the elastic constants of the elasto-plastic porous medium and the phenomenological parameters describing the fluid flow. Note that $[\kappa] = L^2T^{-1}$, whilst ϵ is a pure number. We consider, for simplicity, initial conditions, such as (10.4), satisfying

$$u(x,0) = \frac{A_0}{\ell}\phi\left(\frac{x}{\ell}\right), \tag{10.41}$$

with $u(x,0)$ and the profile $\phi(x)$ normalised as follows:

$$\int_{-\infty}^{\infty} u(x,0)\,dx = A_0; \quad \int_{-\infty}^{\infty} \phi(z)\,dz = 1. \tag{10.42}$$

The dimensionless groups for the problem are

$$\Pi = \frac{u}{A_0}\sqrt{\kappa t}; \quad \Pi_1 = \frac{x}{\sqrt{\kappa t}}; \quad \Pi_2 = \frac{\ell}{\sqrt{\kappa t}}; \quad \Pi_3 = \epsilon, \tag{10.43}$$

and the solution must therefore be of the form

$$\Pi = f(\Pi_1, \Pi_2, \Pi_3). \tag{10.44}$$

---

[6] A discussion of this may be found in Barenblatt's book cited in footnote 3.

This is the same as eqns. (10.13) and (10.14) for the diffusion equation, except that the additional dimensionless group Π_3 is present.

For long times, $\Pi_2 \to 0$, and the considerations of section 10.2 suggest that the asymptotics is determined by the similarity solution: thus we seek a solution of the form

$$u(x,t) = \frac{A_0}{\sqrt{\kappa t}} f(\xi, \epsilon), \quad \xi \equiv x/\sqrt{\kappa t}, \tag{10.45}$$

which is even and vanishes monotonically at infinity. We will denote the positive value of x where $\partial_t u(x,t) = 0$ as $X(t,\epsilon)$, and the corresponding value of ξ as

$$\xi_0 \equiv \frac{X(t,\epsilon)}{\sqrt{\kappa t}}. \tag{10.46}$$

Note that the assumption that the solution is of the form (10.45) implies that ξ_0 is a dimensionless constant, independent of time. Substituting into Barenblatt's equation, we obtain *two* ODEs for the scaling function f, one valid for $\partial_t u < 0$, the other valid for $\partial_t u > 0$:

$$(1+\epsilon)f'' + (\xi f)' = 0, \quad 0 \le \xi \le \xi_0; \tag{10.47}$$

$$f'' + (\xi f)' = 0, \quad \xi_0 \le \xi < \infty. \tag{10.48}$$

Integrating, we obtain

$$(1+\epsilon)f' + (\xi f) = B_1, \quad 0 \le \xi \le \xi_0; \tag{10.49}$$

$$f' + (\xi f) = B_2, \quad \xi_0 \le \xi < \infty, \tag{10.50}$$

where $B_{1,2}$ are constants of integration. The requirement that $f(\xi)$ be even implies that at $\xi = 0$, $f'(\xi) = 0$, and hence $B_1 = 0$. As $\xi \to \infty$, we require that both f and f' tend to zero in order that f remain integrable: hence, $B_2 = 0$. Integrating one more time, we obtain

$$f = \begin{cases} C_1 e^{-\xi^2/2(1+\epsilon)}, & 0 \le \xi \le \xi_0; \\ C_2 e^{-\xi^2/2}, & \xi_0 \le \xi < \infty. \end{cases} \tag{10.51}$$

Now we match f and f' at $\xi = \xi_0$:

$$C_1 e^{-\xi_0^2/2(1+\epsilon)} = C_2 e^{-\xi_0^2/2}; \tag{10.52}$$

$$\frac{C_1}{(1+\epsilon)} e^{-\xi_0^2/2(1+\epsilon)} = C_2 e^{-\xi_0^2/2}. \tag{10.53}$$

The only solution of these equations is the trivial one $C_1 = C_2 = 0$. Hence, there is no non-trivial similarity solution of the form of eqn. (10.45) with continuous derivatives (up to second order).

What then does this imply for the long time behaviour of Barenblatt's equation starting from physical initial conditions with $\ell \neq 0$? Is the problem not well-posed? Do the derivatives develop singularities? In fact, there is a rigorous proof[7] that none of these possibilities occur: the solution to Barenblatt's equation with the stated initial conditions exists, is unique, has continuous derivatives with respect to x to second order, and has a continous derivative with respect to t. We conclude, therefore, that the long time behaviour of Barenblatt's equation is not of the form postulated in eqn. (10.45).

10.3.3 Similarity Solution with an Anomalous Dimension

Even though there is no similarity solution in the conventional sense, it is still possible to find the asymptotic behaviour by *postulating* that as $t \to \infty$, and $\Pi_2 \to 0$, the form of the solution, eqn. (10.44) becomes

$$\Pi \sim \Pi_2^{2\alpha} g(\Pi_1, \Pi_3), \quad \text{as } \Pi_2 \to 0, \tag{10.54}$$

where α is an anomalous dimension, to be determined. We will shortly see how eqn. (10.54) arises from the RG, together with a number of technical assumptions. For now, we follow Barenblatt, and examine the consequences of our hypothesis. Equation (10.54) suggests that we seek a solution to Barenblatt's equation with initial conditions (10.41) and (10.42) of the form

$$u(x,t) = \frac{A_0}{\sqrt{\kappa t}} \frac{\ell^{2\alpha}}{(\kappa t)^\alpha} g\left(\frac{x}{\sqrt{\kappa t}}, \epsilon\right). \tag{10.55}$$

This may seem very strange: if we achieve the limit $\Pi_2 \to 0$ by keeping t fixed, and letting $\ell \to 0$, then $u(x,t)$ tends towards zero or infinity, depending upon the sign of α. The only way to make sense of eqn. (10.55) is to assume that

$$\lim_{\ell \to 0} A_0 \ell^{2\alpha} = \text{constant}. \tag{10.56}$$

In other words, A_0 must depend on ℓ, and in such a way that the product $A_0 \ell^{2\alpha}$ remains finite as $\ell \to 0$! When we discuss renormalisation, we will see that this mathematical requirement has a very physical interpretation: the mass m of the distribution is *not* conserved in Barenblatt's equation, when $\epsilon \neq 0$. The initial condition (10.41) satisfying eqn. (10.56) represents a function more (less) singular than a delta function, for $\epsilon > 0$ ($\epsilon < 0$). Note that the actual value of the constant in eqn. (10.56) is not important, as long as it is not zero or infinity. We will choose it in such a way that

$$g(0, \epsilon) = 1. \tag{10.57}$$

[7] S.L. Kamenomostskaya, *Dokl. Akad. Nauk SSSR* **116**, 18 (1957).

How do we determine the anomalous dimension α? Substituting the hypothesis (10.55) into Barenblatt's equation, we obtain a pair of ODEs for the scaling function g, similar, but not identical to those of eqns. (10.47) and (10.48):

$$(1+\epsilon)g'' + \xi g' + (1+2\alpha)g = 0, \quad 0 \le \xi \le \xi_0; \tag{10.58}$$

$$g'' + \xi g' + (1+2\alpha)g = 0, \quad \xi_0 \le \xi < \infty. \tag{10.59}$$

As before, these equations must be solved subject not only to boundary conditions at the origin and infinity, but also to matching conditions at the point where $\partial_t u(x,t) = 0$. From eqn. (10.55), these matching conditions apply at $\xi = \xi_0$ determined by

$$\xi g' + (1+2\alpha)g(\xi,\epsilon) = 0. \tag{10.60}$$

Thus, eqns. (10.58) and (10.59) must be solved subject to the constraints that

$$g(0,\epsilon) = 1; \quad g'(0,\epsilon) = 0; \quad g(\infty,\epsilon) = 0 \tag{10.61}$$

together with the continuity of $g(\xi_0,\epsilon)$ and $g'(\xi_0,\epsilon)$. There are *five* constraints for *two* second order differential equations.[8] Integrating the pair of ODEs gives *four* constants of integration only. Hence, *in general*, it is not possible to satisfy simultaneously all the constraints. However, there may exist *special* values of α for which all the constraints can be satisfied simultaneously. Thus, the requirement that an overdetermined set of equations have a solution can determine the value (or spectrum of values) of the anomalous dimension α. This sort of condition is exactly what is used to determine energy levels in the time-independent Schrödinger equation, and thus we may think of α as an eigenvalue. However, in contrast to the situation in quantum mechanics, the eigenvalue problem is *non-linear*, because the point ξ_0 is itself unknown at the outset, and is determined by eqn. (10.60). In summary, the anomalous dimension α is determined by requiring that a non-linear eigenvalue problem have a solution: this is sometimes referred to as a **solvability condition**, and α itself is known as a **non-linear eigenvalue**.

Let us now put these considerations into practice. In general, the most effective way of computing the non-linear eigenvalue is numerically, using a **shooting method**.[9] For a given value of ϵ, one proceeds, in principle, by choosing a value for α; then, beginning at $\xi = \infty$ with $g = 0$

---

[8] Equation (10.60) may be taken as a constraint, but then the continuity of $g'(\xi_0)$ is assured automatically; thus there are actually only five constraints in total.

[9] See (*e.g.*) *Numerical Recipes*, W.H. Press, B.P. Flannery, S.A. Teukolsky and W.T. Vetterling (Cambridge University Press, New York, 1986), Section 16.1.

and $g' = 0$, one solves eqn. (10.59) in small increments $\Delta\xi$ back towards $\xi = 0$, monitoring the derivative g'. When condition (10.60) is satisfied, the integration continues, but now using eqn. (10.58); the numerical integration should ensure the continuity of g and g' at ξ_0. In this way, the value $g'(0, \epsilon)$ may be computed. For general α, it will not be zero, thus violating the boundary condition (10.61). One then chooses a slightly different value of α and repeats this process. In this way, one can plot a graph of $g'(0, \epsilon)$ against α, and thence find the value of α where $g'(0, \epsilon) = 0$: this is the value of α corresponding to the given value of ϵ. In practice, one cannot shoot from $\xi = \infty$, and an asymptotic expansion near $\xi = \infty$ of eqn. (10.59) is required to start off the numerical integration. Since eqn. (10.59) is linear, the asymptotic expansion is determined up to an arbitrary multiplicative coefficient, which may be chosen to be of order unity. The choice of this coefficient will of course mean that $g(0, \epsilon)$ is not necessarily unity, but this value was only an arbitrary normalisation in any case, and does not affect the value of the anomalous dimension (*i.e.* the value of α for which $g'(0, \epsilon) = 0$). In fact, for Barenblatt's equation, eqns. (10.58) and (10.59) are simple enough that they may be solved in terms of parabolic cylinder functions, and the values of ξ_0 and α are determined by a pair of transcendental equations (see the exercises at the end of this chapter). However, these transcendental equations must be solved numerically.

The result of this procedure (see exercise 10–2) is that for small ϵ, the solvability condition yields a unique value $\alpha(\epsilon)$ for the anomalous dimension. For sufficiently large values of ϵ, the solutions are apparently not unique, but the eigenvalue with the smallest value for $\alpha(\epsilon)$ corresponds to the stable solution.

In conclusion, although there is no similarity solution of Barenblatt's equation, in the conventional sense, making the hypothesis (10.54) leads to a self-similar solution with a non-trivial exponent or anomalous dimension α. The anomalous dimension is a vestige of the parameters specifying the initial condition, A_0 and ℓ, and the long time behaviour is given by

$$u(x,t) \sim \frac{A}{(\kappa t)^{1/2+\alpha(\epsilon)}} g\left(\frac{x}{\sqrt{\kappa t}}, \epsilon\right), \tag{10.62}$$

where

$$A = B \lim_{\ell \to 0} A_0 \ell^{2\alpha}, \tag{10.63}$$

and B is a constant with value determined by the normalisation $g(0, \epsilon) = 1$. In fact, B can only be determined by solving the actual equation, as will become clear when we discuss this problem from the point of view of the RG.

10.3.4 Intermediate Asymptotics of the Second Kind

We have just seen that the assumption of intermediate asymptotics of the first kind fails for Barenblatt's equation. Of course, Barenblatt's equation is but one example where this occurs, and Barenblatt has given a number of other examples, taken mostly from fluid dynamics, where anomalous dimensions arise and are treated by similar methods.[10] Thus, we are led to formulate the notion of intermediate asymptotics of the second kind, as follows: Given a problem expressed in dimensionless variables $\Pi, \Pi_0, \Pi_1, \ldots, \Pi_n$, with solution

$$\Pi = f(\Pi_0, \Pi_1, \ldots, \Pi_n), \tag{10.64}$$

the asymptotics is said to be of the second kind if, with an appropriate choice of exponents α, α_1, $\alpha_2 \ldots \alpha_n$, the asymptotic behaviour is of the form

$$\Pi \sim \Pi_0^{\alpha} g\left(\frac{\Pi_1}{\Pi_0^{\alpha_1}}, \ldots, \frac{\Pi_n}{\Pi_0^{\alpha_n}}\right), \quad \text{as } \Pi_0 \to 0, \tag{10.65}$$

where g is a scaling function. The exponents $\{\alpha\}$ — the anomalous dimensions — cannot be predicted by dimensional analysis, but must be determined by solving the problem directly, as we saw in the previous section.

Note that in Barenblatt's equation, or in any of the other examples given by Barenblatt, the governing equations are simple PDEs, and contain no stochastic element. Thus, it is not necessary for a problem to be formulated as a partition function (*i.e.* a probability generating function) in order that anomalous dimensions occur. The discussion in chapter 7, not to mention the similarity of eqn. (10.65) with the static scaling hypothesis, eqn. (8.24), may suggest to the reader that the anomalous dimensions in critical phenomena and the anomalous dimensions in intermediate asymptotics of the second kind are identical; this is indeed the case. We will now demonstrate this assertion for Barenblatt's equation by using the RG to show how intermediate asymptotics of the second kind naturally arises (as a consequence of renormalisation), and to compute the anomalous dimensions in an expansion in ϵ.

There is another interesting exercise, which has not, at the time of writing, been accomplished: the identification of the anomalous dimensions in critical phenomena with the exponents appearing in a self-similar solution of the second kind suggests that critical exponents may also be expressible as non-linear eigenvalues. It would be useful to derive exact

---

[10] See footnote 3.

equations for thermodynamic and correlation functions, and seek self-similar solutions of the second kind, thus finding exact relations satisfied by the anomalous dimensions.

10.4 RENORMALISATION

The purpose of this section is to provide a physical explanation of **renormalisation**, in its most general context. Although initially developed in conjunction with perturbative quantum field theory, renormalisation has a direct physical interpretation, conceptually divorced from the context of perturbation theory. Thus, we will see that the RG is essentially non-perturbative.

There are two distinct steps in applying renormalisation and renormalisation group methods to a given problem. The first step is an extension of dimensional analysis, taking into account renormalisation effects. This step shows how the solution to the problem can exhibit asymptotics of the second kind, as expressed by eqn. (10.65). In addition, the exponents found may not be independent, in which case, the RG accounts for the scaling laws between these exponents, as we have seen explicitly already. In the example of the two dimensional Ising model, this development was given in sections 9.2 - 9.4.

The second step is to combine the RG with an approximation scheme, such as perturbation theory, in order to estimate the values of the anomalous dimensions. This second step is exemplified by the explicit calculation of section 9.6 for the two dimensional Ising model, where we used the cumulant expansion to approximate the RG recursion relations. The RG on its own is vacuous, in the sense that it relates coupling constants at different length scales (in the context of critical phenomena), but does not actually calculate any quantities. It must always be combined with an actual calculation of the recursion relations, which is almost always approximate and often involves perturbation theory; thus, it is sometimes said that RG 'improves' (*e.g.*) perturbation theory. We will see this feature explicitly in our treatment of Barenblatt's equation.

The term *renormalisation* is used in two slightly different but related contexts, and it is worth pointing this out explicitly. The first use denotes the way that a certain parameter changes under a specified transformation or operation, an example being the (finite) renormalisation of coupling constants under an RG transformation. The second use refers to the procedure employed to remove divergent terms from perturbation theory.

We will show that the two seemingly different meanings are actually related: it is the fact that parameters become renormalised that enables us to renormalise a theory and render it finite.

There are, then, *two* ingredients to the divergent terms of perturbation theory. The first is intrinsic to the problem at hand, and is the *singular* (but not necessarily *divergent*) behaviour expressed by the correct solution to the problem. For example, we have seen that Barenblatt's equation with (*e.g.*) $\epsilon > 0$ does not possess physical solutions for delta function initial conditions: only a more singular initial condition in the limit $\ell \to 0$ has physical solutions. The second ingredient is the use of an approximation scheme which has 'forgotten' to account for renormalisation. We will shortly see this explicitly, when we construct perturbation theory for Barenblatt's equation, starting from the diffusion equation.

The ideas and techniques discussed in this section are extensively used in statistical mechanics and **field theory** — the generic name given to a physical problem capable of being formulated in terms of a functional integral. However, the formalism is very complicated in these problems, and this tends to obscure the basic physical ideas.[11] Thus, we will study renormalisation in a mathematically simpler context, which is not only interesting in its own right, but which also shows explicitly that renormalisation is not solely the domain of field theory.

10.4.1 Renormalisation and its Physical Interpretation

The crucial difference between the diffusion equation and Barenblatt's equation (with $\epsilon \neq 0$) is that the former is derived from a conservation law for u, whereas the latter is not. To see this, note that in Barenblatt's equation (10.39), the diffusion coefficient κD depends on x:

$$D(x) = \begin{cases} \frac{1}{2} + \frac{\epsilon}{2}\theta\left(X(t,\epsilon) - x\right), & x \geq 0; \\ \frac{1}{2} + \frac{\epsilon}{2}\theta\left(X(t,\epsilon) + x\right), & x \leq 0, \end{cases} \tag{10.66}$$

where θ is the Heaviside function, and $X(t,\epsilon)$ is the *positive* value of x where $\partial_x^2 u(x,t) = 0$. Thus, the rate of change of the mass m, defined in eqn. (10.5) is

$$\partial_t m = \int \kappa D(x)\, \partial_x^2 u(x,t)\, dx \tag{10.67}$$

---

[11] Accounts of renormalisation and the renormalisation group in field theory are given by (*e.g.*) J. Zinn-Justin, *Quantum Field Theory and Critical Phenomena* (Clarendon, Oxford, 1989); J. Collins, *Renormalisation* (Cambridge University Press, New York, 1987); D.J. Amit, *Field Theory, the Renormalisation Group and Critical Phenomena* (World Scientific, Singapore, 1984).

which, upon use of the identity

$$D\partial_x^2 u = \partial_x(D\partial_x u) - (\partial_x D)(\partial_x u) \tag{10.68}$$

and the boundary condition that $\partial_x u \to 0$ as $|x| \to \infty$, becomes

$$\partial_t m = -\kappa \int (\partial_x D)(\partial_x u)\, dx. \tag{10.69}$$

When $\epsilon = 0$, as it is for the diffusion equation, $\partial_x D = 0$, and the mass is conserved, as noted in eqn. (10.6). However, when $\epsilon \neq 0$, the right hand side of eqn. (10.69) does not vanish in general. A similar result holds for the general case of the modified porous medium equation.

Actually, this result is not very surprising — it simply reflects the physics of the situation. The simplest way to see this is to think of the groundwater problem, described by eqn. (10.35). There, the physics dictated that when groundwater exits a pore, a thin wetting layer remains on the walls of the pore. Thus, the quantity of groundwater, which is not trapped in pores and thereby unavailable for transport, continually decreases with time.

This observation has a very important consequence, which is best explained with reference to the diffusion equation. There, the mass m is conserved and $m(t) = m(0)$ for all times $t > 0$. For long times $t \gg \kappa/\ell^2$, the width of the distribution $u(x,t)$ is much larger than ℓ, and intuitively, we expect that the system has 'lost the memory of the initial conditions'. This not only means that the system has lost memory of the initial width ℓ, but also of *the time that has elapsed!* In other words, in the intermediate asymptotic limit $t \to \infty$, it is not possible to discriminate between possible histories of the system, such as (*i*) u was a delta function at $t = 0$, or (*ii*) u was a Gaussian distribution with width ℓ at time $t = 0$, or (*iii*) u was a Gaussian distribution with width ℓ at time $t = 10$. These and many other histories all give rise to the same distribution $u(x)$ at sufficiently long times, *i.e.* in the intermediate asymptotic regime. Nevertheless, given that the governing equation is the diffusion equation, it must always be the case that the mass m at early times was the same as the mass measured in the intermediate asymptotic regime.

A quite different situation prevails for Barenblatt's equation, with $\epsilon \neq 0$. There, the initial conditions are completely undetectable in the intermediate asymptotic regime. The mass $m(0)$ is, in the above sense, *unobservable*, because m changes continuously as a function of time t.

We have encountered a situation similar to this in critical phenomena. There, we considered a physical system, described on the microscopic scale of the lattice spacing a by a Hamiltonian $\mathcal{H}$. We defined an RG transformation, which resulted in a description of the system on a larger length scale $\mathcal{H}_\ell$, and we saw that this process could be continued to larger and larger length scales. We emphasised that this renormalisation process constituted a semi-group, because many different microscopic Hamiltonians flow to the same **fixed point Hamiltonian** $\mathcal{H}^*$, and thus, for a given $\mathcal{H}^*$, there is no unique microscopic Hamiltonian.

As an aside, we remark that the evolution of Hamiltonians $\mathcal{H}$ (or more precisely, probability distributions $\exp \mathcal{H}$) under length scale transformations is analogous to the evolution of the distribution u under transformations in time. The fixed point probability distribution $\exp \mathcal{H}$ is analogous to the distribution u in the intermediate asymptotic regime. We will examine these analogies more closely later on.

In this picture, then, the evolution of the mass m in Barenblatt's equation is analogous to the renormalisation of the coupling constants in statistical mechanics. In Barenblatt's equation, and in non-equilibrium problems in general, the renormalisation occurs in time, whereas in statistical mechanics, the renormalisation occurs in space.[12] Let us now examine the mathematical consequences of these remarks. We begin with a heuristic, but physical derivation of the anomalous dimension α.

10.4.2 Heuristic Calculation of the Anomalous Dimension α

Our starting point is the explicit equation for the rate of change of m, eqn. (10.69). The distribution u is symmetric, and thus

$$\partial_t m = \epsilon\kappa\partial_x u(X(t,\epsilon),t). \tag{10.70}$$

To a first approximation, we may suppose that the removal of mass occurs sufficiently slowly that u retains the form of the distribution that it would have in the case when mass is conserved.[13] Such an approximation would be expected to be valid for small values of ϵ. Thus, we replace A_0 in eqn. (10.8) by $m(t)$ to obtain

$$u(x,t) = \frac{m(t)}{\sqrt{2\pi(\kappa t + \ell^2)}} e^{-x^2/2(\kappa t+\ell^2)}. \tag{10.71}$$

[12] The application of RG methods to Barenblatt's equation, and the possible application to turbulence, crystal growth and other non-equilibrium phenomena is discussed by N.D. Goldenfeld, O. Martin and Y. Oono, *J. Sci. Comp.* **4**, 355 (1989); N.D. Goldenfeld, O. Martin, Y. Oono and F. Liu, *Phys. Rev. Lett.* **64**, 1361 (1990); N. Goldenfeld, O. Martin and Y. Oono, in *Proceedings of the NATO Advanced Research Workshop on Asymptotics Beyond All Orders*, S. Tanveer (ed.) (Plenum, New York, 1992).

[13] I am indebted to J. Goodman for suggesting this simplified argument.

Again in this approximation, the point where $\partial_x^2 u$ changes sign is calculated from eqn. (10.8) to be

$$X(t,\epsilon) = \sqrt{\kappa t + \ell^2}. \tag{10.72}$$

Substituting into eqn. (10.70) we obtain

$$\partial_t m(t) = -\frac{\epsilon \kappa m(t)}{\sqrt{2\pi}} \frac{e^{-1/2}}{(\kappa t + \ell^2)}, \tag{10.73}$$

whose solution is

$$m(t) = m(0) \frac{\ell^{2\alpha}}{(\kappa t + \ell^2)^\alpha}, \tag{10.74}$$

with anomalous dimension

$$\alpha = \frac{\epsilon}{\sqrt{2\pi e}}. \tag{10.75}$$

Thus the long time behaviour is

$$u(x,t) \sim \frac{m(0)\ell^{2\alpha}}{(\kappa t)^{1/2+\alpha}} e^{-x^2/2\kappa t}. \tag{10.76}$$

The renormalisation group makes systematic the above heuristic analysis, and shows that the value of the anomalous dimension is indeed correct to $O(\epsilon)$.

10.4.3 Renormalisation and Dimensional Analysis

Consider Barenblatt's equation, starting with an initial distribution $u(x,0)$, with width ℓ and *initial* mass

$$m_\ell = \int u(x,0)\, dx. \tag{10.77}$$

We have added the subscript ℓ to m to indicate that the mass is associated with a distribution of width ℓ. Now we will redo the dimensional analysis of section 10.3.2, but this time taking into account the renormalisation discussed above. We are interested in the intermediate asymptotic limit $\Pi_2 \equiv \ell/\sqrt{\kappa t} \to 0$, which we choose to achieve here by holding t fixed and letting $\ell \to 0$. In part, this choice is motivated for pedagogical reasons: it is closest to the procedure often used in statistical and quantum field theory![14] Let $m(t)$ be the mass of the distribution $u(x,t)$ at a given

[14] See footnote 3.

time $t > 0$. Our previous argument shows that in fact there is no unique answer to the question: what initial condition gave rise to the observed distribution? For every possible initial condition with width ℓ there is a corresponding mass m_ℓ which will give rise to the observed distribution at time t. In particular, we may even consider the limit of the initial conditions as $\ell \to 0$. The observed mass $m(t)$ must be proportional to m_ℓ, because they have the same units, and hence they may be related by a dimensionless parameter Z:

$$m(t) = Z^{-1} m_\ell. \tag{10.78}$$

We will refer to Z as a **renormalisation constant**. The observed mass m cannot depend on ℓ, and therefore Z must depend upon ℓ. However, Z is dimensionless, and therefore cannot be a function solely of a quantity such as ℓ, which has dimensions of length. To make it dimensionless, we take

$$Z = Z\left(\frac{\ell}{\sqrt{\kappa t}}, \epsilon\right). \tag{10.79}$$

Thus, in terms of the physically observable mass at time t, we have

$$u(x,t) = \frac{Zm}{\sqrt{\kappa t}} f(\xi, \eta, \epsilon), \tag{10.80}$$

where the function f is the *same* function as in eqn. (10.44), and we have defined

$$\xi \equiv \frac{x}{\sqrt{\kappa t}}; \quad \eta \equiv \frac{\ell}{\sqrt{\kappa t}}. \tag{10.81}$$

The distribution $u(x,t)$ is independent of ℓ in the intermediate asymptotic limit, and therefore

$$\ell \frac{du}{d\ell} = 0, \tag{10.82}$$

where the derivative is taken with respect to ℓ, but holding fixed $m(t)$, x, ϵ and κt. The prefactor of ℓ is present for cosmetic reasons, as will become apparent. This equation, expressing the invariance of the physics for long times to changes in the initial conditions, is sometimes called the **bare renormalisation group equation**. The use of the word 'bare' designates parameters present in the original formulation, prior to renormalisation. In the next section, we will discuss how renormalisation removes certain divergences which arise *in perturbation theory*; there we will encounter another renormalisation group equation, but this time the differentiation will be with respect to a parameter μ, introduced during the renormalisation process, which has no direct physical significance.

Evaluating the derivative in (10.82), we find that

$$\ell\left[\frac{\partial Z}{\partial \ell}\frac{m}{\sqrt{\kappa t}}f+\frac{Zm}{\sqrt{\kappa t}}\frac{1}{\sqrt{\kappa t}}\frac{\partial f}{\partial \eta}\right]=0. \tag{10.83}$$

Rearranging gives

$$\frac{d\log Z}{d\log \ell}f+\eta\frac{\partial f}{\partial \eta}=0. \tag{10.84}$$

Now let us take the limit $\ell \to 0$, and *assume* the existence of the limit

$$\lim_{\ell\to 0}\frac{d\log Z}{d\log \ell}=-2\alpha, \tag{10.85}$$

which *defines* the constant α. Our assumption is not trivial, and we will refer to it as the assumption of **renormalisability**. We will discuss and justify this assumption shortly. Collecting together eqn. (10.84) and (10.85), we find

$$\eta\frac{\partial f}{\partial \eta}=2\alpha f \tag{10.86}$$

and thus

$$f(\xi,\eta,\epsilon)=\eta^{2\alpha}g(\xi,\epsilon), \tag{10.87}$$

where g is an as yet undetermined function of ξ and ϵ. Equation (10.85) implies that as $\ell/\sqrt{\kappa t}\to 0$,

$$Z\sim\left(\frac{\ell}{\sqrt{\kappa t}}\right)^{-2\alpha}. \tag{10.88}$$

Hence, we have found that

$$m(t)\sim m(0)\left(\frac{\ell}{\sqrt{\kappa t}}\right)^{2\alpha}\quad \text{as } \ell/\sqrt{\kappa t}\to 0, \tag{10.89}$$

and

$$u(x,t)=\frac{m(t)}{\sqrt{\kappa t}}g(\xi,\epsilon) \tag{10.90}$$

or equivalently

$$u(x,t)=\frac{m_\ell}{\sqrt{\kappa t}}\left(\frac{\ell}{\sqrt{\kappa t}}\right)^{2\alpha}g(\xi,\epsilon). \tag{10.91}$$

Physically, we anticipate that $\partial_t m<0$, and therefore that $\alpha>0$. Thus, we conclude on general grounds that there is an anomalous dimension at long times. Note that we have not yet calculated anything, whereas in the heuristic argument of the previous section, we used the specific approximation (10.71) to estimate the anomalous dimension. If an approximation for f were available, then in conjunction with eqn. (10.84), the anomalous dimension α could be determined. Finally, the observed mass at a fixed time t is finite, and is related to the initial mass by

$$m_\ell = Zm(t) \sim \left(\frac{\ell}{\sqrt{\kappa t}}\right)^{-2\alpha} m(t), \quad \text{as } \ell \to 0. \tag{10.92}$$

Thus, for $m(t)$ fixed at a given time t in the intermediate asymptotic regime, there is a sequence of initial conditions, parametrised by ℓ and m_ℓ, which give rise to the distribution $u(x,t)$, and in this sequence, as we take ℓ smaller and smaller, the mass m_ℓ diverges as $\ell^{-2\alpha}$.

10.4.4 Removal of Divergences and the RG

In the previous section, we discussed the physical renormalisation of the mass in the Barenblatt equation, and showed explicitly that it arose from the fact that the equation is not derivable from a conservation law for u when $\epsilon \neq 0$. We also saw that the initial mass m_ℓ diverged as $\ell \to 0$. In this section, we will examine another aspect of this physics: the fact that perturbation theory about the diffusion equation, an equation which does conserve u, has divergent terms. These divergences are a manifestation of the *singular* nature of the physics we have discussed, and arise in this *form* because of the approximation scheme. However, the *existence* of renormalisation and anomalous dimensions is *not* an artifact of any approximation scheme, and is indeed the essential physics present in the original Barenblatt equation.

We will not analyse the details of perturbation theory in this section, but instead, we will explain how renormalisation introduces anomalous dimensions. Our starting point is again dimensional analysis:

$$u(x,t) = \frac{m_\ell}{\sqrt{\kappa t}} f(\xi, \eta, \epsilon). \tag{10.93}$$

This time, we will renormalise the initial mass m_ℓ, with a renormalisation constant Z, as in eqn. (10.78), but taken as a function of ℓ/μ, where μ is an *arbitrary* length scale. Thus

$$m = Z^{-1} m_\ell; \quad Z = Z(\ell/\mu, \epsilon). \tag{10.94}$$

Here μ is at present unspecified (it may be the radius of the moon or the diameter of a proton!), and we have also included the dependence on the dimensionless parameter ϵ. The *renormalised mass* m is not simply interpreted, which is why we have presented more physical approaches to this topic first. It is, however, a *finite* mass associated with the scale μ, and independent of ℓ.

Why did we choose to introduce a new length scale μ rather than use the natural length scale $\sqrt{\kappa t}$ already present in the problem (as we did in the previous section)? The answer is essentially technical. When we attempt to solve Barenblatt's equation using perturbation theory, we will find that certain integrals diverge as we take the limit $\ell \to 0$. If an integral is finite when a certain parameter in the integral (either the limits or integrand) is finite, but the integral diverges when that parameter is set to zero, then that parameter is said to be a **regularisation parameter**. In statistical mechanics, the lattice spacing is usually a regularisation parameter, whereas in field theory, a regularisation parameter is often artificially introduced in order to control divergences in perturbation theory, using renormalisation. Thus, we may think of the initial width ℓ as the regularisation parameter for Barenblatt's equation.

The divergences in perturbation theory can be eliminated by renormalising the theory, as we will shortly explain. For now, we examine the consequences of renormalising the mass m in Barenblatt's equation. Eqns. (10.93) and (10.94) imply that

$$u(x,t) = \frac{Zm}{\sqrt{\kappa t}} f\left(\frac{x}{\sqrt{\kappa t}}, \frac{\ell}{\sqrt{\kappa t}}, \epsilon\right). \tag{10.95}$$

The function f is singular in the limit $\ell \to 0$; it either tends towards zero or infinity. In perturbation theory, as we will see, this means that the approximation to the function actually diverges. The basic idea of **renormalisation** is to absorb this divergence into a redefinition of the mass m_ℓ. As we have seen, this is perfectly permissible, since m_ℓ is not well-defined physically in the intermediate asymptotic regime. Thus, in writing eqn. (10.95), our intention is to define the renormalisation constant Z in such a way that for small ℓ, the term leading to the divergence in f is cancelled out by a term in Z. Then, proceeding to the limit, the distribution u remains *finite* as $\ell \to 0$. Thus the combination of Z and f in eqn. (10.95) lead to

$$u(x,t) = \frac{m}{\sqrt{\kappa t}} F\left(\frac{x}{\sqrt{\kappa t}}, \frac{\mu}{\sqrt{\kappa t}}, \epsilon\right). \tag{10.96}$$

The function F is given by

$$Z\left(\frac{\ell}{\mu}\right) f\left(\frac{x}{\sqrt{\kappa t}}, \frac{\ell}{\sqrt{\kappa t}}, \epsilon\right) = F\left(\frac{x}{\sqrt{\kappa t}}, \frac{\mu}{\sqrt{\kappa t}}, \epsilon\right) + O(\ell), \quad \text{as } \ell \to 0, \tag{10.97}$$

and contains no singular ℓ dependence, by construction. Equation (10.96) is the statement that the theory is **renormalisable.**

The step, which we have just performed in outline, is known as **renormalisation.** It is by no means obvious that the apparently miraculous cancellation of divergences in f using Z can actually occur; in general, more than one quantity needs to be renormalised in order that all the divergences can be cancelled. If a physical problem can be renormalised with a finite number of renormalisation constants, then it is said to be renormalisable. Not all problems are renormalisable, however. We will discuss the issue of renormalisability later, but for now, note that the Barenblatt equation is guaranteed to be renormalisable, by virtue of the rigorous proof of existence of the solutions that we seek. In other words, if the solution to the initial value problem did not exist, then it would not be possible for us to find a solution by renormalisation, or by any other method. The fact that there does exist a finite solution means that the divergence of the *terms* of perturbation theory are merely artifacts, arising from the fact that the perturbation is around the solution to the diffusion equation, which *does* conserve u, whereas the Barenblatt solution does *not* conserve u.

The renormalisation procedure is not obviously useful, in that it removes the divergences at the expense of introducing the arbitrary length scale μ. Since μ was not present in the formulation of the problem, it is not possible that u can depend upon it. Thus, we have the **renormalisation group equation**

$$\mu\frac{du}{d\mu} = 0, \tag{10.98}$$

where the derivative is taken at fixed values of all the bare parameters m_ℓ, ℓ, ϵ, x, and t. Let

$$\sigma \equiv \frac{\mu}{\sqrt{\kappa t}}. \tag{10.99}$$

Evaluating the derivative in eqn. (10.98), we obtain

$$\frac{dm}{d\mu}F + m\sigma\frac{\partial F}{\partial \sigma} = 0, \tag{10.100}$$

with solution

$$F\left(\frac{x}{\sqrt{\kappa t}}, \frac{\mu}{\sqrt{\kappa t}}, \epsilon\right) = \left(\frac{\mu}{\sqrt{\kappa t}}\right)^{2\alpha} \varphi\left(\frac{x}{\sqrt{\kappa t}}, \epsilon\right). \tag{10.101}$$

Here the function φ is yet to be determined, and α is defined by

$$\frac{d\log m}{d\log \mu} = -2\alpha, \tag{10.102}$$

where m is evaluated from eqn. (10.94) in the limit $\ell \to 0$. Thus, the renormalised mass at scale μ is related to the renormalised mass at scale ρ by

$$m(\mu) = m(\rho)\left(\frac{\rho}{\mu}\right)^{2\alpha}, \tag{10.103}$$

and in the intermediate asymptotic limit, we have

$$u(x,t) = \frac{m(\mu)}{\sqrt{\kappa t}}\left(\frac{\mu}{\sqrt{\kappa t}}\right)^{2\alpha}\varphi\left(\frac{x}{\sqrt{\kappa t}},\epsilon\right) \tag{10.104}$$

Thus, once again we find that as $t \to \infty$, $u \sim t^{-(1/2+\alpha)}$. It is important to note that the function F is supposed to have been explicitly calculated by (*e.g.*) perturbation theory; thus eqn. (10.102) enables the anomalous dimension α to be explicitly calculated.

A slightly different argument than that given above is to re-write eqn. (10.96) as

$$u(x,t) = \frac{Z^{-1}(\ell/\mu)m_\ell}{\sqrt{\kappa t}}F\left(\frac{x}{\sqrt{\kappa t}},\frac{\mu}{\sqrt{\kappa t}},\epsilon\right). \tag{10.105}$$

Then, the renormalisation group equation (10.98) implies that

$$-\frac{d\log Z}{d\log\mu}F + \sigma\frac{\partial F}{\partial\sigma} = 0 \tag{10.106}$$

so that in the limit $\ell \to 0$, assuming renormalisability,

$$\frac{d\log Z}{d\log\mu} = 2\alpha \tag{10.107}$$

and

$$Z \sim \left(\frac{\ell}{\mu}\right)^{-2\alpha}. \tag{10.108}$$

Solving eqn. (10.106), we again find that in the intermediate asymptotic regime

$$u(x,t) = \frac{m_\ell}{\sqrt{\kappa t}}\left(\frac{\ell}{\mu}\right)^{2\alpha}\left(\frac{\mu}{\sqrt{\kappa t}}\right)^{2\alpha}\varphi\left(\frac{x}{\sqrt{\kappa t}},\epsilon\right). \tag{10.109}$$

This concludes our general discussion of renormalisation in the Barenblatt equation.

Let us now turn to the separate question of the validity of perturbation theory. We have explained that the *individual terms* in perturbation theory are divergent in the limit of the regularisation parameter tending to zero. These divergences are the ones that may be absorbed into a redefinition of the phenomenological parameters of the theory through renormalisation. However, the resultant renormalised perturbation expansion may *not* be a convergent series. Indeed, in field theory, the perturbation expansion is invariably in one of the coupling constants, and can be shown to be an asymptotic series at best, by Dyson's argument, given earlier. The divergence of the *summed* perturbation series is an artifact of the expansion itself, and *cannot* be removed by renormalisation, because it has no physical significance. How good an approximation is perturbation theory, then? The answer depends on how the coupling constants change under renormalisation. In quantum electrodynamics, the coupling constant g (in this case the fine structure constant) *increases* at short distances. Thus, the individual terms in the *renormalised* perturbation expansion increase, because each term is proportional to a power of g. Thus, perturbation theory breaks down at short distances in this theory. On the other hand, in quantum chromodynamics, it turns out that the coupling constant g *decreases* at short distances, and thus the perturbation expansion about a non-interacting theory becomes more and more accurate at short distances and high energies. This useful property is known as (ultra-violet) **asymptotic freedom.**

In the Barenblatt equation, the perturbation expansion is in the parameter ϵ, and can be shown to be analytic. In field theory and statistical mechanics, expansions of the anomalous dimensions can be performed in the parameter $\epsilon \equiv d_{uc} - d$, where d is the dimension of space and d_{uc} is the upper critical dimension. This expansion is asymptotic.

10.4.5 Assumption of Renormalisability

Let us pause to discuss the assumption of renormalisability. In eqn. (10.96), the function F is written as having no ℓ dependence, because we have taken the limit $\ell \to 0$. Thus, in the statement of renormalisability, *before* taking the limit $\ell \to 0$, we should really have

$$u(x,t) = \frac{m}{\sqrt{\kappa t}}\tilde{F}\left(\frac{x}{\sqrt{\kappa t}}, \frac{\mu}{\sqrt{\kappa t}}, \frac{\ell}{\sqrt{\kappa t}}, \epsilon\right), \tag{10.110}$$

with $\tilde{F}$ given by

$$Z\left(\frac{\ell}{\mu}\right) f\left(\frac{x}{\sqrt{\kappa t}}, \frac{\ell}{\sqrt{\kappa t}}, \epsilon\right) = \tilde{F}\left(\frac{x}{\sqrt{\kappa t}}, \frac{\mu}{\sqrt{\kappa t}}, \frac{\ell}{\sqrt{\kappa t}}, \epsilon\right). \tag{10.111}$$

The basic idea of renormalisation is to remove the singular dependence on ℓ, and therefore, if a theory is renormalisable, we require that $\tilde{F}$ can be written formally as

$$\tilde{F}\left(\frac{x}{\sqrt{\kappa t}},\frac{\mu}{\sqrt{\kappa t}},\frac{\ell}{\sqrt{\kappa t}},\epsilon\right)=F\left(\frac{x}{\sqrt{\kappa t}},\frac{\mu}{\sqrt{\kappa t}},0,\epsilon\right)+\sum_{n=1}^{\infty}\left(\frac{\ell}{\sqrt{\kappa t}}\right)^{n}A_{n}\left(\frac{x}{\sqrt{\kappa t}},\frac{\mu}{\sqrt{\kappa t}},\epsilon\right),\quad(10.112)$$

which defines the finite functions A_n. The sum represents the corrections to the intermediate asymptotic limit $\ell \to 0$, and gives rise to corrections to scaling.

The statement of renormalisability is that it is possible to construct Z and $\tilde{F}$ satisfying eqn. (10.112) in which the limit $\ell \to 0$ is well-defined. Thus, the existence of the limit

$$\lim_{\ell\to 0}\frac{d\log Z}{d\log\mu}=2\alpha,\qquad(10.113)$$

and hence the validity of the renormalisation group equation are equivalent to the assertion that the theory is renormalisable. This, in turn, is nothing more than the assertion that the Barenblatt equation has a finite solution at long times, with an intermediate asymptotic regime as $\ell/\sqrt{\kappa t} \to 0$. The validity of this assertion follows from the theorem of Kamenomstskaya, referred to earlier.

10.4.6 Renormalisation and Physical Theory

Another way of interpreting the existence of a solution is that a well-defined solution exists in the intermediate asymptotic limit, with a finite mass m, even as $\ell \to 0$. In other words, a description of the system is possible even when the regularisation parameter ℓ has been taken to zero: the phenomenology at late times is insensitive to the behaviour at short times. Moreover, renormalisability implies that a phenomenological description of the system exists, because if the theory were not renormalisable, one would not be able to define a renormalised mass as $\ell \to 0$.

What is a phenomenological description of a physical system? In constructing physical theories, we always use a certain level of description. Thus, in describing the long wavelength behaviour of a magnet, we write down equations for the coarse-grained magnetisation. In describing the motion of a fluid, we write down equations of motion for the velocity

field $\mathbf{v}(\mathbf{r},t)$; however, the velocity field is actually a coarse-grained velocity field, defined by averaging the velocity of many particles in a small volume element in the neighbourhood of the point $\mathbf{r}$. In these and other examples, the variables of interest are always defined with respect to some coarse-graining process. In the two examples above, the coarse-graining was in space, so that phenomena below some scale Λ^{-1} are subsumed somehow into the equations for the coarse-grained variables of interest. Often, Λ^{-1} is a regularisation parameter for the theory, as we have seen in critical phenomena, for example. In non-equilibrium problems, such as the Barenblatt equation, the coarse-graining is in time: no information is available on the system for the initial time when the width of the distribution u was ℓ. Again, ℓ is a regularisation parameter in this example.

In a phenomenological description of a system, we try to describe the behaviour solely in terms of the coarse-grained variables, without reference to the microscopic physics on scales shorter than the coarse-graining length. Inevitably, the description of the coarse-grained variables introduces other parameters: for example, in a magnet, the parameters of the Landau free energy or in a fluid, the coefficient of viscosity. These phenomenological parameters *are* determined by the microscopic physics: for example, the viscosity of a fluid may be calculated, with some approximations, from kinetic theory. A successful phenomenological theory contains only a finite number of such parameters (the smaller the better), and does not attempt to calculate the phenomenological parameters. These are taken to be inputs to the theory.

How does a phenomenological theory change when the microscopic physics is altered, for example by varying the regularisation parameter Λ^{-1}? From the above, the only possible change can be in the *values* of the phenomenological parameters. If new phenomenological parameters have to be introduced whenever the microscopic physics is changed, then the theory is not, by definition, phenomenological. In this sense, then, a successful phenomenological theory should be insensitive to changes in the microscopic physics, although the phenomenological parameters may change.

We conclude from the above discussion that only a renormalisable model can generate a successful phenomenology. Failure of renormalisability would imply that one requires an infinite number of parameters to obtain a well-defined finite description of the physical system of interest, in the limit of small regularisation parameter, and thus, that a phenomenological description of that physical system is not possible.

An interesting corollary of this discussion is that *all* our physical theories are phenomenological. Even the highly successful theories of high energy physics, such as the **electroweak theory** and **quantum chromodynamics**, are phenomenological, precisely because they are renormalisable. They do contain phenomenological parameters, and are perfectly self-consistent, renormalisable theories. Thus, they do not predict their regime of applicability. Only by doing experiments can we determine if there are phenomena which fall outside this description. On the other hand, non-renormalisable theories, such as Fermi's four-fermion theory of the weak interaction, are insensitive to changes in the regularisation parameter only within some range. For example, it turns out that for processes at energies less than about 300 GeV, the four-fermion theory gives a reasonable description, whereas for higher energies, new physics is encountered, which is not described by the theory. The fact that the theory is not renormalisable is an indication that it will have a limited range of validity.

Why do we seek renormalisable theories of physics? Historically, the main reason has been pragmatism: one wants a theory which will always give sensible (*i.e.* finite) results. However, it is also clear that there is no guarantee *a priori* that the predictions of such a theory will always agree with experiment. A renormalisable theory will show no signature of the fact that it may be being used in a physical situation where it is inapplicable.

10.4.7 Renormalisation of the Modified Porous Medium Equation

We now move on from Barenblatt's equation, and briefly discuss the renormalisation[15] of the modified porous medium equation in d dimensions, which we will write in the form

$$\partial_t u(\mathbf{r},t) = D\kappa\Delta_d\, u(\mathbf{r},t)^{1+n}, \tag{10.114}$$

with

$$D = \begin{cases} 1 & \partial_t u < 0; \\ 1-\epsilon & \partial_t u > 0. \end{cases} \tag{10.115}$$

where $\epsilon > 0$.

The new element here is that we will find that there are two anomalous dimensions in the solution, although they are related by a scaling law. This is consistent with the fact that it is still only the mass which becomes renormalised. The intrinsic interest of the porous medium and modified porous medium equations is that they exhibit propagating fronts, in contrast to the diffusion and Barenblatt equations, whose solutions exhibit exponentially decaying tails at spatial infinity.

---

[15] L.-Y. Chen, N.D. Goldenfeld and Y. Oono, *Phys. Rev. A* **44**, 6544 (1991).

We begin by using the results of exercise 10–1 to write down the general form of the solution to the modified porous medium equation:

$$u(r,t) = \frac{m_\ell}{(m_\ell{}^n t)^{d\theta}} f\left(\frac{r}{(m_\ell{}^n t)^\theta}, \frac{\ell}{(m_\ell{}^n t)^\theta}, \epsilon\right). \tag{10.116}$$

The result of exercise 10–1 implies that we need to introduce the renormalised mass $m = Z^{-1}(\ell/\mu, \epsilon)m_\ell$. The statement of renormalisability is, then, that in the limit $\ell \to 0$,

$$u(r,t) = \frac{(Zm)}{[(Zm)^n t]^{d\theta}} f\left(\frac{r}{[(Zm)^n t]^\theta}, \frac{\ell}{[(Zm)^n t]^\theta}, \epsilon\right) \tag{10.117}$$

$$= \frac{m}{(m^n t)^{d\theta}} F\left(\frac{r}{(m^n t)^\theta}, \frac{\mu}{(m^n t)^\theta}, \epsilon\right), \tag{10.118}$$

where F is to be determined. Let the first argument of F be denoted by ξ and the second argument by σ. Then the renormalisation group equation is

$$a\xi\frac{\partial F}{\partial \xi} + b\sigma\frac{\partial F}{\partial \sigma} = cF, \tag{10.119}$$

where

$$a = \gamma n\theta; \quad b = 1 + n\theta\gamma; \quad c = (1 - nd\theta)\gamma, \tag{10.120}$$

and the anomalous dimension is

$$\gamma \equiv \lim_{\ell \to 0} \frac{d\log Z}{d\log \mu}. \tag{10.121}$$

The solution is obtained by the **method of characteristics**, which is explained in the appendix to this chapter, and is

$$F(\xi, \sigma, \epsilon) = \sigma^{c/b}\varphi\left(\frac{\xi}{\sigma^{a/b}}, \epsilon\right), \tag{10.122}$$

where φ is to be determined. Hence, we find that the long time behaviour must be of the form

$$u(r,t) \sim t^{-(d\theta+\alpha)}\varphi\left(\frac{r}{t^{\beta+\theta}}, \epsilon\right), \tag{10.123}$$

with exponents

$$\alpha = \frac{\gamma\theta(1 - nd\theta)}{1 + n\gamma\theta}, \tag{10.124}$$

$$\beta = -\frac{n\gamma\theta^2}{1 + n\gamma\theta}. \tag{10.125}$$

These exponents are not independent, but satisfy a scaling law, reflecting the fact that there is only one anomalous dimension because only one renormalisation constant was required:

$$\alpha + \left(\frac{1 - nd\theta}{n\theta}\right)\beta = 0. \tag{10.126}$$

10.5 PERTURBATION THEORY FOR BARENBLATT'S EQUATION

We now proceed to the formal perturbation theory analysis of Barenblatt's equation in one dimension, written in the form

$$\partial_t u(x,t) = D\kappa\partial_x^2 u(x,t), \tag{10.127}$$

with

$$D = \begin{cases} 1/2 & \partial_t u > 0; \\ (1+\epsilon)/2 & \partial_t u < 0. \end{cases} \tag{10.128}$$

For convenience, we will choose units so that the numerical value of κ is unity, and we will drop it from the equations henceforth.

10.5.1 Formal Solution

Let $X(t,\epsilon)$ be the positive value of x where $\partial_t u(x,t) = 0$. Then, we can write Barenblatt's equation as

$$\left[\partial_t - \frac{1}{2}\partial_x^2\right] u(x,t) = \frac{\epsilon}{2}\theta(X(t,\epsilon) - |x|)\,\partial_x^2 u(x,t), \tag{10.129}$$

with a general solution of the form

$$\begin{aligned} u(x,t) = &\int_{-\infty}^{\infty} dy\, G(x-y,t)\, u(x,0) \\ &+ \frac{\epsilon}{2}\int_0^t ds \int_{-\infty}^{\infty} dy\, G(x-y,t-s)\,\theta\left[-\partial_s u(y,s)\right]\partial_y^2 u(y,s), \end{aligned} \tag{10.130}$$

where G is the Green function

$$G(x,t) \equiv \frac{1}{\sqrt{2\pi t}}e^{-x^2/2t}. \tag{10.131}$$

The idea behind solving Barenblatt's equation by perturbation theory is that when $\epsilon = 0$, the equation is the soluble diffusion equation; therefore, it seems reasonable that, for small ϵ, the solution is in some sense close to that of the diffusion equation. In fact, this point of view is slightly naïve, as we have already seen. Even for small ϵ, there is a *qualitative* difference between the diffusion equation and Barenblatt's equation: one respects a conservation law, whereas the other does not. Nevertheless, we will see that it is the case that the shape of the solution to Barenblatt's equation is *not* qualitatively different from that of the diffusion equation. The term proportional to ϵ in Barenblatt's equation is an example of a **singular perturbation**[16] to a solved problem, where the behaviour for $\epsilon = 0$ is completely different from the behaviour in the limit $\ell \to 0$. A physical example of a singular perturbation is the operator $-(\hbar^2/2m)\partial_x^2$ in Schrödinger's equation for a particle of mass m, where $\hbar$ is Planck's constant: the behaviour of a system when $\hbar = 0$ is quite different from that when $\hbar$ is very small.

10.5.2 Zeroth Order in ϵ

Our perturbation theory proceeds by assuming that the solution can be written in the form

$$u(x,t) = u_0(x,t) + \epsilon u_1(x,t) + \epsilon^2 u_2(x,t) + O(\epsilon^3). \tag{10.132}$$

We assume that $u_n(x,0) = 0$ for $n \geq 1$. The algorithm is to substitute this form of the solution into the formal solution (10.130) and match powers of ϵ on both sides of the equation. For simplicity, we will take as our initial condition the Gaussian profile

$$u_0(x,0) = \frac{m_0}{\sqrt{2\pi\ell^2}} e^{-x^2/2\ell^2}, \tag{10.133}$$

where m_0 is assumed to be a constant. Then the zeroth-order solution is just

$$u_0(x,t) = \frac{m_0}{\sqrt{2\pi(t+\ell^2)}} \exp\left(-\frac{x^2}{2(t+\ell^2)}\right). \tag{10.134}$$

---

[16] See (*e.g.*) C.M. Bender and S.A. Orszag, *Advanced Mathematical Methods for Scientists and Engineers* (McGraw-Hill, New York, 1978); J. Kevorkian and J.D. Cole, *Perturbation Methods in Applied Mathematics* Applied Mathematics Series, vol. 34 (Springer-Verlag, New York, 1981).

10.5.3 First Order in ϵ

To obtain the equation satisfied by the first order correction, we iterate by substituting u_0 into the right hand side of the formal solution (10.130) to give

$$\left[\partial_t - \frac{1}{2}\partial_x^2\right] u_1(x,t) = \frac{\epsilon}{2}\theta(X(t,0) - |x|)\, \partial_x^2 u_0(x,t). \tag{10.135}$$

We can simplify the right hand side by noticing that

$$\frac{1}{2}\partial_x^2 u_0(x,t) = \partial_t u_0(x,t) \tag{10.136}$$

$$= \frac{1}{2}\frac{u_0(x,t)}{t+\ell^2}\left(\frac{x^2}{t+\ell^2} - 1\right). \tag{10.137}$$

Thus, the solution to eqn. (10.135), taking into account the initial condition for u_1, is

$$u_1(x,t) = \frac{1}{2}\int_0^t ds \int_{-X(s,0)}^{X(s,0)} dy\, G(x-y, t-s)\, \frac{u_0(y,s)}{s+\ell^2}\left(\frac{y^2}{t+\ell^2} - 1\right). \tag{10.138}$$

10.5.4 Isolation of the Divergence

This expression for u_1 is divergent as $\ell/\sqrt{t} \to 0$. Here is one way of *isolating* this divergence: we keep t finite and send ℓ to zero. From eqn. (10.137) it is apparent that

$$X(t,0) = \sqrt{t+\ell^2}, \tag{10.139}$$

and thus

$$u_1(x,t) = \frac{m_0}{2}\int_0^t \frac{ds}{s+\ell^2}\int_{-\sqrt{s+\ell^2}}^{\sqrt{s+\ell^2}} dy\, \frac{e^{-(x-y)^2/2(t-s)}}{\sqrt{2\pi(t-s)}}\, \frac{e^{-y^2/2(s+\ell^2)}}{\sqrt{2\pi(s+\ell^2)}}\left(\frac{y^2}{s+\ell^2} - 1\right). \tag{10.140}$$

Now we make the substitution

$$w = \frac{y}{\sqrt{s+\ell^2}}, \tag{10.141}$$

which gives

$$u_1(x,t) = \frac{m_0}{4\pi}\int_0^t \frac{ds}{s+\ell^2}\frac{1}{\sqrt{t-s}}\int_{-1}^{1} ds\, e^{-w^2/2}(w^2-1) \times \exp\left[-\left(w\sqrt{s+\ell^2} - x\right)^2/2(t-s)\right]. \tag{10.142}$$

We are interested in the behaviour at small ℓ. The integral $\int ds/(s+\ell^2)$ looks suspiciously logarithmically divergent (due to the *lower* limit) as $\ell \to 0$, but what about the rest of the integral? In fact, the rest of the integral is regular in ℓ. The only other place where ℓ enters is in the exponential factor, and there is nothing singular about this as $\ell \to 0$. So, we can drop the factor of ℓ from the exponent, knowing that the correction is of $O(\ell)$, and therefore regular as $\ell \to 0$. Now let us examine the behaviour of the integrand at small s, which is the important region as $\ell \to 0$. For $s \to 0$, there is no new singular behaviour other than the logarithmic divergence from the first term in the integrand. Thus, for the purposes of isolating the singularity only, we can set $s = 0$ in the terms of the integrand to the right of $(s+\ell^2)^{-1}$. Finally, then, we write u_1, in the limit $\ell \to 0$, as the sum of a singular part u_1^s and a regular part u_1^r:

$$u_1(x,t) = u_1^s(x,t) + u_1^r(x,t), \tag{10.143}$$

where

$$u_1^s(x,t) = \frac{m_0}{4\pi\sqrt{t}} e^{-x^2/2t} \int_0^t \frac{ds}{s+\ell^2} \int_{-1}^1 dw\, e^{-w^2/2}(w^2-1), \tag{10.144}$$

and

$$u_1^r(x,t) = \frac{m_0}{4\pi} \int_0^t \frac{ds}{s+\ell^2} \int_{-1}^1 dw \left[\frac{e^{-(w\sqrt{s}-x)^2/2(t-s)}}{\sqrt{t-s}} - \frac{e^{-x^2/2t}}{\sqrt{t}} \right] \times e^{-w^2/2}(w^2-1) + O(\ell). \tag{10.145}$$

Now let us evaluate the integrals appearing in u_1^s.

$$\int_{-1}^1 dw\,(w^2-1)\,e^{-w^2/2} = -2e^{-1/2}, \tag{10.146}$$

and thus, as $\ell \to 0$, u_1^s is given by

$$u_1^s = -\,2e^{-1/2}\log\left(\frac{t}{\ell^2}\right)\frac{m_0}{4\pi\sqrt{t}}e^{-x^2/2t} \tag{10.147}$$

$$= -\frac{1}{\sqrt{2\pi e}} u_0(x,t)\log\left(\frac{t}{\ell^2}\right). \tag{10.148}$$

This is the divergence of perturbation theory, advertised in section 10.4.4. Our complete expression *to this order in* ϵ is thus, as $\ell \to 0$,

$$u(x,t) = \frac{m_0}{\sqrt{2\pi t}} e^{-x^2/2t}\left[1 - \frac{\epsilon}{\sqrt{2\pi e}}\log\left(\frac{t}{\ell^2}\right) + O(\epsilon^2)\right] + O(\ell,\epsilon), \tag{10.149}$$

where the term in square brackets is divergent as $\ell \to 0$, and the terms of $O(\ell, \epsilon)$ are regular in this limit. In the remainder of the calculation, it is important to recognise that we have only made a computation accurate to $O(\epsilon)$, and at this stage, we can say nothing about higher orders in perturbation theory.

10.5.5 Perturbative Renormalisation

The next step is to remove the divergence by renormalisation of the mass m_0, as explained in section 10.4. Note that in proceeding blindly from our perturbation theory, we have at no stage had to confront the fact that the mass is not conserved. Only now are we forced to recognise this feature of the physics, by the divergence of perturbation theory.

According to renormalisation theory, we introduce the renormalised mass

$$m = Z^{-1}(\ell/\mu, \epsilon)m_0 \tag{10.150}$$

so that

$$u(x,t) = \frac{Zm}{\sqrt{2\pi t}}e^{-x^2/2t}\left[1 - \frac{\epsilon}{\sqrt{2\pi e}}\log\left(\frac{t}{\ell^2}\right) + O(\epsilon^2)\right] + O(\ell, \epsilon). \tag{10.151}$$

In applying renormalisation to our perturbation theory, the best that we can hope for is to be able to remove the divergences encountered at each order in perturbation theory. Since we have no information about perturbation theory above $O(\epsilon)$, we can only remove the divergences that we have encountered at $O(\epsilon)$. Thus, we assume that we can expand the renormalisation constant Z in a power series in ϵ, and choose the coefficients to remove the divergence in the bare perturbation theory, *i.e.*

$$Z = 1 + \sum_{n=1}^{\infty} a_n(\ell/\mu)\epsilon^n. \tag{10.152}$$

The coefficients a_n are determined order by order in ϵ to make $u(x,t)$ finite. To first order in ϵ, we choose

$$a_1(\ell/\mu) = \frac{1}{\sqrt{2\pi e}}\log\left(\frac{C_1\mu^2}{\ell^2}\right), \tag{10.153}$$

where C_1 is an arbitrary constant. This is the most general form of a_1 which will cancel off the divergence at $O(\epsilon)$, and thus

$$Z = 1 + \frac{2\epsilon}{\sqrt{2\pi e}}\log\left(\frac{C_1\mu}{\ell}\right). \tag{10.154}$$

Indeed, substituting this into eqn. (10.151), we obtain

$$\begin{aligned} u(x,t) = {} & \frac{m}{\sqrt{2\pi t}}e^{-x^2/2t}\left[1 + \frac{\epsilon}{\sqrt{2\pi e}}\log\left(\frac{C_1\mu^2}{\ell^2}\right) + O(\epsilon^2)\right] \\ & \times \left[1 - \frac{\epsilon}{\sqrt{2\pi e}}\log\left(\frac{t}{\ell^2}\right) + O(\epsilon^2)\right] + O(\ell, \epsilon). \end{aligned} \tag{10.155}$$

The cancellation of the divergence occurs because

$$\left[1+\frac{\epsilon}{\sqrt{2\pi e}}\log\left(\frac{C_1\mu^2}{\ell^2}\right)+O(\epsilon^2)\right]\times\left[1-\frac{\epsilon}{\sqrt{2\pi e}}\log\left(\frac{t}{\ell^2}\right)+O(\epsilon^2)\right]$$
$$=1-\frac{\epsilon}{\sqrt{2\pi e}}\left[\log\left(\frac{t}{\ell^2}\right)+\log\left(\frac{\ell^2}{C_1\mu^2}\right)+O(\epsilon)\right]. \quad (10.156)$$

As promised, after renormalisation, in the $\ell \to 0$ limit, the singular dependence on ℓ has been removed.

We can, at this stage, use our general RG considerations, as expressed by eqn. (10.107), to calculate the anomalous dimension α:

$$\alpha = \frac{1}{2}\frac{d\log Z}{d\log\mu} \quad (10.157)$$
$$= \frac{\epsilon/\sqrt{2\pi e}}{1+2\epsilon\log(C_1\mu/\ell)/\sqrt{2\pi e}} \quad (10.158)$$
$$= \frac{\epsilon}{\sqrt{2\pi e}}+O(\epsilon^2). \quad (10.159)$$

Although this is our desired result, let us continue to see what sense we can make of our renormalised expression for u, eqn. (10.155). The formula does not look very promising: it contains an arbitrary length μ and an arbitrary number C_1. However, we can single out one member of this family of solutions by insisting that at the origin at some specified time t^*, the distribution had the value $Q(t^*)$. Then, the corresponding solution u^* is

$$u^*(x,t) = Q(t^*)\frac{u(x,t)}{u(x,0)}, \quad (10.160)$$

which gives the so-called **renormalised perturbation expansion**

$$u^*(x,t) = Q(t^*)\sqrt{\frac{t^*}{t}}e^{-x^2/2t}\left[1-\frac{\epsilon}{\sqrt{2\pi e}}\log\left(\frac{t}{t^*}\right)+O(\epsilon^2)\right]. \quad (10.161)$$

Note that by working only to $O(\epsilon)$, the arbitrary constant C_1 has disappeared, having been pushed to $O(\epsilon^2)$. If we had worked to $O(\epsilon^2)$, we would have been able to push the arbitrary constants to $O(\epsilon^3)$. Renormalisability requires that a unique solution exist to the problem, and so in a renormalisable theory, all constants introduced during the renormalisation procedure C_1, C_2, *etc.* can be pushed to arbitrarily high order in ϵ. This is essentially a statement of perturbative renormalisability.

How useful is the perturbation expansion for α? Dyson's argument does not apply in this case, so there is no reason *a priori* to expect the expansion to be divergent. In fact, it can be shown[17] that α is an analytic function of ϵ.

10.5.6 Renormalised Perturbation Expansion

Of what use is the renormalised perturbation expansion? The arbitrary time t^* is a reflection of the arbitrary length μ introduced in the renormalisation procedure. For times t of order t^*, the expansion may give a reasonable estimate of the behaviour of $u^*(x,t)$. For times t much larger or smaller than t^*, the logarithmic terms in the perturbation series are not small, however, and it may be expected that the perturbation expansion is poor. On the other hand, since t^* is arbitrary, we may adjust it to be close to or equal to any desired time (in the intermediate asymptotic regime); this is equivalent to the fact that u^* cannot depend on t^*. Varying t^* also changes Q, and in such a way as to compensate for the explicit t^* dependence in u^*. Thus, we may write the renormalisation group equation in the form

$$\frac{du^*}{dt^*} = \frac{\partial u^*}{\partial t^*} + \frac{dQ}{dt^*}\frac{\partial u^*}{\partial Q} = 0. \tag{10.162}$$

We can evaluate to $O(\epsilon)$ all the partial derivatives in eqn. (10.162) from the renormalised perturbation expansion and thus deduce how Q varies with t^*:

$$t\frac{dQ}{dt} = -Q\left[\frac{1}{2} + \frac{\epsilon}{\sqrt{2\pi e}} + O(\epsilon^2)\right]. \tag{10.163}$$

Integrating, we obtain

$$Q(t) \sim t^{-1/2+\alpha(\epsilon)}, \tag{10.164}$$

with $\alpha(\epsilon)$ given by eqn. (10.159). The final step is to substitute this functional form into the renormalised perturbation expansion (10.161) and set $t^* = t$, showing that the long time behaviour of u is indeed of the form

$$u(x,t) \sim \frac{e^{-x^2/2t}}{t^{1/2+\alpha(\epsilon)}} + O(\epsilon). \tag{10.165}$$

---

[17] D.G. Aronson and J.-L. Vazquez (manuscript in preparation); other rigorous results are presented by S. Kamin, L.A. Peletier andJ.-L. Vazquez, Institute for Mathematics and its Applications preprint number 817 (1991).

10.5.7 *Origin of Divergence of Perturbation Theory*

We conclude with some remarks about how RG has improved a perturbation expansion, and show that it is the intrinsically singular nature of the Barenblatt equation which leads to the divergence of the terms in perturbation theory. The bare perturbation expansion result, eqn. (10.149), stated

$$u(x,t) = \frac{m_0}{\sqrt{2\pi t}} e^{-x^2/2t}\left[1 - \frac{\epsilon}{\sqrt{2\pi e}}\log\left(\frac{t}{\ell^2}\right) + O(\epsilon^2)\right] + O(\ell,\epsilon). \quad (10.166)$$

The renormalisation group arguments showed that the quantity in square brackets, which we will call $1+\Gamma$, is the first two terms in the expansion

$$1+\Gamma \approx \exp\left[-\frac{\epsilon}{\sqrt{2\pi e}}\log\left(\frac{t}{\ell^2}\right) + O(\epsilon^2)\right] \quad (10.167)$$

$$= \left(\frac{\ell^2}{t}\right)^{\epsilon/\sqrt{2\pi e}+O(\epsilon^2)}. \quad (10.168)$$

In effect, the RG showed that at each order n in perturbation theory, there *must* be a contribution $\Gamma^n/n!$, which when summed, leads to (10.168). If we had used perturbation theory to $O(\epsilon^2)$ we would find a term $\Gamma^2/2!$ plus a correction. This correction would lead, *via* the RG argument to the $O(\epsilon^2)$ term in the anomalous dimension α.

Lastly, note that although the term Γ in the perturbation expansion is *logarithmically divergent* as $\ell \to 0$, the summed expression (10.168) exhibits quite *different* singular behaviour, which is not even divergent when $\epsilon > 0$. This singular behaviour is precisely that of eqn. (10.76), and from the considerations of section 10.4.2, is an intrinsic part of the physics of Barenblatt's equation.

10.6 FIXED POINTS

The RG that we have presented so far in this chapter grew historically from the attempts to renormalise field theories, such as **quantum electrodynamics** and Landau theory. It is not obviously connected with the picture of the RG described in chapter 9. In this section, we will determine the long time behaviour of Barenblatt's equation by defining an appropriate renormalisation group transformation, and seeking fixed points. There are several different ways to do this, but by proper choice of the RG transformation, the fixed point can be chosen to be the self-similar solution governing the intermediate asymptotics. The fixed point structure of Barenblatt's equation is rather simple and uninteresting, compared with the possibilities that occur in critical phenomena, but other equations may present more interesting possibilities.

10.6.1 Similarity Solutions as Fixed Points

Let us define the RG transformation $R_{b,\phi}[u(x,t)]$ on the function $u(x,t)$ at some *arbitrary* time $t \neq 0$. After the transformation has been performed on $u(x,t)$, one obtains the function $u'(x,t)$. One should think of the t argument in these functions as being a label. The RG transformation depends on two parameters b and ϕ, and is specified by the following three steps:

(*i*) Evolve the function $u(x,t)$ forward in time to $t' = bt$, using the desired equation, in this case Barenblatt's equation, represented formally by a time evolution operator $N[u]$. Call the result $u(x,t')$.

(*ii*) Rescale x by defining $x' = b^{\phi}x$.

(*iii*) Rescale u by an amount $Z(b)$ so that $u'(0,t') = u(0,t)$.

Thus, for each x value, an initial time t, and a time evolution operator N, we construct the number $u'(x,t)$ from the initial set $\{u(x,t); -\infty < x < \infty\}$ by

$$\begin{aligned} u'(x,t) &= R_{b,\phi}[u(x,t)] \\ &= \frac{1}{Z(b)}u(b^{\phi}x, bt). \end{aligned} \tag{10.169}$$

The reason that this RG transformation is potentially useful is that the fixed points of the transformation are the self-similar solutions. Let us see this formally. The transformation forms a semi-group:

$$R_{a,\phi}R_{b,\phi} = R_{ab,\phi} \tag{10.170}$$

and this implies that the rescaling factor has the functional form

$$Z(b) = b^{y} \tag{10.171}$$

for some y. Equivalently, the exponent y is defined by

$$y = \frac{d \log Z}{d \log b}. \tag{10.172}$$

For an arbitrary value of ϕ, there may not exist fixed points of the RG transformation. Let us now suppose that we have chosen a value for which a fixed point exists. The fixed point is a function u^* which maps into itself under the RG transformation:

$$u^*(x,t) = b^{-y}u^*(b^{\phi}x, bt). \tag{10.173}$$

Setting $b = 1/t$, we have

$$u^*(x,t) = t^y u^*\left(xt^{-\phi},1\right), \tag{10.174}$$

which is indeed of the self-similar form. Note also the similarity to the static scaling hypothesis.

The RG transformation step (*i*) is the analogue of the coarse-graining in statistical mechanics, and may be implemented approximately in many different ways, including numerically. For purposes of illustration, let us use the renormalised perturbation expansion (10.161) as our approximation. Then we find that

$$Z(b) = b^{-1/2}\left(1 - \frac{\epsilon}{\sqrt{2\pi e}}\log b\right) + O(\epsilon^2) \tag{10.175}$$

and

$$y = -\left(\frac{1}{2} + \frac{\epsilon}{\sqrt{2\pi e}} + O(\epsilon^2)\right). \tag{10.176}$$

Substituting into eqn. (10.174), we find that the fixed point function behaves at long time as

$$u^* \sim t^{-(1/2+\alpha(\epsilon))} f(xt^{-\phi}), \tag{10.177}$$

where the function f is to be determined, as is the value of ϕ. One way to determine ϕ is to consider one iteration of the RG transformation on our usual Gaussian initial condition (10.4). We find

$$u(b^\phi x, bt) = A_0 b^{-1/2}\exp\left(-b^{2\phi}x^2/2bt\right)\left[1 - \frac{\epsilon}{\sqrt{2\pi e}}\log b + O(\epsilon^2)\right]. \tag{10.178}$$

Only if $\phi = 1/2$ can repeated iterations possibly give a non-trivial fixed point: for $\phi > 1/2$, repeated iterations tend to zero for $x \neq 0$, whereas for $\phi < 1/2$ the function tends towards a constant. In conclusion, we have found the same result as that found using the method of the previous section. Although we have presented the RG transformation as a discrete transformation, it is also possible to formulate it as a continuous transformation. The discrete form is the more useful for numerical applications, however, and has proven to be a more efficient way of determining long time behaviour than direct numerical integration of equations of motion.[18]

[18] L.-Y. Chen and N.D. Goldenfeld, unpublished.

10.6.2 Universality in the Approach to Equilibrium

The use of RG methods to study non-equilibrium phenomena is a subject still in its infancy. Nevertheless, it is appropriate to conclude this chapter with a few remarks about the few existing applications. The key point is that whenever an intermediate asymptotic regime is controlled by an RG fixed point, the dynamics there reflects the *local* behaviour of the RG transformation; thus the dynamics in the intermediate asymptotic regime is universal. Different physical systems with different equations of motion may be driven to the same fixed point, and therefore exhibit identical asymptotic behaviour. A corollary of this picture is that for such processes, it is not necessary to solve exactly the equations of motion if one is interested in the universal behaviour. In particular, for computational studies, one may use a **minimal model** of the dynamics, which is in the same universality class as the system of interest, but which is computationally more efficient. In practice, this means making two assumptions: first, that the behaviour of the system of interest does indeed reflect an RG fixed point, and second, that the efficient computational scheme which one may devise is actually in the same universality class of the system of interest. Computationally efficient methods, motivated by RG considerations have been used to study phase separation[19] and related pattern formation problems,[20] block copolymers,[21] crystal growth[22] and the kinetics of the superconducting transition.[23] Algorithms based on rescaling, but not exploiting fixed points, have also been applied to study the finite time blow-up of solutions to such equations as the three-dimensional Euler equations,[24] the non-linear Schrödinger equation[25] and model equations for combustion.[26]

10.7 CONCLUSION

Certain physical problems have asymptotics that cannot be derived from dimensional considerations alone. The RG shows how anomalous

---

[19] A. Shinozaki and Y. Oono, *Phys. Rev. Lett.* **66**, 173 (1991).

[20] See the review by Y. Oono and A. Shinozaki, *Forma* **4**, 75 (1989).

[21] M. Bahiana and Y. Oono, *Phys. Rev. A* **41**, 6763 (1990).

[22] F. Liu and N.D. Goldenfeld, *Phys. Rev. A* **42**, 895 (1990).

[23] F. Liu, M. Mondello and N.D. Goldenfeld, *Phys. Rev. Lett.* **66**, 3071 (1991).

[24] A. Chorin, *Comm. Pure Appl. Math.* **34**, 858 (1981).

[25] M.J. Landman, G.C. Papanicolaou, C. Sulem and P. Sulem, *Phys. Rev. A* **38**, 3837 (1988).

[26] M. Berger and R.V. Kohn, *Comm. Pure Appl. Math.* **41**, 841 (1988).

dimensions arise in these problems, and provides the possibility of computing the anomalous dimensions. The alternative to the RG approach is to guess the form of the asymptotics, in which case, the anomalous dimensions are non-linear eigenvalues of a boundary value problem. The RG not only applies to field theories and statistical mechanics, but to other areas of physics, including non-equilibrium phenomena, which are not formulated as functional integrals, and for which there is apparently no notion of physical scale invariance.

Divergent terms may arise in perturbation theory, whenever one is, in some sense, expanding about the wrong solution. In critical phenomena, the wrong solution is actually mean field theory, whereas in Barenblatt's equation, the wrong solution is that of the diffusion equation. In a renormalisable theory, the divergent terms are artifacts of the perturbation theory, and can be removed by renormalisation. The divergences reflect the singular nature of the problem, which is, however, intrinsic.

APPENDIX 10 - METHOD OF CHARACTERISTICS

This appendix briefly summarises the **method of characteristics**, which is a way to find the general solution of a quasi-linear first order partial differential equation. For simplicity, we will describe the case where a variable u is a function of two independent variables x and y, and satisfies the equation

$$a(x,y,u)u_x + b(x,y,u)u_y = c(x,y,u), \tag{A10.1}$$

where a, b and c are given functions, and as a boundary condition u is specified on a curve C in (x,y), subject to some technical constraints, which we will mention at the end. The notation u_x means $\partial_x u$, *etc.*

The crucial observation is that the solution can be written in three dimensional space as $z = u(x,y)$, which describes a surface known as the **integral surface.** Then the original PDE (A10.1) has a useful geometrical interpretation, as follows. Consider the function

$$F(x,y,z) \equiv u(x,y) - z = 0, \tag{A10.2}$$

for which $dF = 0$, and hence

$$(u_x, u_y, -1) \cdot d\mathbf{r} = 0. \tag{A10.3}$$

This shows that the vector $(u_x, u_y, -1)$ is a normal to the integral surface; but eqn. (A10.1) implies that $(u_x, u_y, -1)$ is also perpendicular to (a,b,c).

Therefore, the vector (a, b, c) is tangential to the surface. Hence the curve $\mathbf{r}(\lambda)$ parameterised by the distance along the curve λ, and given by

$$\frac{d\mathbf{r}}{d\lambda} = (a(x,y,u), b(x,y,u), c(x,y,u)) \tag{A10.4}$$

lies in the integral surface.

Our strategy for constructing the desired solution, the integral surface, is to start on a point x_0, y_0, z_0 on the curve C and solve the so-called **characteristic equations**

$$\frac{dx}{d\lambda} = a(x,y,z); \quad \frac{dy}{d\lambda} = b(x,y,z); \quad \frac{dz}{d\lambda} = c(x,y,z) \tag{A10.5}$$

to obtain the characteristics

$$x = x(x_0, y_0, z_0; \lambda); \quad y = y(x_0, y_0, z_0; \lambda); \quad z = z(x_0, y_0, z_0; \lambda). \tag{A10.6}$$

This characteristic is a curve starting at x_0, y_0, z_0 and lying in the integral surface. Now move to a point on C adjacent to x_0, y_0, z_0 and solve for the new characteristic. In this way, moving along C, we can sweep out the integral surface as a one parameter family of characteristics, the parameter being the one which generates the curve C. Note that it is often useful to write the characteristic equations in the compact form

$$\frac{dx}{a} = \frac{dy}{b} = \frac{dz}{c} = d\lambda. \tag{A10.7}$$

Example 1:

Solve the equation

$$uu_x + u_y = 1 \tag{A10.8}$$

with initial data parameterised by the variable s on a curve C with $x = s$, $y = s$ and $u = s/2$, for $0 \leq s \leq 1$.

Solution: the characteristic equations are, with $z = u$

$$\frac{dx}{d\lambda} = u; \quad \frac{dy}{d\lambda} = 1; \quad \frac{dz}{d\lambda} = 1, \tag{A10.9}$$

with initial conditions $x(0, s) = s$, $y(0, s) = s$ and $z(0, s) = u(0, s) = s/2$. Solve the equation for z and y to obtain

$$z = u = \lambda + s/2; \quad y = \lambda + s. \tag{A10.10}$$

Now solve the equation for x to obtain

$$x = \frac{1}{2}\lambda^2 + \frac{1}{2}s\lambda + s. \tag{A10.11}$$

The solution can now be written down by inverting the equations for x and y to give $\lambda = \lambda(x, y)$ and $s = s(x, y)$, and then substituting these expressions into eqn. A(10.10), yielding $u = u(x, y)$.

Caveats:

Finally, two caveats about the initial data. It should be clear from the geometrical construction of the solution that if the initial data lie on a characteristic, then the solution to the characteristic equations will not sweep out the integral surface. Secondly, the initial data must not be specified on the curve $dy/dx = b/a$; such data would imply that we could not perform the inversion step above and obtain λ and s as functions of x and y.

Example 2:

Solve the quasi-linear eqn. (A10.1) and determine the form of $f(\xi, \eta)$, when a, b and c are constants.

Solution: introduce the **dilation parameter** $t \equiv \exp \lambda$. Then the characteristic equations are

$$\frac{d\xi}{a\xi} = \frac{d\eta}{b\eta} = \frac{df}{cf} = \frac{dt}{t}. \tag{A10.12}$$

The solutions are

$$\xi = t^a \overline{\xi}(s); \quad \eta = t^b \overline{\eta}(s); \quad f = t^c \overline{f}(s), \tag{A10.13}$$

where $\overline{\xi}(s)$, $\overline{\eta}(s)$ and $\overline{f}(s)$ are the values of these quantities at $t = 0$, *i.e.*, on the initial data. To determine $f(\xi, \eta)$, we need to eliminate s and t in favour of ξ and η from the last equality in eqn. (A10.13). The result is

$$f(\xi, \eta) = \eta^{c/b} F\left(\frac{\xi}{\eta^{a/b}}\right), \tag{A10.14}$$

where F is a function to be determined.

EXERCISES

Exercise 10–1

Using dimensional analysis, we will construct a similarity solution for the porous medium equation

$$\partial_t u = \kappa \Delta_d \left(u^{1+n}\right),$$

where Δ_d is the Laplacian in d-dimensions, whose radial part is

$$\frac{1}{r^{d-1}}\partial_r \left(r^{d-1}\partial_r\right),$$

and κ is a constant, which may be taken to have the numerical value 1, without loss of generality.

(a) Show that the radially symmetric similarity solution must be of the form

$$u(r,t) = \frac{Q}{(Q^n t)^{d\theta}} f\left(\frac{r}{(Q^n t)^\theta}\right),$$

where $\theta \equiv (2+nd)^{-1}$, and $Q \equiv \int u(r,0)d^d\mathbf{r}$.

(b) Hence show that

$$f(\xi) = \begin{cases} \left[\frac{n\theta}{2(n+1)}\left(\xi_0^2 - \xi^2\right)\right]^{1/n}, & \xi < \xi_0; \\ 0, & \xi > \xi_0, \end{cases}$$

where $\xi \equiv r/(Q^n t)^\theta$ and ξ_0 is to be determined.

(c) Show that Q is conserved, and hence determine ξ_0.

(d) For $n = 0$, the porous medium equation reduces to the diffusion equation. For $n \neq 0$, the porous medium equation exhibits a propagating front, as we have shown, whereas the diffusion equation has a tail stretching all the way to infinity. Show that in the limit $n \to 0$, the solution in (b) and (c) crosses over smoothly to that of the diffusion equation.

(d) Show that Q is not conserved by the modified porous medium equation.

Exercise 10–2

This question concerns the calculation of the anomalous dimension α of Barenblatt's equation, as discussed in section 10.3.3.

(a) Solve eqns. (10.58) and (10.59), and show that α and ξ_0 are given implicitly by the transcendental equations

$$D_{2\alpha+2}(\xi_0) = 0; \quad F(-\alpha - 1, 1/2, \xi_0^2/2(1+\epsilon)) = 0.$$

Here $D_\alpha(z)$ is the parabolic cylinder function of degree α and $F(a,b,z)$ is the confluent hypergeometric function.[27]

(b) Calculate the anomalous dimension for several values of ϵ ranging from (*e.g.*) 0.001 to (*e.g.*) 0.5, using either the implicit equations derived above or the shooting method discussed in the text.

(c) Find the limiting value of $d\alpha/d\epsilon$ as $\epsilon \to 0$, and compare with the RG value of $1/\sqrt{2\pi e}$. You can also obtain this result analytically from the implicit equations derived in (a).

(d) Show that the analysis of the long time behaviour of the diffusion equation and Barenblatt's equation given in the text for Gaussian initial conditions can be extended to arbitrary initial conditions which are sufficiently localised. Investigate the behaviour for initial conditions with power-law decay at infinity.

Exercise 10-3

This question concerns the long-time behaviour of the modified porous medium equation.

(a) Solve the renormalisation group equation (10.119) for the modified porous medium equation (10.114) and (10.115), using the method of characteristics, and thus derive eqn. (10.122).

(b) Find the long time behaviour of the height $h(r,t)$ of a groundwater mound (r is radial distance from centre of mound), by renormalising the modified porous medium equation for $n = 1$ and $d = 2$, and show that[28]

$$h(r,t) \sim \left[\frac{A}{2(\kappa t)^{1/2+\epsilon/8+O(\epsilon^2)}} - \frac{r^2}{16\kappa t}\right]^+,$$

where the function y^+ is defined to be y when $y > 0$ and 0 when $y < 0$.

Exercise 10-4

This question illustrates how anomalous dimensions may arise even in *linear* problems. We consider the flow of an incompressible fluid past an infinite wedge with opening angle 2α; the wedge is infinitely wide in the z-direction, so that the flow problem may be considered to be two dimensional, in the $x-y$ plane, as shown in the accompanying figure.[29] The flow

---

[27] M. Abramowitz and I.A. Stegun, (eds.) *Handbook of Mathematical Functions* (Dover, New York, 1970).

[28] L.-Y. Chen, N.D. Goldenfeld and Y. Oono, *Phys. Rev. A* **44**, 6544 (1991).

[29] G.I. Barenblatt, *Similarity, Self-Similarity and Intermediate Asymptotics* (Consultants Bureau, New York, 1979); N.D. Goldenfeld and Y. Oono, *Physica A* **177**, 213 (1991).

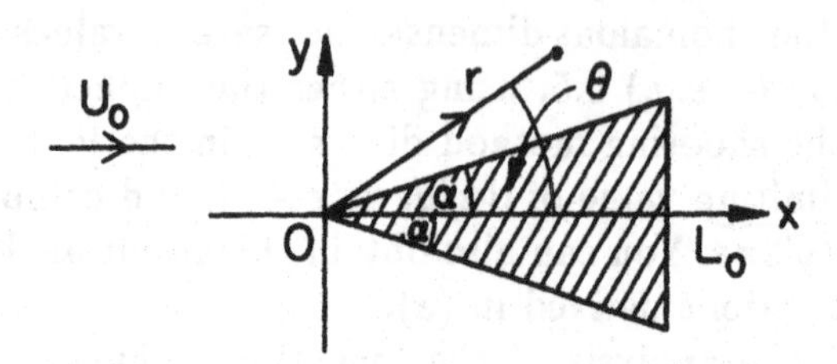

Figure 10.2 Flow of an ideal incompressible fluid past a wedge, as in exercise 10–4.

field $\mathbf{v}$ has a potential ϕ, with $\mathbf{v}(\mathbf{r}) = \nabla\phi(\mathbf{r})$, and due to incompressibility satisfies $\nabla \cdot \mathbf{v} = 0$. Well before impinging on the wedge, the fluid has velocity $\mathbf{v} = (U_0, 0)$.

(a) What are the boundary conditions for ϕ on the sides of the wedge?

(b) ϕ satisfies Laplace's equation. Use cylindrical polar coordinates (r, θ) and dimensional analysis to write down the form of the solution for ϕ. Hence show that it is *not* possible to satisfy the boundary conditions in (a).

(c) Resolve this paradoxical situation, by considering the regularisation where the wedge is allowed to have a finite length L_0, and seeking a solution valid in the region where $r/L_0 \to 0$. Show that a renormalisation of the velocity U_0 at infinity implies that this problem has an anomalous dimension λ, leading to a potential of the form

$$\phi \sim r^{1+\lambda} \cos\left[(\lambda + 1)\theta + \gamma(\alpha)\right].$$

Use the boundary conditions to determine *exactly* λ and γ as functions of α.

(d) Now use perturbation theory to solve the problem. For $\alpha = 0$, we expect $\lambda = 0$. Thus, we try a perturbation theory about $\alpha = 0$, using the complex velocity field $w = v_x - iv_y$. Use conformal mapping to show that the bare complex velocity is

$$w(z) = U_0 + \alpha \left[\frac{U_0}{\pi} \log\left(\frac{z}{L_0}\right) - iU_0\right] + O(\alpha^2).$$

(e) Use the RG to show that the anomalous dimension is

$$\lambda = \frac{\alpha}{\pi} + O(\alpha^2),$$

which should be consistent with your exact result in (c).

CHAPTER 11

Continuous Symmetry

No discussion of phase transitions would be complete without some account of the nature of the correlations in the ordered phase, for systems with a continuous symmetry. This chapter gives an introduction to this important topic, and has two main parts. The first is that when a continuous symmetry is spontaneously broken below a temperature T_c, hence giving rise to **long range order**, transverse correlations (which we define shortly) exhibit power law decay for *all* temperatures $T < T_c$. This result is a form of **Goldstone's theorem**. The second point that we discuss here is the special case of the XY model in two spatial dimensions, where there is a phase transition at a non-zero temperature *in the absence of ordering*: $\langle \mathbf{S} \rangle = 0$ for *all* temperatures! This is the so-called **Kosterlitz-Thouless transition.**

Spontaneous symmetry breaking is ubiquitous, and is accompanied by important phenomena: the acquisition of rigidity, the existence of low energy excitations, and the possibility of topological defects.[1] The physics described in this chapter is relevant to physical systems such as magnets,

[1] See (*e.g.*) P.W. Anderson, *Basic Notions of Condensed Matter Physics* (Benjamin/Cummings, Menlo Park, 1984).

superfluids, superconductors, Heisenberg spin glasses, liquid crystals, rubber, not to mention the phase transitions which may have occured during the early universe.

11.1 CORRELATIONS IN THE ORDERED PHASE

Our discussion of systems with continuous symmetries is based upon the **O(n) model**, which we will treat as the Landau theory for a system with an order parameter **S** with n components ($n > 1$), such as a Heisenberg model. Thus $\mathbf{S} = (S_1, S_2, \ldots, S_n)$ and the statistical mechanics is governed by the effective Hamiltonian

$$-\mathcal{H} = \int d^d\mathbf{r} \left[\frac{1}{2}(\nabla S)^2 + \frac{1}{2}r_0 S^2 + \frac{1}{4}u_0 S^4 - \mathbf{h}\cdot\mathbf{S}\right], \qquad (11.1)$$

where **h** is an external field.[2]

Notation:-

$$(\nabla S)^2 \equiv \sum_{i=1}^{d}\sum_{\alpha=1}^{n}\left[\frac{\partial S_\alpha(\mathbf{r})}{\partial x_i}\right]^2; \qquad (11.2)$$

$$S^2 \equiv \sum_{\alpha=1}^{n}[S_\alpha(\mathbf{r})]^2; \qquad (11.3)$$

$$S^4 \equiv (S^2)^2. \qquad (11.4)$$

The partition function is given by

$$Z = \int D\mathbf{S}\, e^{\mathcal{H}}, \quad DS \equiv \prod_{\alpha=1}^{n} DS_\alpha. \qquad (11.5)$$

The term "O(n)" refers to the fact that in zero external field, $\mathcal{H}$ is invariant when at *each* point in space **r**, the order parameter field **S(r)** is rotated by the *same* angle in n dimensional order parameter space.

The homogeneous part of the Landau free energy density is[3]

$$V(\mathbf{S}) = \frac{1}{2}r_0 S^2 + \frac{1}{4}u_0 S^4, \qquad (11.6)$$

[2] Note that factors of k_BT have been absorbed into the definitions as usual.

[3] Sometimes the homogeneous part of the Landau free energy is called the potential energy and the gradient part is sometimes called the kinetic energy. These terms are misleading in statistical mechanics, and allude to the similarity between the Hamiltonian (11.1) and the Lagrangian of a self-interacting boson field theory.

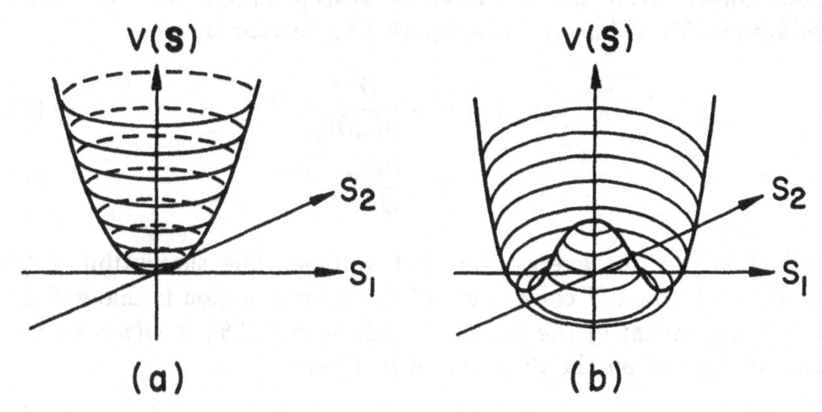

Figure 11.1 Homogeneous part of the Landau free energy for the O(n) model. Sketched for simplicity is the case $n = 2$: (a) $T > T_c$. (b) $T < T_c$.

which has the behaviour shown in figure (11.1), for $\mathbf{h} = \mathbf{0}$. For $r_0 > 0$ there is one minimum at $\mathbf{S} = \mathbf{0}$, corresponding to the high temperature phase of the system which is not ordered, whereas the minimum at $S^2 = -r_0/u_0$ is infinitely degenerate for $r_0 < 0$. The degeneracy corresponds to the directions in order parameter space (*not* real space) in which the system may order below T_c.

11.1.1 The Susceptibility Tensor

In this section, we examine the response function for a system with O(n) symmetry. Suppose the field $\mathbf{h}$ is applied along the direction $\mathbf{n}$ in spin space. Then the spins will order along that direction, and the magnetisation vector is

$$m_\alpha \equiv \langle S_\alpha(x) \rangle = m n_\alpha, \tag{11.7}$$

where m is the magnitude of the magnetisation vector $\mathbf{m}$. We may take $\mathbf{n}$ to be the vector $(1,0,0,\ldots,0)$ and then define the two-point correlation functions

$$\hat{G}_{||}(\mathbf{k}) = \left\langle |S_{1\mathbf{k}}|^2 \right\rangle \tag{11.8}$$

$$V\delta_{\mathbf{k}+\mathbf{k}',0}\delta_{\alpha\beta}\hat{G}_\perp(\mathbf{k}) = \left\langle S_{\alpha\mathbf{k}} S_{\beta\mathbf{k}'} \right\rangle, \quad \alpha, \beta \geq 2. \tag{11.9}$$

The former is the **longitudinal correlation function**, measuring the correlations in the order parameter components parallel to the direction of ordering, whilst the latter is the **transverse correlation function**, measuring the correlations between components of the order parameter which are orthogonal to the direction of ordering. The corresponding susceptibilities are given by the **static susceptibility sum rule**

$$\chi_{||} = \hat{G}_{||}(\mathbf{0}); \quad \chi_\perp = \hat{G}_\perp(\mathbf{0}). \tag{11.10}$$

These **longitudinal** and **transverse** susceptibilities have the following significance. We define the **susceptibility tensor** by

$$\chi_{\alpha\beta} \equiv -\frac{\partial^2 f}{\partial h_\alpha \partial h_\beta}, \tag{11.11}$$

$$\equiv -\frac{\partial m_\alpha}{\partial h_\beta}, \tag{11.12}$$

where f is the free energy per unit volume. The susceptibility tensor describes how the αth component of the magnetisation is changed due to the βth component of the external field. Since $\mathcal{H}\{S\}$ is $O(n)$ symmetric, f cannot depend on the *direction* of $\mathbf{h}$. Hence,

$$f = f(h) \tag{11.13}$$

$$h = |\mathbf{h}| = \left[\sum_{\alpha=1}^{n} h_\alpha^2\right]^{1/2}. \tag{11.14}$$

Thus, differentiating (11.11) and using

$$\frac{\partial}{\partial h_\alpha} = \frac{\partial h}{\partial h_\alpha}\frac{\partial}{\partial h} \tag{11.15}$$

$$= \frac{h_\alpha}{h}\frac{\partial}{\partial h}, \tag{11.16}$$

we get

$$\chi_{\alpha\beta} = -\frac{\partial}{\partial h_\alpha}\frac{h_\beta}{h}\frac{\partial f}{\partial h} \tag{11.17}$$

$$= -\delta_{\alpha\beta}\frac{1}{h}\frac{\partial f}{\partial h} - \frac{h_\alpha h_\beta}{h}\frac{-1}{h^2}\frac{\partial f}{\partial h} - \frac{h_\alpha h_\beta}{h^2}\frac{\partial^2 f}{\partial h^2} \tag{11.18}$$

$$= -\frac{h_\alpha h_\beta}{h^2}\frac{\partial^2 f}{\partial h^2} - \frac{1}{h}\frac{\partial f}{\partial h}\left(\delta_{\alpha\beta} - \frac{h_\alpha h_\beta}{h^2}\right). \tag{11.19}$$

Now, $\mathbf{h} = h\mathbf{n}$ so that

$$\frac{h_\alpha h_\beta}{h^2} = n_\alpha n_\beta. \tag{11.20}$$

Thus

$$\chi_{\alpha\beta} = n_\alpha n_\beta \chi_{\parallel}(h) + \chi_{\perp}(h)(\delta_{\alpha\beta} - n_\alpha n_\beta), \tag{11.21}$$

with the longitudinal susceptibility

$$\chi_{\parallel}(h) = -\frac{\partial^2 f}{\partial h^2}, \tag{11.22}$$

and the transverse susceptibility

$$\chi_{\perp}(h) = -\frac{1}{h}\frac{\partial f}{\partial h} = -\frac{m}{h}. \tag{11.23}$$

Now we can see the physical significance of these quantities: recall that

$$\chi_{\alpha\beta} = -\frac{\partial m_{\alpha}}{\partial h_{\beta}}. \tag{11.24}$$

Equation (11.21) implies that if the probe field $\mathbf{h}$ is in the same direction as $\mathbf{m}$, then the susceptibility is given by $\chi_{\|}$. If the probe field is perpendicular to the direction of $\mathbf{m}$, then the susceptibility is given by $\chi_{\perp}$.

11.1.2 *Excitations for $T < T_c$: Goldstone's Theorem*

The expression for the transverse susceptibility, eqn. (11.23), has profound consequences. For $T < T_c$, the system exhibits spontaneous symmetry breaking when $\mathbf{h} = \mathbf{0}$, and there is **long range order**: $\mathbf{m} \neq \mathbf{0}$. Thus

$$\chi_{\perp}(0)^{-1} = 0. \tag{11.25}$$

Using the static susceptibility sum rule (11.9), this implies that

$$\hat{G}_{\perp}(\mathbf{k} = \mathbf{0})^{-1} = 0 \tag{11.26}$$

In other words, in the transverse direction, the system is *infinitely susceptible* — it requires an infinitesimal amount of field to rotate the direction of magnetisation by a finite, non-zero amount! This is quite reasonable, when we consider the potential $V(\mathbf{S})$ for $r_0 < 0$. There are an infinite number of minima of the effective Hamiltonian, each corresponding to the system spontaneously acquiring a magnetisation along a different direction. It requires no energy to go between these minima, since they are degenerate.

We see that when the O(n) symmetry of $\mathcal{H}$ is spontaneously broken there are two sorts of fluctuation: those *parallel* to the direction in which the system has ordered, and those *perpendicular* to it. These fluctuations cost different amounts of energy: a fluctuation parallel to the direction of ordering (*i.e.* a longitudinal fluctuation) is one which causes the *magnitude* m of $\mathbf{m}$ to change. By inspection of the potential $V(\mathbf{S})$, we see that there is an energy penalty for increasing or decreasing m. On the other hand, fluctuations in the direction perpendicular to $\mathbf{m}$ only change the *direction* of $\mathbf{m}$, and cost no energy, as we have already mentioned.

So far, we have only discussed the case of spatially uniform systems. What happens when we allow for spatial variations of the deviation of the order parameter from the equilibrium value in the ordered state? Now the gradient term in $\mathcal{H}$ must also be considered. Reasoning by continuity, we might expect that longitudinal fluctuations still have high energy, whereas transverse fluctuations will now require a small but non-zero energy, the amount being proportional to $(\nabla S)^2$. Hence, the longer the wavelength of the fluctuation, the smaller the associated free energy. We have made a rather useful deduction: purely because of the ordering, a system with an $O(n)$ symmetry should have an excitation spectrum with modes of arbitrarily low free energy. In fact, there will be one such mode associated with each transverse direction, which makes $n-1$ low energy modes. These modes are called **Goldstone modes.**

Another striking consequence of spontaneous symmetry breaking is the form of the correlation functions below T_c. Let us calculate the two point correlation function for fluctuations about the ordered state, in the Gaussian approximation. The procedure follows that we used to study the Gaussian approximation above T_c:

(*i*) Minimise $-\mathcal{H}$ to find the spatially uniform state.
(*ii*) Calculate the free energy cost of a fluctuation to second order.
(*iii*) Read off the two-point function, using essentially the equipartition of energy, as we did in section 6.3.4.

Step 1

We begin by writing down the potential, using Einstein summation convention, and differentiating to find the mean field solution:

$$V(\mathbf{S}) = \frac{1}{2} r_0 S_\alpha S_\alpha + \frac{1}{4} u_0 S_\alpha S_\alpha S_\beta S_\beta; \tag{11.27}$$

$$\frac{\partial V}{\partial S_\alpha} = r_0 S_\alpha + \frac{1}{4} u_0 (2 S_\alpha (S_\beta S_\beta) \cdot 2) \tag{11.28}$$

$$= S_\alpha (r_0 + u_0 S^2). \tag{11.29}$$

Thus, the mean field solution is, for $r_0 > 0$ (*i.e.* $T > T_c$),

$$S_\alpha = 0 \tag{11.30}$$

and for $r_0 < 0$ (*i.e.* $T < T_c$),

$$\langle S \rangle^2 \equiv m^2 = -r_0 / u_0. \tag{11.31}$$

For $r_0 < 0$, although m^2 is defined, the *direction* of $\mathbf{S}$ is not defined. Therefore, we choose some arbitrary direction of ordering $\mathbf{n} = (1, 0, 0, \ldots, 0)$:

$$\langle \mathbf{S} \rangle = m\mathbf{n}, \tag{11.32}$$

and define the fluctuation $\phi(\mathbf{r})$ by

$$\mathbf{S}(\mathbf{r}) = \mathbf{m} + m\boldsymbol{\phi}(\mathbf{r}). \tag{11.33}$$

Step 2

Now we split up the fluctuation $\boldsymbol{\phi}$ into a part ϕ_1 that is parallel to $\mathbf{m}$ and a part $\boldsymbol{\phi}_\perp$ that is transverse. Thus, we write

$$\mathbf{S} = m\left[\mathbf{n} + \phi_1\mathbf{n} + \boldsymbol{\phi}_\perp\right], \tag{11.34}$$

$$\boldsymbol{\phi}_\perp = (0, \phi_2, \phi_3, \dots \phi_n). \tag{11.35}$$

Note that by including the factor of m in the definition of the fluctuation, eqn. (11.34), $\boldsymbol{\phi}_\perp$ represents the *direction* in order parameter space of the fluctuation.

Now we calculate each term in $V(\mathbf{S})$:

$$S^2/m^2 = 1 + 2\phi_1 + \phi_1^2 + \phi_\perp^2; \tag{11.36}$$

$$S^4/m^4 = 1 + 4\phi_1 + 6\phi_1^2 + 4\phi_1^3 + \phi_1^4 + \phi_\perp^4 + 2\phi_\perp^2 + 4\phi_\perp^2\phi_1 + 2\phi_1^2\phi_\perp^2. \tag{11.37}$$

After substituting into $V(\mathbf{S})$ and some algebra, we find that the quadratic terms are

$$V(\mathbf{S}) = -\frac{1}{4}\frac{r_0^2}{u_0} - r_0 m^2\phi_1^2. \tag{11.38}$$

Note that there are no terms quadratic in $\phi_\perp$. The gradient term is just

$$\frac{1}{2}(\nabla S)^2 = \frac{m^2}{2}(\nabla\phi_1)^2 + \frac{m^2}{2}(\nabla\phi_\perp)^2. \tag{11.39}$$

Finally, the effective Hamiltonian $\mathcal{H}_\phi$ for small fluctuations about the ordered state is given by

$$-\mathcal{H}_\phi\{\phi\} = \frac{m^2}{2}\int d^d\mathbf{r}\,\left[(\nabla\phi_1)^2 + (\nabla\phi_\perp)^2 + (2|r_0|)\phi_1^2\right] + O(\phi_1^3, \phi_1\phi_\perp^2). \tag{11.40}$$

Step 3

Writing this in Fourier space, and identifying the two point correlation functions, we find

$$-\mathcal{H}_\phi\{\phi\} = \frac{1}{V}\sum_{\mathbf{k}}\left[\frac{1}{2}|\hat{\phi}_{1\mathbf{k}}|^2\hat{G}_{\parallel}^0(\mathbf{k})^{-1} + \frac{1}{2}|\hat{\phi}_{\perp\mathbf{k}}|^2\hat{G}_{\perp}^0(\mathbf{k})^{-1}\right], \tag{11.41}$$

with the longitudinal and transverse correlation functions $\hat{G}^0_{\parallel}$ and $\hat{G}^0_{\perp}$ in the Gaussian approximation being

$$\hat{G}^0_{\parallel}(\mathbf{k}) = \frac{m^{-2}}{2|r_0| + k^2} \tag{11.42}$$

and

$$\hat{G}^0_{\perp}(\mathbf{k}) = \frac{m^{-2}}{k^2}. \tag{11.43}$$

The transverse correlation function has no r_0 term in the denominator, and thus the transverse correlations have *power-law* decay for $T < T_c$. The longitudinal correlation function decays *exponentially* below T_c, with a correlation length $\xi_<^{-2} = 2|r_0| = 2\xi_>^{-2}$, which is the result we found in section 5.7.5 for the case $n = 1$. The pole in $\hat{G}^0_{\perp}(\mathbf{k})$ at $k = 0$ is the **Goldstone mode.** We see that for spontaneous symmetry breaking in the O(n) model there is one mode — the longitudinal mode — with a finite correlation length, and $n - 1$ modes — the transverse modes $\phi_\alpha(\alpha = 2 \ldots n)$ — with an infinite correlation length.[4] Other systems with different continuous symmetries may have different numbers of Goldstone modes.

Although our calculation was performed in the Gaussian approximation, the form of eqn. (11.43) is preserved to *all* orders in perturbation theory; this result — Goldstone's theorem — is a direct consequence of the original O(n) symmetry.[5] Goldstone's theorem is generally true when there is a spontaneously broken symmetry, and evasions of the theorem when there are gauge fields present[6] are important in superconductivity and in the electroweak theory, giving rise to a finite electromagnetic penetration depth and intermediate vector bosons respectively. Everyday examples of Goldstone modes include **spin waves** (in magnets) and **phonons** (spontaneous breaking of translational invariance).

---

[4] In quantum field theory, the bare propagator of the boson field ϕ has the form $\hat{G}_0(\mathbf{k}) = 1/(k^2 + m^2)$, where m is the mass of the particle represented by ϕ. Thus m is analogous to ξ^{-1} in statistical mechanics. Goldstone modes are therefore referred to sometimes as *massless* because their propagator is the same as that for massless bosons in quantum field theory.

[5] See (*e.g.*) D.J. Amit, *Field Theory, the Renormalisation Group and Critical Phenomena* (World Scientific, Singapore, 1984), pp. 94-96.

[6] See (*e.g.*) P.W. Anderson, *Basic Notions of Condensed Matter Physics* (Benjamin/Cummings, Menlo Park, 1984); S. Weinberg, *Prog. Theor. Phys. Suppl.* **86**, 43 (1986); H. Wagner, *Z. für Physik* **195**, 273 (1966).

11.1.3 Emergence of Order Parameter Rigidity

Another important consequence of spontaneous symmetry breaking in systems with a continuous symmetry is the emergence of **rigidity**. This is intimately connected with the presence of power law correlations in the transverse response function. Let us examine the transverse contribution to $-\mathcal{H}_\phi$ in real space:

$$-\mathcal{H}_\phi = \text{longitudinal part} + \frac{1}{2}\int d^d\mathbf{r}\, R\,(\nabla\phi_\perp(\mathbf{r}))^2 , \qquad (11.44)$$

where the **stiffness or rigidity** $R = m^2$, within the framework of Landau theory. Recall that $\phi_\perp$ is the direction of a fluctuation about the ordered state. Equation (11.44) shows that *any* spatial variation in the order parameter perpendicular to the direction of ordering raises the energy of the system. Thus, the system exerts a restoring force in response to any attempt to create such a configuration: this is precisely what we mean by rigidity. Within Landau theory, the strength of the response is governed by $R = m^2$, *i.e.* the expectation value of the order parameter. Thus, at least within Landau theory, only when there is long range order and $m \neq 0$, can the phenomenon of rigidity occur.[7] In the high temperature, symmetric phase, the system is not rigid. Nevertheless, the *longitudinal* part of the effective Hamiltonian has the same form in both the high and low temperature phases. It is only by probing the *transverse* response of the system that the state of order can be determined.

Two common examples are the following. Ferromagnets are distinguished from paramagnets by the energy cost of creating a spin wave: the coefficient of $(\nabla\phi_\perp)^2$ is sometimes called the **spin-wave stiffness.** Solids are distinguished from liquids by their ability to support an *infinitesimal* static shear: the analogue of the effective Hamiltonian (11.44) is the coarse-grained free energy of elasticity theory, which for an isotropic solid takes the form[8]

$$F = \frac{1}{2}\int d^d\mathbf{r}\, \left[2\mu u_{ij}^2(\mathbf{r}) + \lambda u_{kk}(\mathbf{r})^2\right] , \qquad (11.45)$$

where $u_{ij}(\mathbf{r})$ is the strain tensor, and λ and μ are the Lamé coefficients. In particular, the existence of the solid state is reflected in the non-zero value of μ, the **static shear modulus.**

---

[7] An exception is the case of certain two-dimensional systems, which undergo the Kosterlitz-Thouless transition; see section 11.2.

[8] L.D. Landau and E.M. Lifshitz, *Theory of Elasticity* (Pergamon, New York, 1986).

11.1.4 Scaling of the Stiffness

Given that the non-zero value of the stiffness is a diagnostic of the the spontaneous breaking of a continuous symmetry, it is of interest to determine how the stiffness rises from zero below the transition temperature. Our Landau theory treatment showed that $R = m^2$, and since $m \sim (-t)^\beta$ for $t < 0$, we conclude that

$$R \sim (-t)^{2\beta}. \tag{11.46}$$

This is, of course not correct, because Landau theory breaks down in the critical region, but we can use scaling to determine the correct relation. The free energy density has units L^{-d}, and therefore scales like $\xi(t)^{-d}$ in the critical region. On the other hand, in physical units where $\phi_\perp$ is dimensionless, the term $(\nabla\phi_\perp)^2$ has dimensions L^{-2}, and thus scales like $\xi(t)^{-2}$. Hence we conclude that

$$R \sim \xi^{2-d} \sim (-t)^{\nu(d-2)} \quad \text{for } t < 0. \tag{11.47}$$

We can use the Josephson scaling law (8.3) together with the scaling laws (8.1):

$$\alpha + 2\beta + \gamma = 2 \tag{11.48}$$

and (8.37):

$$\gamma = \nu(2 - \eta) \tag{11.49}$$

to rewrite eqn. (11.47) as

$$R \sim (-t)^{2\beta - \eta\nu} \quad \text{for } t < 0. \tag{11.50}$$

In Landau theory, $\eta = 0$, and the result (11.50) reduces to eqn. (11.46); but in the critical region, eqn. (11.50) shows that there are corrections to the mean field theory result due to the anomalous dimension η.

11.1.5 Lower Critical Dimension

Now that we have found the fluctuation spectrum about the low temperature ordered state, we can address the issue of the **lower critical dimension** for systems with a continuous symmetry. The basic idea is the same as in the discrete symmetry case: we examine the stability of the ordered state to the fluctuations that are thermally excited. In the present case, we might expect that the ordered state is not very stable compared with the discrete symmetry case, because of the presence of Goldstone modes, and this is indeed correct. In order to decrease the effect of fluctuations, relative to the mean field, it is necessary that the dimensionality

d be greater than two, as we will see. Note that the larger the dimension, the greater the coordination number, and therefore, the more potent is the mean field.

We study the stability of the ordered state by looking at the most dangerous modes, namely the transverse fluctuations. The correlation function for the transverse fluctuations about a putative ordered state is

$$G_{\perp}(\mathbf{r}-\mathbf{r}') = \int_0^{\Lambda} \frac{d^d\mathbf{k}}{(2\pi)^d} \frac{e^{i\mathbf{k}\cdot(\mathbf{r}-\mathbf{r}')}}{k^2}. \tag{11.51}$$

For $d > 2$, the integral is convergent at the lower limit, but for $d \leq 2$, it has an infra-red divergence: these large transverse fluctuations about the ordering direction destroy the long-range order in the system. Thus we conclude that *for all non-zero temperatures,*

$$\langle \mathbf{S} \rangle = 0, \quad d \leq 2. \tag{11.52}$$

Although we have arrived at this conclusion on the basis of the Gaussian approximation, the result, known as the **Mermin-Wagner theorem** can be proven in complete generality[9] Thus, the lower critical dimension for systems with a continuous symmetry is $d = 2$.

11.2 KOSTERLITZ – THOULESS TRANSITION

Exactly at the lower critical dimension, it turns out that the case $n = 2$ and $d = 2$ is special. For this case, it is conventional to introduce the complex field

$$\psi(\mathbf{r}) = S_1(\mathbf{r}) + iS_2(\mathbf{r}) \tag{11.53}$$

and to write the Hamiltonian as

$$-\mathcal{H} = \text{constant} + \int d^2\mathbf{r} \left[\frac{1}{2}|\nabla\psi|^2 + \frac{u_0}{4}\left(|\psi|^2 - \frac{|r_0|}{u_0}\right)^2\right] \tag{11.54}$$

11.2.1 Phase Fluctuations

We argued in section 6.2 that the true T_c is actually below the T_c of Landau theory, because the thermal fluctuations neglected in Landau theory tend to disorder the system. This effect is most severe at or near the lower critical dimension, and thus, in the present case, there will

[9] N.D. Mermin and H. Wagner, *Phys. Rev. Lett.* **17**, 1133 (1966).

be a range of temperatures *below* the mean field transition temperature, and therefore with $r_0 < 0$, but still above the true transition to any ordered state that may exist. In this range, there are both amplitude and direction fluctuations in the order parameter S, but as we have seen above, the amplitude fluctuations are very much suppressed with respect to the direction fluctuations. Thus, to a good approximation, we may take u_0 to be large;[10] but $|r_0|/u_0$ finite, in the temperature range discussed above; setting

$$\psi(\mathbf{r}) = \sqrt{\frac{|r_0|}{u_0}} \exp\left[i\theta(\mathbf{r})\right], \tag{11.55}$$

we obtain the effective Hamiltonian for the **phase fluctuations** $\theta(\mathbf{r})$:

$$-\mathcal{H}_\theta = \frac{K}{2} \int d^2\mathbf{r}\, (\nabla\theta)^2, \tag{11.56}$$

with $K = |r_0|/u_0$. This effective Hamiltonian governs the long wavelength physics for temperatures well below the mean field transition temperature. We will assume that we can neglect the periodicity of θ, which requires that the configuration $\theta(\mathbf{r})+2\pi n$ is equivalent to $\theta(\mathbf{r})$ when n is an integer, and write the partition function as

$$Z = \int_{-\infty}^{\infty} D\theta \exp \mathcal{H}_\theta. \tag{11.57}$$

11.2.2 Phase Correlations

Equation (11.57) represents a model for spin-waves about a putative ordered state. The effective exchange interaction of the spins is

$$J = (k_B T)K. \tag{11.58}$$

By assuming that the fluctuations are small about the ordered state, the extension of the limits to $\pm\infty$ in the partition function is not unreasonable.

---

[10] The effective Hamiltonian (11.54) is sometimes said to describe *soft spins*, because the order parameter has a direction and some variation in the amplitude is allowed. In the limit $u_0 \to \infty$, the amplitude fluctuations about the mean field value are suppressed, and the system is a formal representation of the usual spin system. In this limit, the Hamiltonian is sometimes said to describe *hard spins*.

Let us now look at the order parameter correlation function, which in terms of the phase is proportional to

$$G(\mathbf{r}) \equiv \left\langle e^{i[\theta(\mathbf{r})-\theta(\mathbf{0})]} \right\rangle \tag{11.59}$$

$$= e^{-\frac{1}{2}\langle(\theta(\mathbf{r})-\theta(\mathbf{0}))^2\rangle} \tag{11.60}$$

where, in the second equality, we have used the fact that $\mathcal{H}$ is Gaussian. Then, we have

$$\left\langle (\theta(\mathbf{r}) - \theta(\mathbf{0}))^2 \right\rangle = \frac{2k_BT}{J} \int_0^\Lambda \frac{d^2\mathbf{k}}{(2\pi)^2} \frac{1 - e^{i\mathbf{k}\cdot\mathbf{r}}}{k^2}, \tag{11.61}$$

which follows because

$$\left\langle |\hat{\theta}_\mathbf{k}|^2 \right\rangle = \frac{2k_BT}{Jk^2} \tag{11.62}$$

from eqn. (11.57). Using the integral representation of the Bessel function J_0, eqn. (11.61) becomes

$$\left\langle (\theta(\mathbf{r}) - \theta(\mathbf{0}))^2 \right\rangle = \frac{2k_BT}{J} \int_0^\Lambda \frac{dk}{2\pi} \frac{1 - J_0(kr)}{k}, \tag{11.63}$$

which leads to the asymptotic result

$$\left\langle (\theta(\mathbf{r}) - \theta(\mathbf{0}))^2 \right\rangle = \frac{k_BT}{\pi J} \log \frac{r}{\Lambda^{-1}} \quad \text{for } r \gg \Lambda^{-1}. \tag{11.64}$$

Thus, we find that the correlation function itself is given by

$$G(r) = r^{-\eta}; \quad \eta = \frac{k_BT}{2\pi J}. \tag{11.65}$$

We have reached a rather startling conclusion. Since $\left\langle (\theta(r) - \theta(0))^2 \right\rangle \sim \log r$, there is no long range order: the angular deviation between spins increases as the separation increases. Furthermore, the order parameter correlation function decays *algebraically* to zero, with an exponent η which is not universal, but depends on T and J. In conclusion, we can view the spin-wave model as having a line of critical points from $T = 0$ to $T = \infty$.

The spin-wave approximation is, at best, valid at low temperature. At high temperature, we would expect that the system is a true paramagnet, with exponentially decaying correlations. Therefore, we conclude that the true phase diagram has power law correlations at low temperatures, but no long range order; and exponentially decaying correlations with no long range order at high temperatures. Hence, there must be a phase transition at some intermediate temperature T_{KT}, known as the Kosterlitz-Thouless transition temperature. The line of critical points, which we have just derived, actually only exists for $0 \le T \le T_{\mathrm{KT}} < T_c$, where T_c is the transition temperature in the original Landau theory.

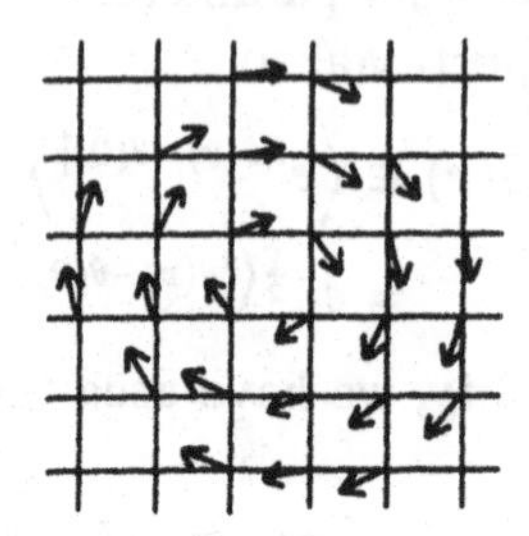

Figure 11.2 A vortex with unit winding number. The arrows denote the direction of the order parameter, and their magnitude denotes the amplitude. The order parameter is defined on a square lattice.

11.2.3 Vortex Unbinding

How does the putative transition occur at T_{KT}? Berezinskii,[11] and Kosterlitz and Thouless[12] proposed that the transition was associated with the unbinding of **vortices**. Vortices are singular spin configurations which we have neglected in our calculation of the long wavelength behaviour of the spin-wave model.

A vortex configuration is shown in figure (11.2). It has the property that when integrated around any closed path

$$\oint d\mathbf{r} \cdot \nabla\theta = 2\pi n, \tag{11.66}$$

where n, a positive or negative integer is called the **winding number.** This is easy to understand: since $d\theta = \nabla\theta \cdot d\mathbf{r}$, the line integral in (11.66) is just the total change in phase along the path. By tracing around a closed path in the figure, one can check that if the path encloses the vortex centre, the winding number is 1; otherwise, the line integral is 0. The winding number is closely related to the familiar concept of the Burgers' vector associated with a dislocation in a solid.

What energy is associated with a vortex? To answer this, we need to know the field configuration of a vortex, which we can obtain from eqn. (11.66): we simply take as a contour a circle of radius $r \gg a$ around the

---

[11] V.L. Berezinskii, *Sov. Phys. JETP* **34**, 610 (1972) [*Zh. Eksp. Teor. Fiz.* **61**, 1144 (1972)].

[12] J.M. Kosterlitz and D.J. Thouless, *J. Phys. C* **5**, L124 (1972); *ibid.* **6**, 1181 (1973).

vortex. Hence we find that $\nabla\theta \sim 1/r$ and so each vortex has energy

$$E_1 = \frac{1}{2}J\int(\nabla\theta)^2 d^2\mathbf{r} = \pi J \log\left(L/\Lambda^{-1}\right), \tag{11.67}$$

where L is the linear dimension of the system. For a single vortex in an infinite system, the energy is infinite. But for a *pair* of vortices with opposite winding numbers, the energy is

$$E_{\text{pair}}(r) \approx 2\pi J \log\left(r/\Lambda^{-1}\right). \tag{11.68}$$

where r is the vortex separation.

Kosterlitz and Thouless suggested that at low temperature, the vortices are bound in pairs, but at high T, they become unbound. We can estimate how this happens by making a crude version of the energy-entropy argument. At low temperature, the energy cost of a single vortex is E_1. The entropy of a single vortex in a system of size L is given approximately by

$$S = k_B \log(L/\Lambda^{-1})^2 \tag{11.69}$$

because $(L/\Lambda^{-1})^2$ is the number of possible lattice sites that the core could occupy. Thus the free energy of an isolated vortex is

$$F = E_1 - TS = (\pi J - 2k_B T)\log\left(L/\Lambda^{-1}\right). \tag{11.70}$$

At low temperature, the free energy cost of creating a vortex diverges as $L \to \infty$. But at high temperature, it is favorable to create isolated vortices. Thus, the unbinding occurs at a transition temperature

$$T_c = \frac{\pi J}{2k_B}. \tag{11.71}$$

This is the essence of the Kosterlitz-Thouless theory. The simple picture presented here neglects interactions between vortices, but a much more detailed RG theory of the transition confirms the essential correctness of these arguments.[13]

---

[13] J.M. Kosterlitz and D. Thouless, *J. Phys. C* **6**, 1181 (1973); *ibid.* **7**, 1046 (1974); a comprehensive review of the extensive applications of the Kosterlitz-Thouless transition is given by D. Nelson in *Phase Transitions and Critical Phenomena*, vol. 7, C. Domb and M.S. Green (eds.) (Academic, New York, 1983).

11.2.4 Universal Jump in the Stiffness

What happens to the stiffness or rigidity R at the Kosterlitz-Thouless transition? Landau theory predicts that the stiffness is $R = |r_0|/u_0 \neq 0$, despite the fact that it also predicts that the phase fluctuations imply that m is zero at *all* temperatures. Note that within Landau theory, the stiffness R happens to be the same as the effective exchange interaction J between spins in the XY model version of the theory.

The RG theory of the Kosterlitz-Thouless transition shows that these conclusions are qualitatively correct. However, the stiffness R becomes smaller than J for $T > 0$ and vanishes *discontinuously* at T_{KT}, with

$$R(T_{KT}^{-}) = \frac{2}{\pi} k_B T_{KT}. \tag{11.72}$$

Thus, the *ratio*

$$\frac{R(T_{KT}^{-})}{k_B T_{KT}} = \frac{2}{\pi}, \tag{11.73}$$

which is a universal number. This prediction has been verified experimentally for the superfluid transition of thin helium films[14] and for the roughening transition of equilibrium crystal surfaces.[15]

The RG theory also predicts that the exponent η is given by a modified form of eqn. (11.65):

$$\eta = \frac{k_B T}{2\pi R}. \tag{11.74}$$

At the transition itself, eqns. (11.73) and (11.74) imply that

$$\eta(T_{KT}) = 1/4. \tag{11.75}$$

[14] I. Rudnick, *Phys. Rev. Lett.* **40**, 1454 (1978); D.J. Bishop and J. Reppy, *Phys. Rev. Lett.* **40**, 1727 (1978).
[15] P.E. Wolf, F. Gallet, S. Balibar and E. Rolley, *J. de Physique* **46**, 1987 (1985).

CHAPTER 12

Critical Phenomena Near Four Dimensions

In this chapter, we discuss a very popular method of studying critical phenomena analytically: the so-called ϵ-expansion. We saw in chapter 7 that for dimensions less than four, naïve perturbation theory in the quartic coupling constant u_0 breaks down in the critical region, because the dimensionless expansion parameter

$$\bar{u}_0 = u_0 \bar{a}^{-\epsilon/2} t^{-\epsilon/2} \tag{12.1}$$

diverges as $t \to 0$. Here $\bar{a}$ is a positive constant and

$$\epsilon = 4 - d. \tag{12.2}$$

Thus, for $d < 4$, the nth term in a perturbation expansion in u_0 will diverge when $n > 2/\epsilon$. This divergence is equivalent to the power law infra-red divergence of the integrals in perturbation theory, explained in section 6.4, and for many years, prevented progress in understanding critical phenomena.

There are (at least) two ways to work around this problem. The first is the **momentum shell RG**[1] which is closely related to the real space RG

---

[1] **The use of the term 'momentum' is historical and derives from the analogy with quantum field theory.**

that we used in chapter 9 to study the two dimensional Ising model. The basic idea is that to construct the RG recursion relations, it is only ever necessary to perform a partial trace over the *short wavelength* degrees of freedom, on scales less than the block size. In the momentum space RG, this partial trace is performed on the *Fourier components* of the order parameter, so that the range of integration is always $\Lambda/\ell < |\mathbf{k}| < \Lambda$, where Λ is the coarse-graining length, which serves as an ultra-violet cut-off to the integrals, and $k = |\mathbf{k}|$ is the wavenumber. In *this* range of integration, there can be no infra-red divergences, because k is restricted to be away from zero, and the RG recursion relations can be calculated in perturbation theory in u_0. The second step of the RG is to find fixed points of the recursion relations. We will see that in addition to a fixed point representing Landau theory with $u_0^* = 0$, another fixed point exists, with a value for u_0^* which is in general large and outside the regime of validity of perturbation theory. However, if we regard the spatial dimension d as a variable, then for $\epsilon \ll 1$, an expansion of the recursion relation can also be made in powers of ϵ! Near four dimensions, there is a fixed point for $u_0^* = O(\epsilon) \neq 0$, which is accessible by perturbation theory. This fixed point governs the critical behaviour, and near four dimensions, the global flow diagram, exponents *etc.* can be constructed. As an additional bonus, it turns out that the numerical results obtained by setting $\epsilon = 1$ in the formula obtained are reasonably accurate. The ϵ expansion is divergent, but assumed to be Borel summable and thus the numerical results using this method can be made rather accurate.[2]

The second way to work around the infra-red problem is to use naïve perturbation theory, but to perform a *double* expansion in both u_0 and ϵ. The basic point here is that we can write

$$t^{-\epsilon/2} = \exp\left[-\frac{\epsilon}{2}\log t\right] \tag{12.3}$$

$$= \sum_{n=0}^{\infty}\left(-\frac{\epsilon}{2}\right)^n (\log t)^n, \tag{12.4}$$

suggesting that in this double expansion, even *below* four dimensions, the perturbation expansion will only contain *logarithmic divergences.* These

[2] J. Zinn-Justin, *Quantum Field Theory and Critical Phenomena* (Clarendon, Oxford, 1989), Chapter 25. A third way to calculate critical exponents is explained in this reference: perturbation expansion above T_c at *fixed* dimension. This perturbation expansion is divergent, as is the ϵ-expansion, but has been rigorously proved to be Borel summable.

logarithmic divergences can be renormalised using the techniques described in chapter 10, applied to field theory[3] and the critical behaviour then follows. Here, we shall not follow this approach, often referred to as the **field theoretic approach.**

12.1 BASIC IDEA OF THE EPSILON-EXPANSION

The starting point is the effective Hamiltonian of Landau theory for the Ising uiversality class, which we will write in terms of the scalar order parameter $S(\mathbf{r})$, in order to emphasise the similarities with your treatment of the RG for the Ising model:

$$-\mathcal{H}\{S\} = \int d^d\mathbf{r}\left[\frac{1}{2}(\nabla S)^2 + \frac{1}{2}r_0 S^2 + \frac{1}{4}u_0 S^4 - h_0 S\right]. \tag{12.5}$$

The method described in this chapter can be applied to the effective Hamiltonian for any universality class, with order parameters more complicated than the scalar one here. When we refer to Landau theory below, we explicitly mean the effective Hamiltonian (12.5). Note that exercise 3–3 shows that, using the Hubbard-Stratonovich transformation, spin models such as the Ising model can be mapped into an effective Hamiltonian of the form of (12.5), but with additional terms higher order in S. These terms are irrelevant at the Gaussian fixed point, as shown in exercise 7–1.

The basic idea is that if we start with this Landau theory and apply (in some as yet unspecified way) a RG transformation, the coupling constants r_0, u_0, h_0 will flow towards fixed point values r^*, u^*, h^*. For $d > 4$, we have already argued that the Gaussian approximation gives the correct exponents. Thus, we might expect that at the critical fixed point corresponding to the Gaussian theory

$$r^* = h^* = u^* = 0. \tag{12.6}$$

This is the so-called **Gaussian fixed point.** For $d < 4$, it will turn out that this becomes unstable to another fixed point, the so-called **Wilson-Fisher fixed point.** By this we mean that the Gaussian fixed point acquires a new relevant variable for $d < 4$, which sends the RG flows towards the Wilson-Fisher fixed point. For $\epsilon \ll 1$, the Wilson-Fisher fixed point is near the Gaussian fixed point, which is why perturbation theory can be successfully used. We can see why this might be the case by the following heuristic argument.

---

[3] See the exposition by D.J. Amit, *Field Theory, the Renormalisation Group and Critical Phenomena* (World Scientific, Singapore, 1984).

The coupling constant has dimensions $[u_0] = L^{d-4}$, from eqn. (7.8). Hence, if we neglect the contribution to the recursion relation from other coupling constants, under a RG transformation R_ℓ, we have

$$u_0 \to u_0' = \ell^{4-d} u_0 = \ell^\epsilon u_0. \tag{12.7}$$

We can write this recursion relation in differential form as

$$\frac{du_s}{ds} = \epsilon u_s, \tag{12.8}$$

where $s \equiv \log \ell$. This recursion relation would predict that for $\epsilon > 0$, the quartic coupling constant grows unboundedly under iteration of the RG recursion relation. Of course, we know that we cannot ignore the effects of other coupling constants, so this recursion relation is not correct. Now, suppose that the correct recursion relation is actually

$$\frac{du_s}{ds} = \epsilon u_s - A u_s^2 + O(u_s^3), \tag{12.9}$$

where A is some constant of order unity. Then, for $A < 0$, the only fixed point consistent with $u_0 \geq 0$ would be $u_0^* = 0$. On the other hand, if A turned out to be positive, then

$$u_0^* = \frac{\epsilon}{A} \tag{12.10}$$

is a possible fixed point, in addition to the Gaussian fixed point $u_0^* = 0$. The fact that there might be a non-trivial fixed point at $u^* = O(\epsilon)$ suggests that perturbation theory in ϵ might be successful, and this is indeed what we shall show in this chapter.

A critical fixed point with $u_0^* < 0$ is not physically acceptable, because the Landau free energy density would have no lower bound, and therefore could always be minimised by taking $|S| \to \infty$.

12.2 RG FOR THE GAUSSIAN MODEL

As a prelude to our RG analysis of the Landau theory for the Ising universality class, it is convenient to consider the **Gaussian model**, which is defined by the Landau free energy (12.5), with $u_0 = 0$. Of course, without the quartic coupling, it is impossible to have an ordered phase. Nevertheless the Gaussian model should be adequate if we restrict ourselves to $T > T_c$, where it has the merit that it is exactly soluble (as discussed in chapter 6). Furthermore, according to the Ginzburg criterion, this model should be valid for $d > 4$ and $T > T_c$.

Our starting effective Hamiltonian is the Landau theory, which we write in Fourier space (see section 6.3.3) as

$$-\mathcal{H} = \frac{1}{V} \sum_{0<|\mathbf{k}|<\Lambda} \frac{1}{2} |\hat{S}_{\mathbf{k}}|^2 (r_0 + k^2). \tag{12.11}$$

We define a RG transformation by performing a partial trace over (or "integrating out") the degrees of freedom with wavenumber

$$\frac{\Lambda}{\ell} < |\mathbf{k}| < \Lambda \quad \text{for } \ell > 1. \tag{12.12}$$

This will leave us with an effective Hamiltonian for degrees of freedom with $0 < |\mathbf{k}| < \Lambda/\ell$, and we will not quite be able to simply read-off the recursion relations for the coupling constant. To return the Hamiltonian to its original form, we need a second step: a *rescaling* of space, so that the new Hamiltonian is defined in terms of the original degrees of freedom. The procedure followed below is similar to the real space RG for the Ising model of section 9.6, and the reader may wish to review this before proceeding.

12.2.1 Integrating Out the Short Wavelength Degrees of Freedom

We define the long and short wavelength components of $S(\mathbf{r})$ respectively by

$$\hat{S}'_\ell(\mathbf{k}) = \hat{S}_{\mathbf{k}} \quad \text{for } 0 < |\mathbf{k}| < \frac{\Lambda}{\ell}, \tag{12.13}$$

and

$$\hat{\sigma}_\ell(\mathbf{k}) = \hat{S}_{\mathbf{k}} \quad \text{for } \frac{\Lambda}{\ell} < |\mathbf{k}| < \Lambda. \tag{12.14}$$

In real space

$$S(\mathbf{r}) = \int_0^{\Lambda/\ell} \frac{d^d\mathbf{k}}{(2\pi)^d} \hat{S}_{\mathbf{k}} e^{i\mathbf{k}\cdot\mathbf{r}} + \int_{\Lambda/\ell}^{\Lambda} \frac{d^d\mathbf{k}}{(2\pi)^d} \hat{S}_{\mathbf{k}} e^{i\mathbf{k}\cdot\mathbf{r}} \tag{12.15}$$

$$\equiv S'_\ell(\mathbf{r}) + \sigma_\ell(\mathbf{r}). \tag{12.16}$$

To emphasise the analogy with the real space RG: S'_ℓ corresponds to the block spins, whilst σ_ℓ corresponds to the microscopic degrees of freedom within a block.

The effective Hamiltonian can be split into long and short wavelength parts:

$$-\mathcal{H} = \frac{1}{V}\sum_{k<\Lambda/\ell}\frac{1}{2}|\hat{S}'_\ell|^2(r_0+k^2) + \frac{1}{V}\sum_{\Lambda/\ell<|\mathbf{k}|<\Lambda}\frac{1}{2}|\hat{\sigma}_\ell|^2(r_0+k^2) \quad (12.17)$$

$$= -\mathcal{H}_s\{\hat{S}_\ell(\mathbf{k})\} - \mathcal{H}_\sigma\{\hat{\sigma}_\ell(\mathbf{k})\}. \quad (12.18)$$

The partition function is simply a function of r_0:

$$Z(r_0) = \int DS\, e^{\mathcal{H}} \quad (12.19)$$

$$= \int \prod_{0<|\mathbf{k}|<\Lambda} d\hat{S}_\mathbf{k}\, e^{\mathcal{H}\{\hat{S}_k\}} \quad (12.20)$$

$$= \int \prod_{0<|\mathbf{k}|<\Lambda/\ell} d\hat{S}_\ell(\mathbf{k}) \prod_{\Lambda/\ell<|\mathbf{k}|<\Lambda} d\hat{\sigma}_\ell(\mathbf{k})\, e^{\mathcal{H}_S+\mathcal{H}_\sigma} \quad (12.21)$$

$$= Z_S \cdot Z_\sigma. \quad (12.22)$$

Z_σ is the analogue of the term $Z_0(K)^M$ in the real space RG of section 9.6, and has the explicit form

$$Z_\sigma = \exp\left[\frac{1}{2}\sum_{\Lambda/\ell<|\mathbf{k}|<\Lambda}\log\frac{2\pi V}{r_0+k^2}\right] \quad (12.23)$$

We will ignore this regular, multiplicative prefactor of Z_s for the purposes of computing critical exponents, since it will not enter the recursion relations for r_0 and u_0; it will, however, affect the free energy. Now let us study Z_S.

$$Z_S = \int D\hat{S}'_\ell \exp\left[-\frac{1}{V}\sum_{|\mathbf{k}|<\Lambda/\ell}\frac{1}{2}(r_0+k^2)|\hat{S}'_\ell(k)|^2\right]. \quad (12.24)$$

This is almost of the form of eqn. (12.11) but the momentum integral is cut-off at Λ/ℓ rather than Λ. We want to be able to read off the recursion relations for the coupling constants, and so we first need to put the Hamiltonian $\mathcal{H}'$ into the same form as $\mathcal{H}$. This can be easily achieved by rescaling the momentum and fields, which are, after all, only dummy variables.

12.2.2 Rescaling of Fields and Momenta

Define

$$\mathbf{k}_\ell \equiv \ell \mathbf{k} \tag{12.25}$$

and

$$\hat{S}_\ell(\mathbf{k}_\ell) \equiv z^{-1} \hat{S}'_\ell(\mathbf{k}). \tag{12.26}$$

The field $\hat{S}'_\ell$ has been rescaled to $\hat{S}_\ell$ by a factor z^{-1}, which is sometimes called a **wave function renormalisation**. We need to introduce z, so that the coefficient of the $(\nabla S_\ell)^2$ term in $\mathcal{H}'$ will be $1/2$, just as it was in $\mathcal{H}$. This adjustment is necessary in order that there is the possibility of a non-trivial fixed point: the necessity to determine z corresponds, in the description of the fixed points of Barenblatt's equation in chapter 10, to the necessity to determine $\phi = 1/2$.

Substituting in these definitions, we have

$$Z_S(r_0) = \int D\hat{S}_\ell \exp\left[-\frac{1}{2}\int_0^\Lambda \frac{d^d\mathbf{k}_\ell}{(2\pi)^d}\ell^{-d}\left(r_0 + \frac{\mathbf{k}_\ell^2}{\ell^2}\right)|\hat{S}_\ell(\mathbf{k}_\ell)|^2 z^2\right]. \tag{12.27}$$

The coefficient of $\mathbf{k}_\ell^2$ is $-\ell^{-(2+d)}z^2/2$, so to ensure that this coefficient is actually equal to $1/2$, we choose

$$z = \ell^{1+d/2}. \tag{12.28}$$

Then

$$Z_S(r_0) = \int D\hat{S}_\ell \exp\left[-\frac{1}{2}\int_0^\Lambda \frac{d^d\mathbf{k}_\ell}{(2\pi)^d}(r_\ell + \mathbf{k}_\ell^2)|\hat{S}_\ell(\mathbf{k})|^2\right], \tag{12.29}$$

with

$$r_\ell = \ell^{-d} z^2 r_0 = \ell^2 r_0. \tag{12.30}$$

This is our desired RG recursion relation.

12.2.3 Analysis of Recursion Relation

A very useful way to analyse RG recursion relations is to write them as a differential RG transformation, and we will follow this procedure here, for purposes of illustration rather than necessity. We consider making a change $\ell \to \ell + \delta\ell$: then

$$r_{\ell+\delta\ell} = (\ell + \delta\ell)^2 r_0 \tag{12.31}$$

leading to

$$\frac{dr_\ell}{d\ell} = \frac{2}{\ell} r_\ell. \tag{12.32}$$

As usual, we make the substitution $s = \log \ell$ to give

$$\frac{dr_s}{ds} = 2r_s. \tag{12.33}$$

The fixed point is $r_s^* = 0$ and the corresponding relevant eigenvalues are

$$\Lambda_r = \ell^2; \quad y_r = 2. \tag{12.34}$$

Note that y_r can be directly read off from the differential form of the RG recursion relations.

12.2.4 Critical Exponents

As $r_0 \propto t$, the exponent y_r is simply y_t: hence we can read off the exponent ν.

$$\nu = \frac{1}{y_t} = \frac{1}{2}. \tag{12.35}$$

What happens if we include an external field? A uniform field simply couples to the $\mathbf{k} = 0$ component of $\hat{S}_\mathbf{k}$, and the Hamiltonian becomes

$$\mathcal{H} \to \mathcal{H} + H \int d^d\mathbf{r}\, S(\mathbf{r}) = \mathcal{H} + H\hat{S}_0. \tag{12.36}$$

This term is unaffected by integrating out the short wavelength degrees of freedom in the RG transformation, but is affected by the wave function renormalisation. It becomes

$$H\hat{S}'_\ell(0) = z\hat{S}_\ell(0)H \tag{12.37}$$

$$= H\ell^{1+d/2}\hat{S}_\ell(0) \tag{12.38}$$

$$= H_\ell \hat{S}_\ell(0). \tag{12.39}$$

Thus we obtain the recursion relation for the field:

$$H_\ell = \ell^{1+d/2} H, \tag{12.40}$$

which can be written in the differential form

$$\frac{dH_s}{ds} = \left(1 + \frac{d}{2}\right) H_s. \tag{12.41}$$

Hence, the RG eigenvalue y_h has the value

$$y_h = 1 + \frac{d}{2}, \tag{12.42}$$

and the values for the critical exponents can now be obtained from the scaling laws. Using the relation $d\nu = 2 - \alpha$, we find that

$$\alpha = 2 - d/2. \tag{12.43}$$

The gap exponent $\Delta = y_h/y_t = (1 + d/2)/2$, and thus, using the scaling law (9.35), we obtain

$$\eta = 0. \tag{12.44}$$

Substitution into the result of exercise 9–1 (a) immediately gives

$$\delta = \frac{d+2}{d-2}. \tag{12.45}$$

From the scaling law (8.26), we find that

$$\beta = \frac{d-2}{4}. \tag{12.46}$$

Lastly, the susceptibilty exponent is obtained from the Rushbrooke scaling law (8.29):

$$\gamma = 1. \tag{12.47}$$

If we now compare our RG results with the values for the exponents that we already obtained from the Gaussian approximation for the fluctuations about Landau theory, we find an unwelcome surprise: they do not agree! The values for β and δ are incorrect. Why? The reader is encouraged to contemplate the wreckage of our theory and to try to find the error, before reading on.

12.2.5 A Dangerous Irrelevant Variable in Landau Theory

The exponents β and δ are both defined with respect to thermodynamic properties below or at T_c. This suggests that our error is connected with the fact that the Gaussian model is not well defined for $T < T_c$, and so should have nothing to say about these exponents, as we have already anticipated.

However, it is an instructive exercise to see what goes wrong when one blindly tries to apply the Gaussian model below T_c, and to see how to deduce the correct exponents in the Gaussian approximation from RG.

After all, we saw in exercise 6–1 that below T_c, the Landau free energy can be expanded about the degenerate minima, and the fluctuation part has precisely the same form as the effective Hamiltonian we have been using in this section. Furthermore, the Ginzburg criterion shows that for $d > 4$ we can neglect the interaction between the fluctuations.

As you might have guessed, the error made in proceeding blindly has to do with how u_0 transforms under RG transformations. The RG tells us that, suppressing non-essential variables, the singular part of the free energy density transforms as

$$f_s(t, h, u_0) = \ell^{-d} f_s(t\ell^{y_t}, h\ell^{y_h}, u_0\ell^{y_u}) \tag{12.48}$$

where we have used t rather than r_0. The scaling of the magnetisation M, *i.e.* $\langle S(\mathbf{r})\rangle$, is given by

$$M(t, h, u_0) \equiv -\frac{1}{k_B T}\frac{\partial f_s}{\partial h} = \ell^{-d+y_h} M(t\ell^{y_t}, h\ell^{y_h}, u_0\ell^{y_u}). \tag{12.49}$$

Setting $h = 0$, and replacing $\ell = t^{-1/y_t}$ we get the scaling law

$$M(t, 0, u_0) = t^{-(y_h-d)/y_t} M(1, 0, u_0 t^{-y_u/y_t}). \tag{12.50}$$

The Gaussian model gave $y_t = 2y_h = 1 + d/2$. What about y_u? We expect u_0 to be irrelevant at the Gaussian fixed point for $d > 4$, based upon the Ginzburg criterion. Thus, one might guess that $y_u \propto 4 - d$, and in fact this will be shown below to be correct:

$$y_u = 4 - d \tag{12.51}$$

and u_0 becomes relevant for $d < 4$.

Now let us re-examine our argument leading to the erroneous critical exponents. What we did was effectively to set $u_0 = 0$ in eqn. (12.50), thus deriving

$$M(t, 0, u_0) \sim t^{\beta} = t^{-(y_h-d)/y_t} M(1, 0, 0). \tag{12.52}$$

which leads to eqn. (12.46), using the values for y_h and y_t. However, this step is incorrect. Landau theory gives $M = \sqrt{-r_0/u_0}$, and so

$$M(t, 0, u_0) \propto u_0^{-1/2}. \tag{12.53}$$

Thus we cannot set $u_0 = 0$ in the scaling law (12.50). The quartic coupling u_0 is an example of a **dangerous irrelevant variable**.

How can we deduce the correct values for β and δ? What we must do is to use the information in eqn. (12.53) about the scaling function to take the $u_0 \to 0$ limit of the scaling law (12.50): writing the scaling law as

$$M(t,0,u_0) = t^{-(y_h-d)/y_h} F_M\left(\frac{u_0}{t^{y_u/y_t}}\right) \tag{12.54}$$

where the scaling function F_M has the behaviour

$$F_M(x) \propto (-x)^{-1/2} \quad \text{for } x < 0, \tag{12.55}$$

we obtain

$$M(t,0,u_0) \sim t^{-(y_h-d)/y_t+y_u/2y_t} \sim t^{\beta}, \quad \text{for } u_0 \neq 0. \tag{12.56}$$

Thus

$$\beta = -\frac{1}{2}\left(1-\frac{d}{2}\right)+\frac{(4-d)}{4} = \frac{1}{2}. \tag{12.57}$$

Similarly we can obtain the correct value for δ by using the fact that for $t = 0$

$$M(0,h,u_0) \propto u_0^{-1/3} \tag{12.58}$$

which follows from eqn (5.30). Setting $t = 0$ in eqn. (12.49) and making the obvious choice of ℓ to obtain the scaling law we get:

$$M(0,h,u_0) = h^{-(y_h-d)/y_h} M(0,1,u_0 h^{-y_u/y_h}). \tag{12.59}$$

Now, using eqn. (12.58) we obtain

$$\frac{1}{\delta} = -\frac{(y_h-d)}{y_h} + \frac{1}{3}\left(\frac{y_u}{y_h}\right), \tag{12.60}$$

and thus $\delta = 3$, which is the correct answer for the Gaussian approximation.

To summarise: for $d > 4$, u_0 is an irrelevant variable. The RG calculation for the exponents y_t and y_h was performed in the Gaussian model, neglecting u_0. In order to obtain the critical exponents, we needed to use the RG in conjunction with information obtained from an approximation scheme (in this case mean field theory). Even though u_0 is an irrelevant variable at the Gaussian fixed point, it is dangerous for $T < T_c$.

12.3 RG BEYOND THE GAUSSIAN MODEL

In this section we will perform the RG calculation of the critical exponents in the complete Landau theory, near $d = 4$. The calculation is quite complicated because there will be many terms to count when we integrate over the short wavelength degrees of freedom – the $\hat{\sigma}_\ell(\mathbf{k})$ – to obtain the renormalised effective Hamiltonian. During the gory details, bear in mind that, conceptually, we are doing nothing more than what we did in the previous section and in section 9.6. The calculation is more complicated here than in section 9.6, because it is performed in Fourier space, and is systematic to the first order in ϵ.

12.3.1 Setting Up Perturbation Theory

Our starting point is the effective Hamiltonian

$$-\mathcal{H}\{S\} = \int d^d\mathbf{r}\left[\frac{1}{2}(\nabla S)^2 + \frac{1}{2}r_0 S^2 + \frac{1}{4}u_0 S^4 - h_0 S\right], \qquad (12.61)$$

As before $0 < |\mathbf{k}| < \Lambda$. In working in Fourier space, will sometimes use the notation

$$\int_{\mathbf{k}} \equiv \int_0^\Lambda \frac{d^d\mathbf{k}}{(2\pi)^d} \equiv \int_0^\Lambda \bar{d}\mathbf{k}, \qquad (12.62)$$

and we will set the external field h_0 to zero.

The next task is to write the quartic term in momentum space.

$$\int d^d\mathbf{r}\, S(\mathbf{r})^4 = \int d^d\mathbf{r} \int_{\mathbf{k}_1\mathbf{k}_2\mathbf{k}_3\mathbf{k}_4} \hat{S}_{\mathbf{k}_1}\hat{S}_{\mathbf{k}_2}\hat{S}_{\mathbf{k}_3}\hat{S}_{\mathbf{k}_4} e^{i\sum_i \mathbf{k}_i\cdot\mathbf{r}}. \qquad (12.63)$$

Using the definitions of Fourier transforms in section 5.7.2, this becomes

$$\int d^d\mathbf{r}\, S(\mathbf{r})^4 = \int_{\mathbf{k}_1\ldots\mathbf{k}_4} (2\pi)^d\delta(\mathbf{k}_1 + \cdots + \mathbf{k}_4)\hat{S}_{\mathbf{k}_1}\ldots\hat{S}_{\mathbf{k}_4}. \qquad (12.64)$$

Finally, the effective Hamiltonian in Fourier space is

$$-\mathcal{H}\{S\} = \frac{1}{2}\int_{\mathbf{k}}(r_0 + k^2)|\hat{S}_{\mathbf{k}}|^2 + \frac{1}{4}u_0\int_{\mathbf{k}_1\ldots\mathbf{k}_4}\hat{S}_{\mathbf{k}_1}\ldots\hat{S}_{\mathbf{k}_4} \times (2\pi)^d\,\delta(\mathbf{k}_1 + \cdots + \mathbf{k}_4). \qquad (12.65)$$

Now we split up the field into short and long wavelength components as in eqns. (12.13) and (12.14). The difficulty of dealing with the interaction term u_0 is now painfully obvious. The quadratic part of $-\mathcal{H}\{S\}$ does not couple $\hat{S}'_\ell$ and $\hat{\sigma}_\ell$, but the quartic term does. Thus, we will write the effective Hamiltonian as

$$-\mathcal{H}\{S\} = -\mathcal{H}_{S'}\{\hat{S}'_\ell\} - \mathcal{H}_\sigma\{\hat{\sigma}_\ell\} - V(\hat{S}'_\ell, \hat{\sigma}_\ell) \qquad (12.66)$$

Here $\mathcal{H}_S$ and $\mathcal{H}_\sigma$ have the same meaning as in (12.17). Substitution of the definitions (12.13) and (12.14) into the quartic term of the effective Hamiltonian (12.65) gives the explicit form of $V(S'_\ell, \sigma_\ell)$, which we shall evaluate later.

Let us now go through and sketch the outline of the calculation we are about to do, using as our model, the calculation of section 12.2. The partition function is

$$Z(r_0, u_0) = \int DS\, e^{\mathcal{H}} \tag{12.67}$$

$$= \int DS\, e^{\mathcal{H}_{S'} + \mathcal{H}_r + V} \tag{12.68}$$

$$= \int DS'_\ell\, e^{\mathcal{H}_{S'}} \int D\sigma_\ell\, e^{\mathcal{H}_\sigma + V}. \tag{12.69}$$

After integration over σ_ℓ, wave-function renormalisation of S'_ℓ and rescaling of $\mathbf{k}$, we shall obtain

$$Z(r_0, u_0) = \int DS_\ell\, e^{\mathcal{H}_\ell}, \tag{12.70}$$

where $\mathcal{H}_\ell$ is the effective Hamiltonian for the coarse-grained order parameter S_ℓ. We can write down a formal expression for $\mathcal{H}_\ell$, just as we did in the real space RG of section 9.6. We define the average

$$\langle A(S'_\ell) \rangle_0 \equiv \frac{\int D\sigma_\ell\, e^{\mathcal{H}_\sigma} A(S'_\ell, \sigma_\ell)}{\int D\sigma_\ell\, e^{\mathcal{H}_\sigma}}, \tag{12.71}$$

which integrates out over the short wavelength degrees of freedom. Then we can write the partition function as

$$Z = Z_\sigma(r_0) \int DS'_\ell\, e^{\mathcal{H}_{S'}} \langle e^V \rangle_0, \tag{12.72}$$

where

$$Z_\sigma(r_0) = \int D\sigma_\ell\, e^{\mathcal{H}_\sigma} \tag{12.73}$$

is a non-singular contribution to the partition function, involving only Fourier components $\Lambda/\ell < |\mathbf{k}| < \Lambda$. As before, we shall ignore this term for purposes of computing the critical exponents. The partition function for the long wavelength degrees of freedom can be approximated using the **cumulant expansion**, and we obtain

$$Z = \int DS'_\ell\, e^{\mathcal{H}_{S'}} \langle e^V \rangle_0$$

$$= \int DS'_\ell\, e^{\mathcal{H}_{S'}} e^{\langle V \rangle_0 + \frac{1}{2}\left[\langle V^2 \rangle_0 - \langle V \rangle_0^2\right] + O(V^3)}. \tag{12.74}$$

As in the discussion of the Gaussian theory, we shall need to make a wave function renormalisation of S'_ℓ:

$$\hat{S}'_\ell(\mathbf{k}) = z\hat{S}_\ell(\mathbf{k}_\ell); \tag{12.75}$$

$$\mathbf{k}_\ell = \ell\mathbf{k}, \tag{12.76}$$

giving the renormalised Hamiltonian

$$\mathcal{H}_\ell\{\hat{S}_\ell(\mathbf{k}_\ell)\} = \mathcal{H}_{S'}\{z\hat{S}_\ell\} + \langle V\rangle_0 + \frac{1}{2}\left[\langle V^2\rangle_0 - \langle V\rangle_0^2\right] + O(V^3). \tag{12.77}$$

To read off the coupling constant recursion relations, we write the renormalised Hamiltonian in the form

$$\begin{aligned}-\mathcal{H}_\ell\{S_\ell\} = & \frac{1}{2}\int_{\mathbf{k}} u_{2,\ell}(\mathbf{k})|\hat{S}_\ell(\mathbf{k})|^2 \\ & + \frac{1}{4}\int_{\mathbf{k}_1\ldots\mathbf{k}_4}(2\pi)^d\delta\left(\sum_{i=1}^{4}\mathbf{k}_i\right)u_{4,\ell}(\mathbf{k}_1\ldots\mathbf{k}_4)\hat{S}_{\mathbf{k}_1}\cdots\hat{S}_{\mathbf{k}_4} \\ & + O(\hat{S}^6).\end{aligned} \tag{12.78}$$

The coefficients $u_{2,\ell}(\mathbf{k}), u_{4,\ell}(\mathbf{k}_1\cdots\mathbf{k}_4)$ are the renormalised versions of r_0+k^2 and u_0 respectively. Irrelevant terms (at the Gaussian fixed point) – in fact higher order gradient terms – are generated and appear as the wavenumber dependence of $u_{2,\ell}$ and $u_{4,\ell}$. There will also be terms of order $\hat{S}^6$ *etc.*, giving rise to coupling constants such as $u_{6,\ell}, u_{8,\ell}$, all of which are functions of momenta, in general. Fortunately, we shall find that to $O(\epsilon)$ these complication do not arise. That is to say, once we have generated recursion relations and found the critical fixed point, the following estimates will be valid:

$$\begin{aligned} r^* &= O(\epsilon) \\ u^* &= O(\epsilon) \\ u_2^*(\mathbf{k}) &= r^* + k^2 + O(\epsilon^2) \\ u_4^*(\mathbf{k}) &= u^* + O(\epsilon^2) \\ u_6^*, u_8^*, \ldots &= O(\epsilon^2) \end{aligned} \tag{12.79}$$

It is quite involved to check that these estimates are correct, and we will not be particularly persuasive here.[4]

---

[4] For details, consult the review article by K.G. Wilson and J. Kogut, *Phys. Rep.* C 12, 75 (1974).

12.3.2 Calculation of $\langle V \rangle_0$: Strategy

In this section, we will sketch the *strategy* of the calculation of $\langle V \rangle_0$, to prepare the reader for the algebraic horrors to follow. We wish to compute

$$V = \frac{1}{4} u_0 \int_{\mathbf{k}_1 \mathbf{k}_2 \mathbf{k}_3 \mathbf{k}_4} (2\pi)^d \delta(\mathbf{k}_1 + \mathbf{k}_2 + \mathbf{k}_3 + \mathbf{k}_4) \hat{S}_{\mathbf{k}_1} \hat{S}_{\mathbf{k}_2} \hat{S}_{\mathbf{k}_3} \hat{S}_{\mathbf{k}_4} \tag{12.80}$$

where

$$\int_{\mathbf{k}} \hat{S}_{\mathbf{k}} = \int_0^{\Lambda/\ell} \frac{d^d\mathbf{k}}{(2\pi)^d} \hat{S}'_\ell(\mathbf{k}) + \int_{\Lambda/\ell}^{\Lambda} \frac{d^d\mathbf{k}}{(2\pi)^d} \hat{\sigma}_\ell(\mathbf{k}). \tag{12.81}$$

The strategy is to substitute eqn. (12.81) into the expression (12.80), and then perform the average $\langle \cdots \rangle_0$. We will see later how to *calculate* the terms that we get from this procedure. Our first step is to *generate* these terms. Then, we will show how, in principle, we can calculate the renormalised coupling constants $u_{2,\ell}$, $u_{4,\ell}$ *etc.*

To get the basic idea, think symbolically of (12.81) as saying

$$\hat{S} = \hat{S}'_\ell + \hat{\sigma}_\ell. \tag{12.82}$$

Substituting into $S(\mathbf{k}_1)S(\mathbf{k}_2)S(\mathbf{k}_3)S(\mathbf{k}_4)$ will give terms like

$$S'S'S'S' + 4S'\sigma S'S' + 6\sigma\sigma S'S' + \cdots \tag{12.83}$$

We will need to be careful and keep track of the arguments of the S' and σ fields. Once we have expanded V in this way, we shall put $\langle \cdots \rangle_0$ around the expression for V. Now $\langle \cdots \rangle_0$ only averages over $\hat{\sigma}_\ell$, and therefore, we will be able to factor the S' out of the average. For example, a term such as

$$\begin{aligned} \frac{u_0}{4} \int_0^{\Lambda/\ell} đ\mathbf{k}_1 \int_0^{\Lambda/\ell} đ\mathbf{k}_2 \int_{\Lambda/\ell}^{\Lambda} đ\mathbf{k}_3 \int_{\Lambda/\ell}^{\Lambda} đ\mathbf{k}_4 \, (2\pi)^d \delta(\mathbf{k}_1 + \mathbf{k}_2 + \mathbf{k}_3 + \mathbf{k}_4) \\ \times \left\langle \hat{S}'_\ell(\mathbf{k}_1) \hat{S}'_\ell(\mathbf{k}_2) \hat{\sigma}_\ell(\mathbf{k}_3) \hat{\sigma}_\ell(\mathbf{k}_4) \right\rangle_0 \end{aligned} \tag{12.84}$$

will become

$$\begin{aligned} \frac{u_0}{4} \int_0^{\Lambda/\ell} đ\mathbf{k}_1 \int_0^{\Lambda/\ell} đ\mathbf{k}_2 \int_{\Lambda/\ell}^{\Lambda} đ\mathbf{k}_3 \int_{\Lambda/\ell}^{\Lambda} đ\mathbf{k}_4 \, (2\pi)^d \delta(\mathbf{k}_1 + \mathbf{k}_2 + \mathbf{k}_3 + \mathbf{k}_4) \\ \times \hat{S}'_\ell(\mathbf{k}_1) \hat{S}'_\ell(\mathbf{k}_2) \langle \hat{\sigma}_\ell(\mathbf{k}_3) \hat{\sigma}_\ell(\mathbf{k}_4) \rangle_0 . \end{aligned} \tag{12.85}$$

The quantity $\langle\hat{\sigma}_\ell(\mathbf{k}_3)\hat{\sigma}_\ell(\mathbf{k}_4)\rangle_0$ is easily calculable, as are the other terms generated by this procedure:

$$\langle\hat{\sigma}_\ell(\mathbf{k})\rangle_0\,; \quad \langle\hat{\sigma}_\ell(\mathbf{k}_1)\hat{\sigma}_\ell(\mathbf{k}_2)\hat{\sigma}_\ell(\mathbf{k}_3)\rangle_0\,; \quad \langle\hat{\sigma}_\ell(\mathbf{k}_1)\cdots\hat{\sigma}_\ell(\mathbf{k}_4)\rangle_0\,. \tag{12.86}$$

Once we have obtained an expression like (12.85), what do we then do with it? Suppose we now perform the integrals over $\mathbf{k}_3$ and $\mathbf{k}_4$, after we have calculated $\langle\hat{\sigma}_\ell(\mathbf{k}_3)\hat{\sigma}_\ell(\mathbf{k}_4)\rangle_0$ (which is, after all, just some function of $\mathbf{k}_3$ and $\mathbf{k}_4$.) Then we will be left with a term

$$\frac{u_0}{4}\int_0^{\Lambda/\ell} đ\mathbf{k}_1 \int_0^{\Lambda/\ell} đ\mathbf{k}_2\, \hat{S}'_\ell(\mathbf{k}_1)\hat{S}'_\ell(\mathbf{k}_2) \times (\text{some function of } \mathbf{k}_1, \mathbf{k}_2). \tag{12.87}$$

After we have re-scaled the momentum and performed a wave function renormalisation, this term will be of the form

$$\int_{\mathbf{k}} u_{2,\ell}(k)|\hat{S}_\ell(\mathbf{k})|^2. \tag{12.88}$$

Hence, the term (12.84) in the expansion of $\langle V\rangle_0$ ends up making a contribution to the renormalised coupling constant $u_{2,\ell}$ in the renormalised Hamiltonian. Proceeding in similar fashion for all the terms in $\langle V\rangle_0$, $\langle V^2\rangle_0 - \langle V\rangle_0^2$, we can, in principle, calculate the renormalised coupling constants in a systematic way.

We are almost ready to put this program into effect. Having seen that we are going to end up with correlation functions of $\hat{\sigma}_\ell(\mathbf{k})$, computed according to eqn. (12.71), let us now discuss how they are evaluated.

12.3.3 Correlation Functions of $\hat{\sigma}_\ell(\mathbf{k})$: Wick's Theorem

These correlation functions are rather easy: we basically did the work in section 6.3.4 when we computed $\langle\eta_{\mathbf{k}}\eta'_{\mathbf{k}}\rangle$ in the Gaussian approximation. Recall that

$$-\mathcal{H}_\sigma = \frac{1}{2}\int_{\Lambda/\ell}^{\Lambda} đk\,(r_0+k^2)|\hat{\sigma}_\ell(\mathbf{k})|^2 \tag{12.89}$$

$$= \frac{1}{V}\sum_{\mathbf{k}}{}' \frac{1}{2}(r_0+k^2)|\hat{\sigma}_\ell(\mathbf{k})|^2 \tag{12.90}$$

where we have defined

$$\sum_{\mathbf{k}}{}' \equiv \sum_{\Lambda/\ell<|\mathbf{k}|<\Lambda}. \tag{12.91}$$

The form of $\mathcal{H}_\sigma$ implies that we can follow completely the discussion of section 6.3.4 for the Gaussian theory. Thus

$$\langle \hat{\sigma}_\ell(\mathbf{k}_1) \ldots \hat{\sigma}_\ell(\mathbf{k}_m) \rangle_0 = \frac{\int_{-\infty}^{\infty} d\hat{\sigma}(\mathbf{k}_1) \ldots d\hat{\sigma}(\mathbf{k}_m) [\hat{\sigma}(\mathbf{k}_1) \ldots \hat{\sigma}(\mathbf{k}_m)] e^{\mathcal{H}_\sigma}}{\int_{-\infty}^{\infty} d\hat{\sigma}(\mathbf{k}_1) \ldots d\hat{\sigma}(\mathbf{k}_m) e^{\mathcal{H}_\sigma}}. \tag{12.92}$$

Following the arguments of section 6.3.4, where we calculated the correlation function for the Gaussian approximation, we get

$$\langle \hat{\sigma}_\ell(\mathbf{k}_1) \cdots \hat{\sigma}_\ell(\mathbf{k}_m) \rangle_0 = 0, \quad \text{for } m \text{ odd.} \tag{12.93}$$

When $m = 2$, we have simply the two-point correlation function result of section 6.3.4

$$\langle \hat{\sigma}_\ell(\mathbf{k}_1) \hat{\sigma}_\ell(\mathbf{k}_2) \rangle_0 = \delta_{\mathbf{k}_1 + \mathbf{k}_2, 0} V G_0(\mathbf{k}_1), \tag{12.94}$$

where the **propagator** is defined by

$$G_0(\mathbf{k}) \equiv \frac{1}{r_0 + k^2}. \tag{12.95}$$

In the infinite volume limit, we obtain (using the results of section 5.7.2)

$$\langle \hat{\sigma}_\ell(\mathbf{k}_1) \hat{\sigma}_\ell(\mathbf{k}_2) \rangle_0 = (2\pi)^2 \delta(\mathbf{k}_1 + \mathbf{k}_2) G_0(\mathbf{k}_1). \tag{12.96}$$

To calculate the four-, six-, eight-, *etc.* point functions, consider again the numerator of eqn. (12.92). Since $\mathcal{H}_\sigma$ may be written as the sum of Hamiltonians, one for each mode labelled by $\mathbf{k}$, the integral always factorises. If there is a variable of integration $\hat{\sigma}(\mathbf{k})$ which appears in the integrand raised to an odd power, the whole numerator vanishes, because $\exp \mathcal{H}_\sigma$ is an even function of the $\hat{\sigma}(\mathbf{k})$ variables. By virtue of eqns. (12.94) or equivalently (12.96), the only non-zero contributions to the numerator arise when *pairs* of wavevector labels sum to zero. Every possible way to pair up the $\mathbf{k}$ labels of the variables of integration gives a contribution. To check this assertion, the reader is urged to do or at least contemplate exercise 12–1, after noting that the wavevectors are simply labels for the σ_ℓ fields. This conclusion is best written out explicitly; for $m = 4$, we obtain

$$\begin{aligned} \langle \hat{\sigma}_\ell(\mathbf{k}_1) \hat{\sigma}_\ell(\mathbf{k}_2) \hat{\sigma}_\ell(\mathbf{k}_3) \hat{\sigma}_\ell(\mathbf{k}_4) \rangle_0 = \\ (2\pi)^d \delta(\mathbf{k}_1 + \mathbf{k}_2) G_0(\mathbf{k}_1) \, (2\pi)^d \delta(\mathbf{k}_3 + \mathbf{k}_4) G_0(\mathbf{k}_3) \\ + (2\pi)^d \delta(\mathbf{k}_1 + \mathbf{k}_3) G_0(\mathbf{k}_1) \, (2\pi)^d \delta(\mathbf{k}_2 + \mathbf{k}_4) G_0(\mathbf{k}_2) \\ + (2\pi)^d \delta(\mathbf{k}_1 + \mathbf{k}_4) G_0(\mathbf{k}_1) \, (2\pi)^d \delta(\mathbf{k}_2 + \mathbf{k}_3) G_0(\mathbf{k}_3). \end{aligned} \tag{12.97}$$

The reader is urged to write down the corresponding expression for the six-point function. These results, which are simple consequences of the properties of Gaussian integrals, are known as **Wick's theorem**.

12.3.4 Evaluation of $\langle V\rangle_0$

Now we are ready to do the calculation of $\langle V\rangle_0$ in earnest. According to the considerations of section 12.3.2 and eqn. (12.83), we can write $\langle V\rangle_0$ as a sum of terms: $\langle V\rangle_0^{SSSS}$, $\langle V\rangle_0^{SSS\sigma}$, $\langle V\rangle_0^{SS\sigma\sigma}$, *etc.*, where the superscript indicates symbolically the number of $\hat{S}$ and $\hat{\sigma}$ fields in the term. The terms with an odd number of $\hat{\sigma}$ fields will be zero by Wick's theorem. Then $\langle V\rangle_0^{SSSS}$ will contribute to $u_{4,\ell}$, $\langle V\rangle_0^{SS\sigma\sigma}$ will contribute to $u_{2,\ell}$, and $\langle V\rangle_0^{\sigma\sigma\sigma\sigma}$ will contribute to the regular part of the free energy. We will denote the contribution of $\langle V^m\rangle_0$ to these quantities by the superscript m. Let us now calculate each of these contributions explicitly.

We begin with

$$\langle V\rangle_0^{SSSS} = \frac{1}{4}u_0\int_0^{\Lambda/\ell} dk_1\cdots dk_4\,(2\pi)^d\delta(\mathbf{k}_1+\cdots+\mathbf{k}_4)\hat{S}'_\ell(\mathbf{k}_1)\cdots S'_\ell(\mathbf{k}_4), \tag{12.98}$$

where we have used $\langle 1\rangle_0 = 1$. The next step is to renormalise:

$$\mathbf{k}_\ell = \ell\mathbf{k} \tag{12.99}$$

$$\hat{S}_\ell(\mathbf{k}_\ell) = z^{-1}\hat{S}'_\ell(\mathbf{k}). \tag{12.100}$$

Using the identity

$$\delta(\mathbf{k}_\ell/\ell) = \ell^d\delta(\mathbf{k}_\ell), \tag{12.101}$$

we obtain the $O(V^1)$ contribution to $u_{4,\ell}$, which we denote by $u_{4,\ell}^{(1)}$:

$$\frac{1}{4}u_{4,\ell}^{(1)} = \frac{1}{4}u_0 z^4\ell^{-3d}. \tag{12.102}$$

Next we consider $\langle V\rangle_0^{SS\sigma\sigma}$. This involves two $\hat{\sigma}$ fields, and thus, after integrating over these variables and using Wick's theorem (12.96), we obtain

$$\begin{aligned}\langle V\rangle_0^{SS\sigma\sigma} &= \frac{1}{4}u_0\int_0^{\Lambda/\ell} dk_1\,dk_2\,\hat{S}'_\ell(\mathbf{k}_1)\hat{S}'_\ell(\mathbf{k}_2)(2\pi)^d\delta(\mathbf{k}_1+\mathbf{k}_2)\\ &\times\int_{\Lambda/\ell}^{\Lambda} dk_3\frac{1}{r_0+{\mathbf{k}_3}^2}.\end{aligned} \tag{12.103}$$

Here, the integral over $\mathbf{k}_3$ is simply a function of Λ, ℓ and r_0, so it is unaffected by the rescaling and renormalisation step, which we perform next, to give:

$$\langle V\rangle_0^{SS\sigma\sigma} = \frac{1}{4}u_0\ell^{-2d}\ell^d z^2\int_0^{\Lambda} dk_\ell\,\hat{S}_\ell(\mathbf{k}_\ell)\hat{S}_\ell(-\mathbf{k}_\ell)\int_{\Lambda/\ell}^{\Lambda} dk\frac{1}{r_0+k^2}. \tag{12.104}$$

There are six such terms contributing to $u^{(1)}_{2,\ell}$. There is another contribution to $u^{(1)}_{2,\ell}$ which comes from $\mathcal{H}_{S'}$, and which is the term we got in the Gaussian model analysis of section 12.2. Collecting together all the terms for the quadratic part of $\mathcal{H}_\ell$, we have

$$\frac{1}{2}\int_0^\Lambda đ\mathbf{k}_\ell\, z^2\ell^{-d}\left(r_0+\frac{k_\ell^2}{\ell^2}\right)|\hat{S}_\ell(\mathbf{k}_\ell)|^2 + \frac{6}{4}u_0\ell^{-d}z^2\int_0^\Lambda đ\mathbf{k}_\ell\,|\hat{S}_\ell(\mathbf{k}_\ell)|^2\int_{\Lambda/\ell}^\Lambda đ\mathbf{k}\,\frac{1}{r_0+k^2}. \tag{12.105}$$

Thus, we can read off

$$u^{(1)}_{2,\ell} = z^2\ell^{-d}\left[r_0+\frac{\mathbf{k}_\ell^2}{\ell^2}+3u_0\int_{\Lambda/\ell}^\Lambda \frac{đ\mathbf{k}}{r_0+k^2}\right]. \tag{12.106}$$

As before, we will choose z so that the coefficient of $\mathbf{k}_\ell^2$ is just 1/2. This gives $z=\ell^{1+d/2}$ as before. In summary, to $O(V^1)$ we obtain the renormalised coupling constants

$$u^{(1)}_{2,\ell} = r_0\ell^2+\mathbf{k}_\ell^2+3u_0\ell^2\int_{\Lambda/\ell}^\Lambda \frac{đ\mathbf{k}}{(r_0+k^2)}, \tag{12.107}$$

$$u^{(1)}_{4,\ell} = u_0\ell^{4-d} = u_0\ell^\epsilon \tag{12.108}$$

As anticipated in the general comments of section 12.1, $u'\sim u_0\ell^\epsilon$.

To find a fixed point which is non-trivial, we must go at least to $O(V^2)$. This is a rather involved calculation, and to carry it out in a manner capable of systematic improvement, we need to become more proficient at perturbation theory. Thus, we now break off our calculation of the RG recursion relations to make a digression on the topic of Feynman diagrams. The reader who is not interested in the calculational details, or for whom this is the first exposure to RG for Landau theory, is advised to skip the next section, and to proceed directly to the RG analysis of the recursion relations, given in section 12.5. In any event, the reader should bear in mind that the apparent complexity of RG calculations is usually due to the technical details of obtaining the recursion relations, and *not* to the RG procedure itself.

12.4 FEYNMAN DIAGRAMS

The terms in the perturbation series rapidly get quite complicated, as the reader has probably noticed. Feynman developed a useful shorthand for the terms, which is to write down diagrams indicating how different fields are paired up. There is one-to-one correspondence between each **Feynman diagram** or **Feynman graph** and a term in the perturbation series. This correspondence is not unique: there are many ways to write down diagrams for the terms, and people have their own preferences. Once one has figured out this correspondence (the Feynman rules), then one only has to be able to draw all the appropriate diagrams to see what one should compute. Another use of diagrams is to communicate efficiently what calculation one has done by writing down the diagram rather than a long expression: Feynman rules usually only differ by combinatorial factors, so the structure of the integrals is readily communicated by the diagram.[5]

The presentation here is rather pedestrian, and the reader will be able to streamline it once the main points have been grasped; however, to benefit from this section, the active participation of the reader is required. So, with pencils at the ready, we proceed by writing down diagrams for the first order calculation that we have already done.

12.4.1 Feynman Diagrams to O(V)

Consider

$$\langle V\rangle_0^{SSSS} = \frac{u_0}{4}\int_0^{\Lambda/\ell} đ\mathbf{k}_1 \ldots đ\mathbf{k}_4\,(2\pi)^d\delta(\mathbf{k}_1+\cdots\mathbf{k}_4)\hat{S}'_\ell(\mathbf{k}_1)\cdots\hat{S}'_\ell(\mathbf{k}_4). \tag{12.109}$$

We write this as

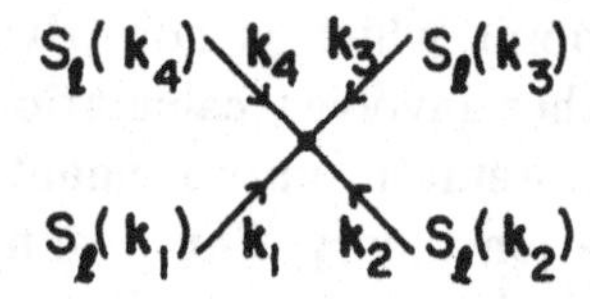

$S_\ell(k_4)$ k_4 k_3 $S_\ell(k_3)$

$S_\ell(k_1)$ k_1 k_2 $S_\ell(k_2)$

Each **external leg** of the diagram represents a $\hat{S}'_\ell$ field, with the $\mathbf{k}_i$ labels deployed in any arbitrary fashion. External legs have one end dangling, *i.e.* not connected to any other part of the diagram, and the momentum $\mathbf{k}_i$ associated with each leg should be thought of as flowing into the leg from

[5] Strictly speaking, Feynman diagrams describe terms in perturbation theory for *field theory*, where there is no separation into short and long wavelength degrees of freedom. The rules for a perturbation calculation in field theory are of course different from those described below.

the outside. We will see why in a moment. The central vertex represents the factor

$$\bullet = \frac{u_0}{4}(2\pi)^d\,\delta(\mathbf{k}_1 + \mathbf{k}_2 + \mathbf{k}_3 + \mathbf{k}_4). \tag{12.110}$$

In general, each vertex is connected to 4 legs, because the interactions are quartic in this model, and the sum of the momenta flowing into each vertex must sum to zero. Momentum is conserved. Finally, an integral $\int d\!\!\!{}^-\mathbf{k}_i$ over each $\mathbf{k}_i$ in the external legs is performed, with $0 < |\mathbf{k}_i| < \Lambda/\ell$. The reader should check that the diagram, when decoded, does correspond to the algebraic expression (12.109).

Let us now consider $\langle V\rangle_0^{SS\sigma\sigma}$. We begin by following the procedure above. We write down a vertex, and label the external legs; however, this time, two of the external legs are labelled by $\hat{\sigma}_\ell$ fields. Now we need to add a new rule, to describe the averaging over the $\hat{\sigma}_\ell$ fields. Once we have averaged over the $\hat{\sigma}_\ell$ fields, they disappear from the description completely, and we are left with a function of their momenta, as given by eqn. (12.96). We represent this sequence of operations as follows: draw a dotted line between the two $\hat{\sigma}_\ell$ fields, sitting on the ends of the external legs. This represents the intention to perform the average. The average itself ties together the two $\hat{\sigma}_\ell$ fields, to create an **internal line**, characterised by the **internal momentum k**.

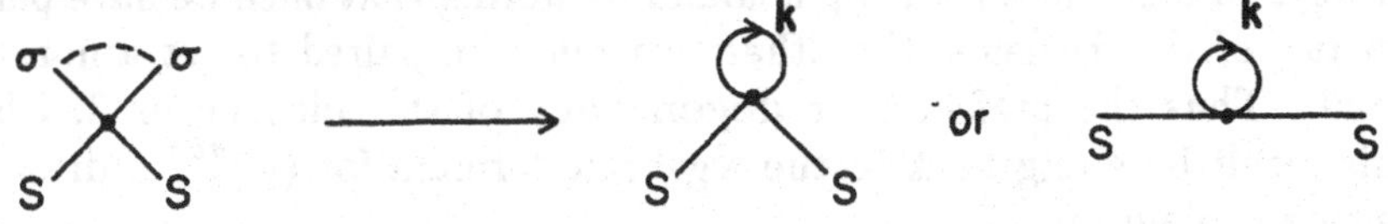

Each internal line carries momentum **k**, and represents a factor $G_0(\mathbf{k})$. The external lines represent $\hat{S}'_\ell(\mathbf{k}_1)$ and $\hat{S}'_\ell(\mathbf{k}_2)$. Having constructed the diagram, drawn in the momenta (and their directions), conserved momentum at the vertices, we now integrate over the internal and external momenta. The external momenta are as above, whereas the $\int d\!\!\!{}^-\mathbf{k}$ of the internal momenta is performed for $\Lambda/\ell < |\mathbf{k}| < \Lambda$. Thus

$$\begin{aligned}\underline{\;\bigcirc\;} &= \frac{u_0}{4}\int_0^{\Lambda/\ell} d\!\!\!{}^-\mathbf{k}_1 d\!\!\!{}^-\mathbf{k}_2\,(2\pi)^d\delta(\mathbf{k}_1+\mathbf{k}_2)\hat{S}'_\ell(\mathbf{k}_1)\hat{S}'_\ell(\mathbf{k}_2)\int_{\Lambda/\ell}^{\Lambda}\frac{d\!\!\!{}^-\mathbf{k}}{r_0+k^2}\\ &= \frac{u_0}{4}\int_0^{\Lambda/\ell} d\!\!\!{}^-\mathbf{k}_1|\hat{S}'_\ell(\mathbf{k}_1)|^2\int_{\Lambda/\ell}^{\Lambda}\frac{d\!\!\!{}^-\mathbf{k}}{r_0+k^2}.\end{aligned} \tag{12.111}$$

Thus we see that our results for the renormalised coupling constants (12.106) and (12.102) can be written as

$$u_{2,\ell}^{(1)} = z^2\ell^{-d}\left[r_0\ell^2 + \mathbf{k}_\ell^2 + 12\ \underline{\;\bigcirc\;}\ \right] \tag{12.112}$$

$$u_{4,\ell}^{(1)} = z^4 \ell^{-3d} \left[\times \right] . \tag{12.113}$$

In writing the results this way, we used our known expressions to construct the diagrams and the Feynman rules as we went along. In particular, the prefactors of diagrams in the recursion relations were obtained from section 12.3.

Let us conclude this section by writing down $\langle V \rangle_0$ in these diagrams, but *not* refering to our earlier results. The strategy is to draw all the diagrams, then work out their multiplicative prefactors. To $O(V^1)$, we simply write down a vertex with four external legs. We can write down 5 separate **primitive diagrams** with 0,1,2,3 or 4 $\hat{\sigma}_\ell$ fields, the remainder being $\hat{S}'_\ell$ fields. Those with odd numbers of $\hat{\sigma}_\ell$ fields vanish by Wick's theorem. The diagram with all $\hat{S}'_\ell$ fields only occurs once.

The diagram with two $\hat{\sigma}_\ell$ fields is the one with the prefactor 12 in eqn. (12.112). Its prefactor here (which we now calculate without using our knowledge of eqn. (12.112)) is given by noticing that once we have deployed the $\hat{\sigma}_\ell$ fields on the external legs (with the associated momenta), there is only *one* way to pair up the $\hat{\sigma}_\ell$ fields. The number of deployments is $4!/2!2! = 6$.

Finally, there is the primitive diagram with four $\hat{\sigma}_\ell$ fields. The number of ways to pair these fields up is found by noting that once we have paired up *two* of the $\hat{\sigma}_\ell$ fields, the other two *must* be paired to get a non-zero result. Thus the prefactor or **degeneracy** of this diagram is 3. Check this result by going back to the algebraic formula for $\langle V \rangle_0^{\sigma\sigma\sigma\sigma}$; draw the resultant diagram.

The value of the diagram is

$$\begin{aligned} \infty &= \frac{u_0}{4} \int_{\Lambda/\ell}^{\Lambda} d\!\!{}^-\mathbf{k}_1 \cdots d\mathbf{k}_4 \, (2\pi)^d \delta(\mathbf{k}_1 + \cdots + \mathbf{k}_4) \\ &\quad \times \langle \hat{\sigma}_\ell(\mathbf{k}_1) \cdots \hat{\sigma}_\ell(\mathbf{k}_4) \rangle_0 \\ &= \frac{u_0}{4} \left(\int_{\Lambda/\ell}^{\Lambda} d\!\!{}^-k \, G_0(\mathbf{k}) \right)^2 . \end{aligned} \tag{12.114}$$

In conclusion, our final expression for $\langle V \rangle_0$ is

$$\langle V \rangle_0 = \times \; + 6 \; \multimap \; + 3 \; \infty \; . \tag{12.115}$$

12.4.2 Feynman Diagrams for $\langle V^2\rangle_0 - \langle V\rangle_0^2$

Let us begin by considering $\langle V^2\rangle_0$. Recall that

$$V = \frac{u_0}{4}\int_{\mathbf{k}_1\ldots\mathbf{k}_4}(2\pi)^d\delta(\mathbf{k}_1+\cdots+\mathbf{k}_4)\hat{S}(\mathbf{k}_1)\ldots\hat{S}(\mathbf{k}_4), \tag{12.116}$$

with

$$\int_{\mathbf{k}}\hat{S}(\mathbf{k}) = \int_0^{\Lambda/\ell} d\!\!{}^-\mathbf{k}\,\hat{S}'_\ell(\mathbf{k}) + \int_{\Lambda/\ell}^{\Lambda} d\!\!{}^-\mathbf{k}\,\hat{\sigma}_\ell(\mathbf{k}). \tag{12.117}$$

Combining eqns. (12.116) and (12.117) to form V^2 generates many terms. Before averaging, each term can be constructed from a primitive diagram, which is made up of *two* vertices and their associated four external legs placed next to each other. Each external leg may be either a $\hat{S}'_\ell$ field or a $\hat{\sigma}_\ell$ field. We saw in the previous section that there are 5 different ways to deploy $\hat{S}'_\ell$ and $\hat{\sigma}_\ell$ fields on the legs of a single vertex, with 0,1,2,3 or 4 $\hat{\sigma}_\ell$ fields. Thus, when we juxtapose two vertices, and allow each to be decorated in 5 ways, there are $5 \times 5 = 25$ total primitive diagrams. Then, our algorithm requires us to form *all possible pairs* of $\hat{\sigma}_\ell$ fields and perform the averaging, to form the final diagrams with their internal lines. The labour of this task is significantly reduced, when we realise that in many of these diagrams, $\hat{\sigma}_\ell$ fields are paired only with another $\hat{\sigma}_\ell$ field decorating the *same* vertex. Thus, the *value* of the completed diagram is actually a *product* of the values of two separate diagrams. For example, there is the contribution to $\langle V^2\rangle_0$ by

When the completed diagram is actually a product of other diagrams, it is referred to as a **disconnected diagram**. Note that the diagram above is equal to the square of the first term in eqn. (12.115) for $\langle V\rangle_0$. Thus, when we compute the **cumulant**

$$\langle V^2\rangle_c \equiv \langle V^2\rangle_0 - \langle V\rangle_0^2, \tag{12.118}$$

this disconnected diagram is actually cancelled! Exercise 12–2 invites the reader to verify that this cancellation is not an isolated instance: *all* of the disconnected diagrams in $\langle V^2\rangle_0$ are cancelled by diagrams in $\langle V\rangle_0^2$. Thus, in the cumulant $\langle V^2\rangle_c$, the only non-cancelling diagrams are those where the primitive diagrams are connected. The fact that the disconnected

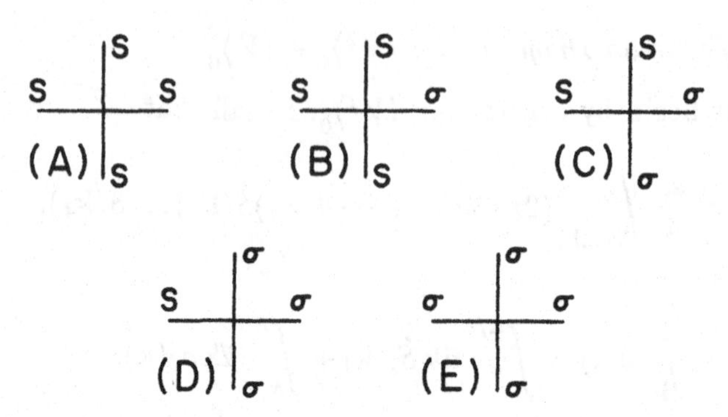

Figure 12.1 The five possible vertices and their decorations, from which are constructed the Feynman diagrams for the Ising universality class Landau theory.

diagrams do not contribute to $\langle e^V \rangle_0$ is an example of the **linked cluster theorem**, which is proved in the appendix. The subscript "c" can be thought of as denoting "cumulant" or "connected".

Let us return now to the computation of $\langle V^2 \rangle_c$. We begin by writing down the final result, and then discussing the derivation, for those interested in mastering diagrammatic perturbation theory.

$$\frac{1}{2}\langle V^2\rangle_c = 8 \quad + 36 \quad + 48 \quad + 48 \quad + 72 \quad + 72 \quad + 36 \quad + 12 \tag{12.119}$$

The derivation is best presented in two steps: first work out the possible diagrams, second calculate the combinatoric factors.

Step 1

The starting point is that we are working to $O(V^2)$, and so when we construct the primitive diagrams, we need two vertices, each with its four external legs. Each leg may be associated with a $\hat{\sigma}_\ell$ field or a $\hat{S}'_\ell$ field. Each end of a line can be either an S' or a $\hat{\sigma}_\ell$ field. We will pair up all the ends in all possible ways to create diagrams with 2 vertices; there are five possible vertices, as shown in figure (12.1), and we have to form all possible pairs by juxtaposition (A-A, A-B, A-C, *etc.*). Then we must tie together the $\hat{\sigma}_\ell$ fields by averaging and using Wick's theorem. The labour is reduced by only considering diagrams that will end up being connected.

We know A-A is cancelled out by the linked cluster theorem. A cannot form a connected diagram with any other vertex, because it has no free $\hat{\sigma}_\ell$ fields.

Consider vertex E. It can pair with itself in two ways to form E-E(a) and E-E(b)

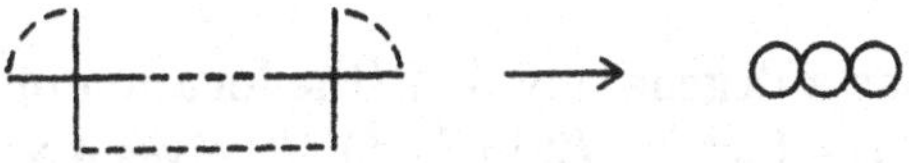

(12.120)

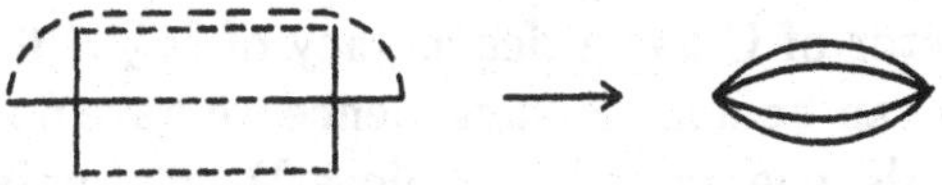

(12.121)

plus one other diagram that is cancelled out by the linked cluster theorem. (Exercise: draw this diagram.) E cannot pair with B because that would leave a free $\hat{\sigma}_\ell$ field which would average to zero. Likewise with D. Pairing with A can only produce disconnected diagrams which are cancelled out by the linked-cluster theorem. We are left with E-C. Ignoring the disconnected diagrams, E-C gives

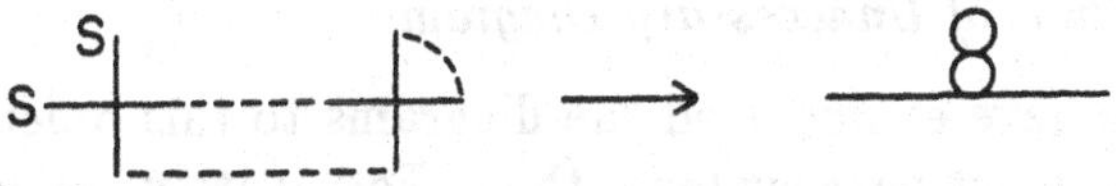

(12.122)

Now consider vertex C. It can pair with itself to form C-C

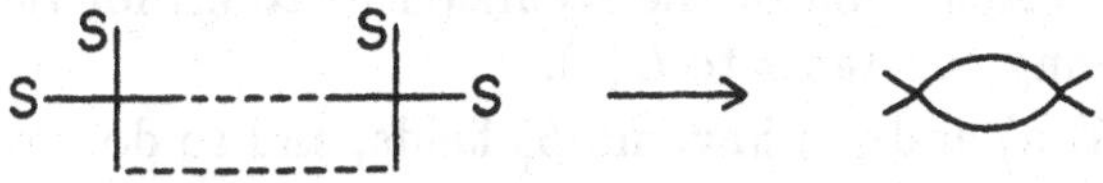

(12.123)

ignoring disconnected diagrams. C cannot pair with B or D because of a free $\hat{\sigma}_\ell$- field being generated. Vertex D can pair with B to give D-B

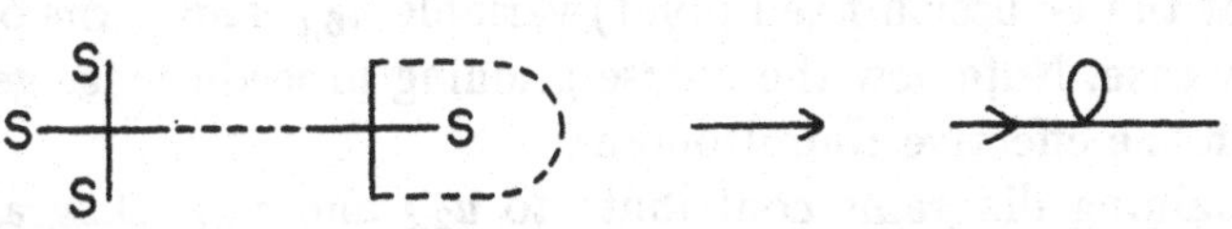

(12.124)

and with itself to give D-D, for which there are *two* resultant connected diagrams (a) and (b):

(12.125)

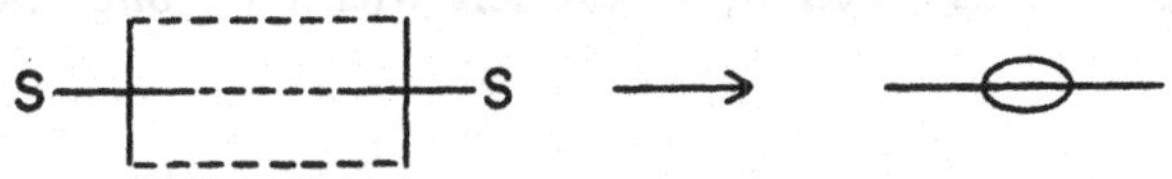

(12.126)

Finally, B can also pair with itself to form B-B

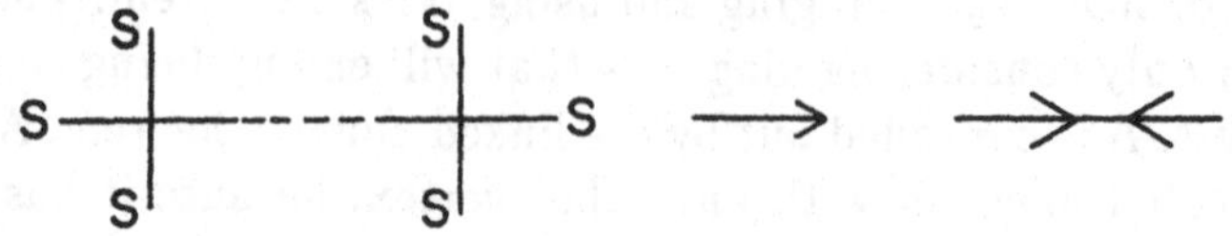

S S S S S S S S → →← (12.127)

Step 2

Now we work out the degeneracy for each graph in the expression for $\langle V^2\rangle_c/2$.

B-B: Each B has a degeneracy of 4 (the location of the $\hat{\sigma}_\ell$ field). Thus, including the factor 1/2 in $\langle V^2\rangle_c/2$ gives $4\times 4\times(1/2)=8$.

C-C: Each factor of C has a degeneracy of ${}^4C_2=6$. The pairing of the $\hat{\sigma}_\ell$ fields can be done 2 ways: hence we have $6\times 6\times 2\times(1/2)=36$.

B-D: B has a degeneracy of 4, as does D. The pairing of the $\hat{\sigma}_\ell$ fields can be done in 3 ways, because there are 3 $\hat{\sigma}_\ell$ fields on the D to pair with the $\hat{\sigma}_\ell$ on the B. Finally, note that this graph does *not* equal its mirror image; it comes from B-D and D-B. Thus we get $4\times 4\times 3\times 2\times(1/2)=48$.

The remaining degeneracies are left as exercises.

12.4.3 Elimination of Unnecessary Diagrams

Now that we have enumerated the diagrams to this order in perturbation theory, we must next evaluate them. Fortunately, we will see that most of the diagrams are not necessary for our present purpose, which, you may recall, is the calculation of the recursion relations for the quadratic and quartic coupling constants to $O(\epsilon)$.

Diagrams E-E(a) and (b) have no $\hat{S}'_\ell$ fields, and so do not contribute to the recursion relations for the coefficients of powers of $\hat{S}'_\ell$. Diagram B-B has six $\hat{S}'_\ell$ fields, and therefore will represent a contribution to the irrelevant (at the Gaussian fixed point) variable $u_{6,\ell}$, and turns out to be $O(\epsilon^2)$ in any case. Note how the coarse-graining procedure has generated new terms in the effective Hamiltonian.

The remaining diagrams contribute to $u_{2,\ell}$ and $u_{4,\ell}$. Diagrams D-D and E-C contribute to $u_{2,\ell}$, but are second order in u_0, and therefore, at the fixed point $u^*=O(\epsilon)$, they will contribute to r at $O(\epsilon^2)$. Thus, we ignore them for our $O(\epsilon)$ calculation. Diagrams B-D and D-D are actually zero: the reader is urged to calculate the value of these diagrams, by translating them into integrals, and checking that they vanish.

Now that the dust has settled, we are left with only *one* diagram to calculate: C-C.

12.5 THE RG RECURSION RELATIONS

The conclusion of our diagrammatic analysis is that the recursion relations read:

$$u^{(2)}_{2,\ell} = z^2\ell^{-d}\left[(r_0 + k^2) + 12 \;\underline{\;\bigcirc\;}\;\right]; \tag{12.128}$$

$$u^{(2)}_{4,\ell} = z^4\ell^{-3d}\left[\;\times\; - 36 \;\rangle\!\bigcirc\!\langle\;\right]. \tag{12.129}$$

The diagram from C-C has the value

$$\left(\frac{u_0}{4}\right)^2 \int_0^{\Lambda/\ell} \bar{d}\mathbf{k}_1 \cdots \bar{d}\mathbf{k}_4\, (2\pi)^d\, \delta(\mathbf{k}_1 + \cdots + \mathbf{k}_4)\, \hat{S}'_\ell(\mathbf{k}_1) \cdots \hat{S}'_\ell(\mathbf{k}_4)$$
$$\times \int_{\Lambda/\ell}^{\Lambda} \frac{\bar{d}\mathbf{k}_5}{r_0 + \mathbf{k}_5^2} \frac{1}{r_0 + (\mathbf{k}_1 + \mathbf{k}_2 - \mathbf{k}_5)^2}. \tag{12.130}$$

Note that this diagram leads to a form of the quartic coupling which is *not* diagonal in real space: the coupling between the $\hat{S}'_\ell$ fields is not simply a delta function, as it was in the original Hamiltonian (12.65), but has some wavevector dependence. We assert that the wavevector dependence is not important for the $O(\epsilon)$ calculation; thus we expand the last integral in (12.130) about $\mathbf{k}_1 = \mathbf{k}_2 = \mathbf{0}$, and ignore the corrections.[6]

The next step is to perform the momentum rescaling and the wave-function renormalisation, just as we did in section 12.3. Thus, our recursion relations become

$$r' = \ell^2 \left[r_0 + 3u_0 I_1\right], \tag{12.131}$$
$$u' = u_0 \ell^{\epsilon} \left[1 - 9u_0 I_2\right], \tag{12.132}$$

where

$$I_1 \equiv \int_{\Lambda/\ell}^{\Lambda} \frac{\bar{d}\mathbf{k}}{r_0 + k^2} \tag{12.133}$$

and

$$I_2 \equiv \int_{\Lambda/\ell}^{\Lambda} \frac{\bar{d}\mathbf{k}}{(r_0 + k^2)^2}. \tag{12.134}$$

These are our RG recursion relations, to $O(u_0^2)$. Higher orders of u_0 in the recursion relation for r' are not needed, because they will turn out to

---

[6] For details, consult the review article by K.G. Wilson and J. Kogut, *Phys. Rep.* C **12**, 75 (1974).

be $O(\epsilon^2)$, but they are needed in the recursion relation for u' in order to obtain a non-trivial fixed point.

The calculation we have just performed becomes incredibly laborious for higher orders in ϵ and u_0. The present method is used in practice for $O(\epsilon)$ calculations, and is usually the quickest way to get the phase diagram, upper and lower critical dimensionalities, *etc.* However, for accurate computation of the critical exponents, the most streamlined technique is to use the field-theoretic approach mentioned in the introduction of this chapter.

Finally: a guide for the perplexed. Conceptually, we have done nothing more than the simple calculation of the recursion relation for the two dimensional Ising model using a real space renormalisation group, given in sections 9.6.2 and 9.6.3. The present calculation, however, is a more controlled approximation scheme.

12.5.1 Feynman Diagrams For Small $\epsilon = 4 - d$

To analyze the recursion relations (12.131) and (12.132), we evaluate I_1 and I_2 in $\epsilon = 4 - d$ dimensions. Thus, we treat spatial dimensionality as a continuous variable, and assume that it is legitimate to expand the integrals as a power series about four dimensions (*i.e.* $\epsilon = 0$). Such an expansion is formal, and although divergent, is assumed to be Borel summable.[7] It is a very strong assumption that a given property of a physical system varies smoothly with the spatial dimensionality, and it is almost certainly incorrect in some cases. For example, topological characteristics, such as defect structures, entanglement of polymer loops, *etc.* are sensitive to the spatial dimensionality, and the existence of the Kosterlitz-Thouless transition is a vivid reminder of this. Nevertheless, in systems where there are no such complications, expansions in spatial dimensionality have proven to be very fruitful for calculating critical exponents and thermodynamic functions in the critical regime.

Disregarding the potential complications, then, we expand

$$I_{1,2}(\epsilon) = I_{1,2}(0) + \epsilon I'_{1,2}(0) + O(\epsilon^2). \tag{12.135}$$

Since I_1 and I_2 appear as coefficients of u_0 in the recursion relations, we do not even need to evaluate them beyond *zeroth* order in ϵ, for the purpose of finding the fixed point to $O(\epsilon)$. However, it is worth explaining how to continue these integrals in continuous dimensions, as this is necessary for higher orders in ϵ.

---

[7] See footnote 2.

There are two classes of integral which typically arise from a Feynman diagram calculation: spherically symmetric integrals and uniaxially symmetric integrals.

Spherically Symmetric Integrals

Consider the spherically symmetric integral of some function f:

$$I = \int \frac{d^d\mathbf{q}}{(2\pi)^d} f(q^2) = \frac{S_d}{(2\pi)^d} \int dq\, q^{d-1} f(q^2), \tag{12.136}$$

where S_d is the surface area of a unit sphere in d-dimensions, evaluated in exercise 12–2(g) as

$$S_d = \frac{2\pi^{d/2}}{\Gamma(d/2)}. \tag{12.137}$$

In this formula, note that d need not necessarily be an integer. Now write $d = 4 - \epsilon$: the radial integral has one dummy variable, and can be performed (in principle) to yield an answer which can only depend on ϵ.

Uniaxially Symmetric Integrals

Choose the axis of symmetry so that $\mathbf{q} \cdot \hat{\mathbf{z}} = \cos\theta$. Then the integral can be written in the form

$$I = \int \frac{d^d\mathbf{q}}{(2\pi)^d} f(q^2, \theta) = C \int dq q^{d-1} \int_0^\pi d\theta (\sin\theta)^{d-2} f(q^2, \theta) \tag{12.138}$$

where we can determine the constant C by looking at the special case when f is independent of θ.

$$C = \frac{S_d}{(2\pi)^d} \frac{1}{\int_0^\pi d\theta\, (\sin\theta)^{d-2}} = \frac{1}{(2\pi)^d} S_{d-1}. \tag{12.139}$$

The radial integral can then be evaluated to give a function of ϵ as before.

Returning to the problem at hand, we only need lowest order results in ϵ for the integrals $I_{1,2}$. Since we anticipate that the fixed point value $r^* = O(\epsilon)$, we also expand the integrand in powers of r_0, which is legitimate because the integrals $I_{1,2}$ are well-behaved near $r_0 = 0$. Therefore I_1 is given by

$$\begin{aligned}
\int_{\Lambda/\ell}^{\Lambda} \frac{d^d\mathbf{k}}{(2\pi)^d} \frac{1}{r_0 + k^2} &= \int_{\Lambda/\ell}^{\Lambda} \frac{d^d\mathbf{k}}{k^2} \left[1 - \frac{r_0}{k^2} + O(r_0^2)\right] \\
&= \int_{\Lambda/\ell}^{\Lambda} \frac{d^d\mathbf{k}}{(2\pi)^d} \frac{1}{k^2} - r_0 \int_{\Lambda/\ell}^{\Lambda} \frac{d^d\mathbf{k}}{(2\pi)^d} \frac{1}{k^4} + O(r_0^2, \epsilon) \\
&= K_4 \int_{\Lambda/\ell}^{\Lambda} k dk - r_0 K_4 \int_{\Lambda/\ell}^{\Lambda} \frac{dk}{k} + O(r_0^2, \epsilon) \\
&= \frac{K_4 \Lambda^2}{2} \left[1 - \frac{1}{\ell^2}\right] - r_0 K_4 \log \ell + O(r_0^2, \epsilon).
\end{aligned} \tag{12.140}$$

Here, the constant K_4 is defined by

$$K_d \equiv \frac{S_d}{(2\pi)^d}, \tag{12.141}$$

with $K_4 = 1/8\pi^2$. In a similar way, we evaluate I_2 as

$$\int_{\Lambda/\ell}^{\Lambda} \frac{d^d\mathbf{k}}{(2\pi)^d} \frac{1}{(r_0 + k^2)^2} = \int_{\Lambda/\ell}^{\Lambda} \frac{d^d\mathbf{k}}{(2\pi)^d} \frac{1}{k^4} + O(r_0, \epsilon) \tag{12.142}$$

$$= K_4 \log \ell + O(r_0, \epsilon). \tag{12.143}$$

12.5.2 Recursion Relations to $O(\epsilon)$

The recursion relations become

$$r' = \ell^2 \left[r_0 + \frac{3}{4\pi^2}\left(\frac{u_0}{4}\right)\Lambda^2\left(1 - \frac{1}{\ell^2}\right) - \frac{3}{2\pi^2}\left(\frac{u_0}{4}\right) r_0 \log \ell\right] + O(\epsilon u_0, r_0^2 u_0), \tag{12.144}$$

$$\frac{u'}{4} = \ell^\epsilon \left[\frac{u_0}{4} - \frac{9}{2\pi^2}\left(\frac{u_0}{4}\right)^2 \log \ell\right] + O(\epsilon^3, r_0 \epsilon^2). \tag{12.145}$$

Writing

$$\ell^\epsilon = e^{\epsilon \log \ell} = 1 + \epsilon \log \ell + \frac{1}{2!}\epsilon^2 (\log \ell)^2 + O(\epsilon^3), \tag{12.146}$$

we finally get the recursion relations to this order of accuracy in a form suitable for use:

$$r' = \ell^2 \left[r_0 + \frac{3}{4\pi^2}\frac{u_0}{4}\Lambda^2\left(1 - \frac{1}{\ell^2}\right) - \frac{3}{2\pi^2}\frac{u_0}{4} r_0 \log \ell\right]; \tag{12.147}$$

$$\frac{u'}{4} = \frac{u_0}{4} + \frac{u_0}{4}\left[\epsilon - \frac{9}{2\pi^2}\frac{u_0}{4}\right] \log \ell. \tag{12.148}$$

12.5.3 Fixed Points to $O(\epsilon)$

The Gaussian fixed point is a solution, as expected, with

$$r^* = u^* = 0. \tag{12.149}$$

However, now there is a new fixed point, the one which we have worked so hard to find:

$$\frac{u^*}{4} = \frac{2\pi^2}{9}\epsilon. \tag{12.150}$$

Substituting into the recursion relation (12.147) we find

$$r^* = -\frac{\epsilon\Lambda^2}{6}. \tag{12.151}$$

The non-trivial fixed point is sometimes called the **Wilson-Fisher fixed point.** Note that ℓ cancelled out of the equation for the fixed point, as it must do: the fixed point on a trajectory should not depend on the size of the steps one makes along that trajectory. Also note that r^* depends explicitly on the cut-off Λ. However, our general considerations predict that the critical exponents will not depend on the cut-off, as we will see explicitly below.

Now we need to linearise the RG at these fixed points to calculate the exponents. At the Gaussian fixed point, the linearised RG transformation is

$$\mathbf{M} = \begin{pmatrix} \partial r'/\partial r & \partial r'/\partial u \\ \partial u'/\partial r & \partial u'/\partial u \end{pmatrix} = \begin{pmatrix} \ell^2 & 3\Lambda^2(\ell^2-1)/4\pi^2 \\ 0 & \ell^\epsilon \end{pmatrix}. \tag{12.152}$$

At the Wilson-Fisher fixed point, we find

$$\mathbf{M} = \begin{pmatrix} \ell^{2-\epsilon/3} & 3\Lambda^2(\ell^2-1)/4\pi^2 \\ 0 & \ell^{-\epsilon} \end{pmatrix}. \tag{12.153}$$

The eigenvalues of the linearised RG transformation, which we will call Λ_t and Λ_2 are:

Gaussian fixed point

$$\Lambda_t = \ell^2 \Rightarrow y_t = 2. \tag{12.154}$$
$$\Lambda_2 = \ell^\epsilon \Rightarrow y_2 = \epsilon. \tag{12.155}$$

Wilson-Fisher fixed point

$$\Lambda_t = \ell^{(2-\epsilon/3)} = \ell^{2(1-\epsilon/6)} \Rightarrow y_t = 2(1-\epsilon/6). \tag{12.156}$$
$$\Lambda_2 = \ell^{-\epsilon} \Rightarrow y_2 = -\epsilon. \tag{12.157}$$

This result for y_2 was announced earlier in eqn. (12.51).

12.5.4 RG Flows and Exponents

We also need the right eigenvectors $\mathbf{v}_{1,2}$, satisfying

$$\mathbf{M}\mathbf{v} = \Lambda \mathbf{v}, \tag{12.158}$$

in order to examine the flows. We use the coupling constant space labelled by r and $u/4$, for a system with quadratic coupling constant r and quartic coupling constant u:

$$\mathbf{v} = (r, u/4), \tag{12.159}$$

At the Gaussian fixed point, we find

$$\mathbf{v}_1 = (1,0); \quad \mathbf{v}_2 = \left(-\frac{3\Lambda^2}{4\pi^2}, 1\right). \tag{12.160}$$

The corresponding result for the Wilson-Fisher fixed point is left as an exercise for the reader. Note that $\mathbf{v}_1$ is not perpendicular to $\mathbf{v}_2$, which is why we did not label the eigenvalues Λ_t and Λ_u, as one might have wished. Note also that the position $(r^*_{\mathrm{WF}}, u^*_{\mathrm{WF}}/4)$ of the Wilson-Fisher fixed point is in the direction of $\mathbf{v}_2$ from the Gaussian fixed point:

$$\begin{aligned}(r^*_{\mathrm{WF}}, u^*_{\mathrm{WF}}/4) &= \left(-\frac{\epsilon\Lambda^2}{6}, \frac{2\pi^2\epsilon}{9}\right) \\ &= \frac{2\pi^2\epsilon}{9}\mathbf{v}_2. \end{aligned} \tag{12.161}$$

Thus, the Wilson-Fisher fixed point lies along $\mathbf{v}_2$, a distance $O(\epsilon)$ away from the Gaussian fixed point. According to this result, then, for $\epsilon < 0$ (*i.e.* $d > 4$), the Wilson-Fisher fixed point has a *negative* value of the quartic coupling constant, which is unphysical. Thus, one expects that the Gaussian fixed point controls the critical behaviour. For $\epsilon > 0$ (*i.e.* $d < 4$), the Wilson-Fisher fixed point is in a physical region of parameter space, whilst in the special case $\epsilon = 0$ (*i.e.* $d = 4$), the two fixed points coincide.

Let us look at the consequences for the RG flows.

Flows for $d > 4$

The linearised RG transformations near the Gaussian fixed point, as reflected in eqns. (12.154) and (12.155), show that the fixed point is stable along the $\mathbf{v}_2$ direction, but unstable in the r direction (*i.e.* to t). Thus u is an irrelevant variable, and the critical exponent $\nu = 1/y_t = 1/2$.

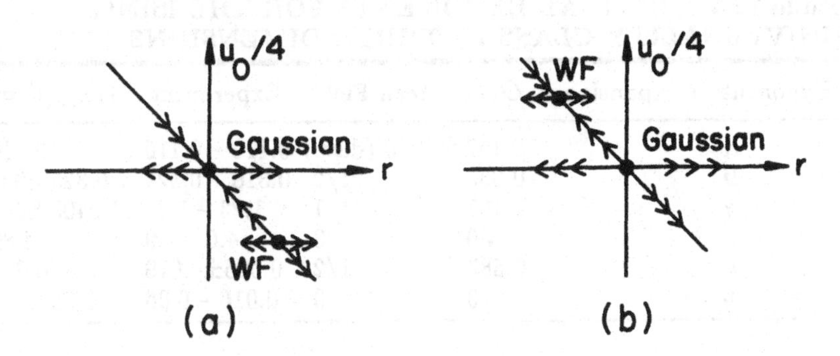

Figure 12.2 RG flows near four dimensions: (a) $d > 4$. (b) $d < 4$.

Flows for $d < 4$

Now $\epsilon > 0$, and the Gaussian fixed point is unstable along both $\mathbf{v}_1$ and $\mathbf{v}_2$. Along $\mathbf{v}_2$, the flow is towards the Wilson-Fisher fixed point, now in a physical part of parameter space. The flows approach the Gaussian fixed point, are repelled along $\mathbf{v}_2$ towards the Wilson-Fisher fixed point, and then are repelled along the r (*i.e.* t) direction to the high or low temperature sink. Thus, there is a crossover, with the asymptotic critical behaviour governed by the Wilson-Fisher fixed point. Thus,

$$\nu = \frac{1}{y_t} = \frac{1}{2}\left(1 + \frac{\epsilon}{6}\right) + O(\epsilon^2). \tag{12.162}$$

The crossover exponent is

$$\phi = \frac{|y_2|}{y_t} = \frac{\epsilon}{2} \tag{12.163}$$

as announced in exercise 9–3.

Let us conclude by calculating the values of the critical exponents, using the scaling laws.

$$\gamma = 1 + \epsilon/6 \tag{12.164}$$
$$\eta = 0 \tag{12.165}$$
$$\alpha = \epsilon/6 \tag{12.166}$$
$$\beta = \frac{1}{2} - \frac{\epsilon}{6} \tag{12.167}$$
$$\delta = 3 + \epsilon. \tag{12.168}$$

The fact that $\eta = 0$ to this order in ϵ simply means that $\eta = O(\epsilon^2)$. This reflects the fact that the wave function renormalisation was

$$z = \ell^{1+d/2}. \tag{12.169}$$

Table 12.1 CRITICAL EXPONENTS FOR THE ISING UNIVERSALITY CLASS IN THREE DIMENSIONS

Exponent	ϵ-expansion to $O(\epsilon)$	Mean Field	Experiment	Ising ($d = 3$)
α	0.167	0 (disc.)	0.110 – 0.116	0.110(5)
β	0.333	1/2	0.316 – 0.327	0.325±0.0015
γ	1.167	1	1.23 – 1.25	1.2405±0.0015
δ	4.0	3	4.6 – 4.9	4.82(4)
ν	0.583	1/2	0.625±0.010	0.630(2)
η	0	0	0.016 – 0.06	0.032±0.003

At second order in ϵ, the wave function renormalisation would acquire an anomalous dimension, which, as we have disussed in section 7.2, is equivalent to a non-zero value of η. As a crude estimate of the critical exponents in three dimensions, we take the liberty of setting $\epsilon = 1$ in the results above. The results, shown in table (12.1), are closer to the correct results than those of mean field theory.

12.6 CONCLUSION

In this chapter, we showed that systematic RG calculations are possible near the upper critical dimension, and yield semi-quantitative results for the critical behaviour, and indeed the phase diagram itself. The ϵ-expansion has proved to be invaluable in this respect. It is also worth mentioning that many other expansion techniques have been devised, to capitalise on small parameters in a given problem. These include ϵ-expansions about the lower critical dimension and expansion in $1/n$, where n is the number of components of the order parameter. In addition, many numerical methods have also been developed to implement the renormalisation group.

A full survey of the advanced development of the RG and the range of problems to which it has been applied is beyond the scope of this book, although a paper trail has been laid in the footnotes. The interested reader will find many of these developments mentioned in the Nobel lecture of K.G. Wilson.[8]

[8] K.G. Wilson, *Rev. Mod. Phys.* **55**, 583 (1983).

APPENDIX 12 - THE LINKED CLUSTER THEOREM

The cancellation of disconnected diagrams in the second-order perturbation theory calculation of the renormalised Hamiltonian is an example of the **linked cluster theorem**, which applies not only in the present context, but in field theory, many-body theory and the cluster expansion of non-ideal gases, where it was first formulated by J. Meyer.

Theorem

$$\langle e^V \rangle_0 = \exp\left[\sum_i W(C_i)\right], \tag{A12.1}$$

where $W(C_i)$ is the value of the ith *connected* diagram C_i, including the degeneracy of the diagram.

Proof

The left hand side may be written

$$\langle e^V \rangle_0 = \sum_{m=0}^{\infty} \frac{1}{m!} \langle V^m \rangle_0 . \tag{A12.2}$$

V^m has m vertices. When we write down the terms in V^m, we draw all possible primitive diagrams with m vertices, decorated in all possible ways by $\hat{\sigma}_\ell$ and $\hat{S}'_\ell$ fields. Consider one of these primitive diagrams with m vertices and $\hat{S}'_\ell$ and $\hat{\sigma}_\ell$ fields deployed in some fashion. If we now average just this one primitive diagram, we will tie together the $\hat{\sigma}_\ell$ fields in all possible ways to generate several diagrams, some of which may be fully connected, the remainder being disconnected – *i.e.* the juxtaposition or product of several connected diagrams. There are $m!$ orderings of the vertices and thus $\langle V^m \rangle_0 / m!$ is the sum of all connected and disconnected diagrams with m vertices. Each diagram appears only once, by virtue of the $m!$. Now perform the sum over m: the left hand side is the sum of *all* diagrams, connected and disconnected.

Now let us consider a diagram made up of n connected diagrams. We will call such a diagram (obviously disconnected when $n > 1$) a n-diagram. Let

$$S_n \equiv \text{sum of all } n\text{-diagrams}. \tag{A12.3}$$

To construct S_n, think of an n-diagram as being made up of n boxes in a line, each of which may contain a connected diagram. Let C_i be the ith connected diagram, labelled in some arbitrary manner. Then the value of an n-diagram with (*e.g.*) C_7 in box 1, C_9 in box 2, C_{137} in box 3, *etc.* is $W(C_7)W(C_9)W(C_{137})\ldots$. We can make a different n-diagram by putting (*e.g.*) C_2 in box 1, but otherwise leaving all the other boxes the same. The value of the new diagram would be $W(C_2)W(C_9)W(C_{137})\ldots$. Alternatively, we could have substituted (*e.g.*) C_4 for C_9, in which case, the diagram would now be $W(C_7)W(C_4)W(C_{137})\ldots$. Since the value of the n-diagram is the product of the values of the component connected diagrams, the substitutions can be performed in each box independently. For given contents of boxes $2, 3, \ldots, n$ we can substitute each connected diagram into box 1; this gives a value

$$(W(C_1) + W(C_2) + W(C_3) + \cdots) \times (\text{rest of } n\text{-diagram}). \tag{A12.4}$$

Likewise, for given contents of boxes $1, 3, 4, \ldots, n$, we can substitute each connected diagram into box 2 *etc.* In this way, we can generate all the n-diagrams; however, this process generates each n-diagram $n!$ times, because the *order* of the connected diagrams in each n-diagram is immaterial. Thus,

$$S_n = \frac{(W(C_1) + W(C_2) + W(C_3) + \cdots)^n}{n!}. \tag{A12.5}$$

Finally, the left hand side is $\sum_n S_n$. Thus

$$\langle e^V \rangle_0 = \sum_n S_n \tag{A12.6}$$

$$= \sum_n \frac{(\sum_i W(C_i))^n}{n!} \tag{A12.7}$$

$$= \exp\left[\sum_i W(C_i)\right]. \tag{A12.8}$$

Q.E.D.

EXERCISES

Exercise 12–1

This question is about Gaussian integrals; by doing it, you will be able to follow the demonstration of **Wick's theorem** given in the text.

(a) Prove that

$$\langle x_q x_r \rangle \equiv \frac{\int_{-\infty}^{\infty} d^n x \, x_q x_r e^{-\frac{1}{2}A_{ij}x_i x_j}}{\int_{-\infty}^{\infty} d^n x \, e^{-\frac{1}{2}A_{ij}x_i x_j}} = A^{-1}_{qr}$$

where A is a real symmetric $n \times n$ matrix, and Einstein summation convention is used.

(b) Using the same notation as above, prove that

$$\langle x_a x_b x_c x_d \rangle = \langle x_a x_b \rangle \langle x_c x_d \rangle + \langle x_a x_d \rangle \langle x_b x_c \rangle + \langle x_a x_c \rangle \langle x_b x_d \rangle .$$

Exercise 12–2

These exercises cover some technical aspects of Feynman diagram calculations.

(a) Check that all the disconnected diagrams in $\langle V^2 \rangle_0$ are cancelled by $\langle V \rangle_0^2$.

(b) Calculate the third order cumulant in terms of the first and second cumulants. Verify explicitly that the disconnected diagrams for S^4 Landau theory are cancelled (do a reasonable fraction of them at least).

(c) Convince yourself of the result that if $\mathcal{H} = \mathcal{H}_0 + \mathcal{H}_1$, then the two-point function is given by $G(\mathbf{r} - \mathbf{r}') = \langle S(\mathbf{r}) S(\mathbf{r}') e^{\mathcal{H}_1} \rangle_c$ where the expectation value is with respect to $\mathcal{H}_0$ and the subscript c stands for connected part.

(d) Complete the calculation of the degeneracies for the Feynman diagrams of $\langle V^2 \rangle_c / 2$.

(e) Calculate the diagrams designated B-D and D-D in the text, and show that they vanish.

(f) Let p_i be the ith prime number. Show that

$$\prod_i (1 + p_i) = 1 + 2 + 3 + 4 + 5 + 6 + \cdots = \sum_{n=1}^{\infty} n.$$

Of course, this result is purely formal; the series is divergent. Use the idea of your demonstration to show that

$$\prod_i (1 + p_i^{-s}) = \sum_{n=1}^{\infty} \frac{1}{n^s}, \quad \text{Re } s > 1.$$

Explain the relevance of this question to the linked cluster theorem.

(g) Find the surface area of a unit sphere in d dimensions by evaluating the integral $\int d^d \mathbf{q} e^{-q^2}$ in two different ways: first in Cartesian co-ordinates and second in spherical polar co-ordinates. You will need the result that the Gamma function is defined by

$$\Gamma(z) \equiv \int_0^{\infty} t^{z-1} e^{-t} \, dt, \quad \text{Re } z > 0.$$

Exercise 12–3

This question asks you to analyse the RG recursion relations for the Landau theory of the Ising universality class, using the differential RG.

(a) Starting from the recursion relations for S^4 Landau theory, derive their differential form:

$$\frac{dr}{ds} = 2r + \frac{Au}{1+r}$$

$$\frac{du}{ds} = u\left[\epsilon - \frac{Bu}{(1+r)^2}\right],$$

where $A = 12K_4$, $B = 36K_4$, $u \equiv u_0/4$ and $K_4 = 1/8\pi^2$.

(b) Find the fixed points and the linearised RG in their vicinity. Sketch the flows for $\epsilon > 0$ and $\epsilon < 0$. Try and be as complete as you can. Show that the temperature-like scaling field is a linear combination of r and u.

(c) For the remainder of this question, we study the marginal case $\epsilon = 0$. Show that

$$\xi \sim t^{-1/2}(\log t)^{1/6}.$$

(d) Consider the scaling form of the free energy density, and show that a simple-minded approach gives

$$f(t,u) \sim t^2(-\log t)^{-2/3}$$

and find the form of the specific heat, C. Explain briefly why this answer might be wrong. In fact, the correct result is that[9]

$$C \sim (-\log t)^{1/3}.$$

Try and derive this yourself.

[9] See D. Nelson, *Phys. Rev. B* **9**, 3504 (1975); J. Rudnick and D. Nelson, *Phys. Rev. B* **13**, 2208 (1976).

Index

Printed in the United States
by Baker & Taylor Publisher Services

Printed in the United States
by Baker & Taylor Publisher Services